LINEAR ACCELERATOR AND BEAM OPTICS CODES

AIP CONFERENCE PROCEEDINGS 177

RITA G. LERNER
SERIES EDITOR

LINEAR ACCELERATOR AND BEAM OPTICS CODES
LA JOLLA INSTITUTE 1988

EDITOR:
CHARLES R. EMINHIZER
PHYSICAL DYNAMICS, INC.

AMERICAN INSTITUTE OF PHYSICS   NEW YORK 1988

Authorization to photocopy items for internal or personal use, beyond the free copying permitted under the 1978 US Copyright Law (see statement below), is granted by the American Institute of Physics for users registered with the Copyright Clearance Center (CCC) Transactional Reporting Service, provided that the base fee of $3.00 per copy is paid directly to CCC, 27 Congress St., Salem, MA 01970. For those organizations that have been granted a photocopy license by CCC, a separate system of payment has been arranged. The fee code for users of the Transactional Reporting Service is: 0094-243X/87 $3.00.

Copyright 1988 American Institute of Physics.

Individual readers of this volume and non-profit libraries, acting for them, are permitted to make fair use of the material in it, such as copying an article for use in teaching or research. Permission is granted to quote from this volume in scientific work with the customary acknowledgment of the source. To reprint a figure, table or other excerpt requires the consent of one of the original authors and notification to AIP. Republication or systematic or multiple reproduction of any material in this volume is permitted only under license from AIP. Address inquiries to Series Editor, AIP Conference Proceedings, AIP, 335 E. 45th St., New York, NY 10017.

L.C. Catalog Card No. 88-46074
ISBN 0-88318-377-3
DOE CONF-880175

Printed in the United States of America.

Contents

Preface vii

PARMILA—An Overview 1
    G. P. Boicourt
PARMTEQ—A Beam Dynamics Code for the RFQ Linear Accelerator 22
    K. R. Crandall and T. P. Wangler
Trace, an Interactive Beam Dynamics Program 29
    K. R. Crandall
LTRACK—Beam-Transport Calculation Including Wake Field Effects 37
    K. C. D. Chan and R. K. Cooper
Electron Ray Tracing Programs for Gun Design and Beam Transport 45
    W. B. Herrmannsfeldt
Status of the "Path" Magnetic Optics Design Code 59
    R. J. Kashuba, R. J. Schmitt, and P. F. Meads, Jr.
Principles of GIOS and COSY 74
    H. Wollnik, B. Hartmann, and M. Berz
Some Particle Beam Computer Programs Adapted from Plasma Physics Research 86
    B. B. Godfrey
Recent Applications of Superfish 105
    R. L. Gluckstern and F. Neri
An Interactive Interface to the Beam Optics Code "Transport" 109
    R. R. Silbar
Status and Future of the 3D Mafia Group of Codes 117
    F. Ebeling, R. Klatt, F. Krawcyk, E. Lawinsky, T. Weiland, and S. G. Wipf
Magnus-3D: Accelerator Magnet Calculations in 3 Dimensions 131
    S. Pissanetzky
Three-Dimensional Electromagnetic Particle Codes and Applications to Accelerators 137
    A. Mankofsky
Numerical Limits on P.I.C. Simulation of Low Emittance Transport 161
    I. Haber and H. Rudd
Hollow Beam Dynamics in the DESY Wake Field Accelerator 173
    P. Schütt, W. Bialowons, F.-J. Decker, F. Ebelig, R. Wanzenberg, T. Weiland, and X. Chengde
Magnetic Optics Design 181
    E. A. Heighway
Beam Dynamics of the RF Electron Gun of the BNL Accelerator Test Facility 204
    K. T. McDonald
Optics Code Development at Los Alamos 220
    C. T. Mottershead and W. P. Lysenko
Nonlinear Beam Dynamics in a Funnel for Combining Two Intense Ion Beams 228
    J. H. Whealton, R. J. Raridon, K. E. Rothe, W. R. Becraft, and T. L. Owens

Lie Algebraic Methods for Charged Particle Optics 261
   A. Dragt

Self-Consistent Transfer Maps for Beams with Space Charge Using Lie Algebraic Methods 265
   R. D. Ryne

Differential Algebraic Treatment of Beam Dynamics to Very High Orders Including Applications to Spacecharge 275
   M. Berz

Analytical Solution of Single Particle Equations of Motion to High Orders: The Formula Manipulator Hamilton and the Fifth Order Code Cosy 5.0 301
   M. Berz and H. Wollnik

Algorithms for the Self-Consistent Simulation of High Power Klystrons 313
   K. Eppley

Moment Invariants for Particle Beams 323
   W. P. Lysenko and M. S. Overley

The Correction of Aberrations in Beams Filling Elliptical Phase-Space Areas .. 336
   H. Wollnik

Summary of the Working Group on Magnet Codes 345
   S. Pissanetzky

Summary for the Working Group on Linac Codes 347
   T. P. Wangler and J. Staples

Summary of the Working Group on Wakefield Codes 351
   K. C. D. Chan and R. Cooper

Summary of the Working Group on Transport Codes 352
   C. R. Eminhizer

List of Partcipants 355

Preface

On January 19–21, 1988, the Workshop on Linear Accelerator and Beam Optics Codes was held at the Kona Kai Beach and Tennis Resort on Shelter Island in San Diego, California. The purpose of the workshop was to bring together researchers from universities, national laboratories, and private industry who are developing and using computer codes for accelerator and beam optics design. The workshop was attended by 133 participants: 71 from national laboratories, 34 from private industry, 14 from academic institutions, 10 from foreign countries and 4 from government agencies.

The development of computer codes for accelerator and beam optics system design has paralleled the tremendous progress in this technology. With accelerators costing hundreds of millions to billions of dollars, the development of computer codes to design, simulate, and control these machines has become indispensable. Simulation codes have permitted design studies that would have been prohibitively expensive otherwise. In striving to produce high-energy and high-current beams with very low emittance, the need for sophisticated codes has grown. This workshop was held to promote the development and interchange of new ideas, to introduce new researchers in this area, to broaden the knowledge of those engaged in research in this area, and to discuss, to stimulate, to recommend, and to promote code development.

The agenda covered a wide range of computer codes including those used to design magnets, accelerating structures such as RFQ, linacs and low- and high-energy beam transport systems. Methods such as particle-in-cell (PIC), Lie algebra, and differential algebra were presented. Codes used in a wide range of applications such as the design of rf cavities, power supplies, magnets, ion sources, accelerator simulation, and control codes were discussed. Codes used to study special effects such as wake fields, space charge and higher order nonlinear effects were included.

We are grateful, for support of the workshop and publication of these proceedings, to the United States Department of Defense and Department of Energy and the La Jolla Institute. In particular, we thank Dr. Michael Lavan, Director of the Directed Energy Weapons Directorate of the United States Army Strategic Defense Command in Huntsville, Dr. David Sutter of the Office of Energy Research at DOE in Germantown, and Dr. Adolf Hochstim, President of La Jolla Institute.

We gratefully acknowledge the contributions of Major Richard Abbott of USA/SDC in Huntsville and Professor Alex Dragt of the University of Maryland, who were instrumental in initiating and promoting this workshop. A special thanks to the members of the organizing committee: Drs. David Carey (FNL), Richard Cooper (LANL), George Gillespie (PDI), and William Hermannsfeldt (SLAC). We also thank the session and workshop chairmen for a job well done.

A very special thanks to the local conference organizers: Grace Pitts (LJI) and Rosalie Rocher (LJI). Grace Pitts made arrangements and coordinated meeting activities with the Kona Kai Beach and Tennis Resort, handled participant registration, and provided administrative assistance during, before and after the meeting. Rosalie Rocher prepared the workshop agenda, prepared the participant lists, and assisted the editor in the processing of the manuscripts.

PARMILA — AN OVERVIEW*

G. P. Boicourt
Los Alamos National Laboratory, Los Alamos, NM 87545, USA

ABSTRACT

A tutorial overview of the drift tube linac beam dynamics code, PARMILA, is presented. Its structure and operation are described and its origins and applications are reviewed briefly. A major application, the evaluation of the effects of errors in construction and deviations from design operating conditions, and the extent to which these effects can be addressed by PARMILA are discussed briefly. The user friendliness, accuracy, options in the code, and improvements to the code are described. References to this material are provided for the benefit of the reader.

WHAT IS PARMILA?

The name PARMILA is an acronym for "Phase And Radial Motion In Linear Accelerators." I suggest that it should be remembered by "Phase And Radial Motion in Ion Linear Accelerators," because there now exists a PARMELA code that stands for "Phase And Radial Motion in Electron Linear Accelerators."

PARMILA is a multiparticle simulation code that generates a drift-tube linac, generates an input beam, and moves the beam through the linac. It contains detailed nonlinear modeling with the accelerator's longitudinal coordinate as the independent variable. It is written in FORTRAN, and is considered essentially a second-order code. It is modularized and uses the chain-matrix method to pass collections of particles through the system under study. The code is six-dimensional except for the space-charge calculation. For the space-charge calculation, the ion beam is treated as if it were a function of radius and longitudinal position only. Each particle is assumed to be a ring of charge that contributes to the electric field on a two-dimensional $R - Z$ mesh. This is sometimes termed a 2-1/2 dimensional treatment. The mesh can be held fixed for an entire run or may be readjusted at fixed intervals set by the user.

APPLICATIONS OF PARMILA

PARMILA is used to design and study the beam dynamics of drift tube linacs (DTL's) and low- and high-energy beam-transport lines (LEBTs and HEBTs). A major use is to study the effect of errors, that is, deviations from design values, on the transported or accelerated beam. Also, it has been used to

---

* Work supported by the U.S. Department of Energy, Office of High Energy and Nuclear Physics.

predict beam spill. In this connection, it is difficult to determine its accuracy. At present no adequate beam spill prediction method exists, so even an order of magnitude estimate is valuable. Therefore, the results produced by PARMILA, however limited, should be appreciated.

HISTORY

In the discussion following Don Swenson's presentation,[1] Fred Mills indicates that the starting point for PARMILA was a report by W. K. H. Panofsky.[2] Besides himself, he credits Phil Morton,[1,3] Don Swenson,[1,4] and Don Young,[1,4] with extending Panofsky's work. PARMILA, in its original form, was written by Don Swenson at MURA (Midwestern Universities Research Association) about 1963.[4] It was brought to Los Alamos by him about 1964. At Los Alamos, it was expanded and improved. The major contributors were K. R. Crandall, R. A. Jameson, R. S. Mills, J. E. Stovall, and D. A. Swenson. It is still being modified and continues to grow as models are improved and capabilities are expanded.

VERSIONS

PARMILA has been distributed worldwide over the course of time, and it is likely that the code has experienced many modifications producing many versions. This paper describes the Los Alamos Group AT-6 version. The closest other version, Los Alamos Group AT-1, was the source of the AT-6 version in 1979. While developing the AT-6 version, there was a strong attempt to preserve the modeling of the AT-1 version, but not the coding. The motivation for developing this new version was to model the FMIT (Fusion Materials Irradiation Test) accelerator, which was to accelerate a 100-mA deuteron beam to 35 MeV, to operate cw for 20 years and to allow hands-on maintenance. Because of this maintenance requirement, beam spill became one of the most important design considerations. A PARMILA version was needed that could handle large numbers of pseudo-particles efficiently in order to estimate the amount and location of any beam spill.

The AT-6 version has been optimized for a CRAY environment. The code was restructured to allow efficient vectorization, and we estimate now that the present code is about 10 times faster than the initial version. Some of the recoding will transfer without problems to other machines, but the vector merge operations would have to be changed for use on a nonvector machine.

Other differences of the AT-6 version include its ability to run larger numbers of particles and to simulate more error conditions, but it has fewer output capabilities than the original version. Its common structure has been extensively revised, and some new capabilities have been added. For instance, it can handle separated quadrupole singlets contained in a common drift tube including individual quadrupole alignment and strength errors.

PRINCIPAL TASKS PERFORMED BY PARMILA

The principal tasks performed by PARMILA can be grouped under the following headings.
- The generation of a drift-tube linac
- The generation of an input beam
- Beam dynamics through one or more of the following:
  ▷ LEBT line
  ▷ the DTL
  ▷ HEBT line
- The dumping of results to various files for study and post processing.

The user directs the work of PARMILA by entering labels followed by appropriate data in the input file. Table I lists the labels that PARMILA presently recognizes.

Table I. Task direction labels currently recognized by PARMILA

ADJUST	BEGIN	CHANGE	CHARGE	ELIMIT	END
ERROR	ERSPQD	INPUT	LINAC	LINOUT	OPTCON
OUTPUT	PHLAW	PHSHFT	RUN	SCHEFF	SFDATA
SPQUAD	START	STEER	STOP	TANK	TITLE
TNKERR	TRANS1	TRANS2			

DTL GENERATION

From an rf standpoint, a DTL is a series of cells all of which oscillate at the same frequency in the TM_{010} mode. PARMILA must be supplied with data for a sample set of cells in the accelerator energy range. These data depend on the general cell geometry chosen for the DTL and are obtained from runs of an rf cavity code such as SUPERFISH.[5,6] The geometrical assumptions used to make the rf cavity code runs determine all but three of the individual cell parameters. Figure 1 shows a schematic of a typical DTL cell. Some of the geometrical values that are assumed for the rf code runs are tank diameter D, drift-tube diameter SD, bore radius RH, etc. The three values PARMILA must determine are the cell length L_n, the gap length G_n, and the gap center displacement Y_n. PARMILA calculates these values in sequence. Each set of three values is calculated iteratively as described in Refs. 7 and 8.

Synchronous phase is set according to information input before DTL generation. Quadrupole lengths are set according to information following a TANK label. Quadrupole gradients are calculated from a straight line in the stability diagram. The user must supply the slope and intercept of the line [4,9,10] as part of the data following a TANK label.

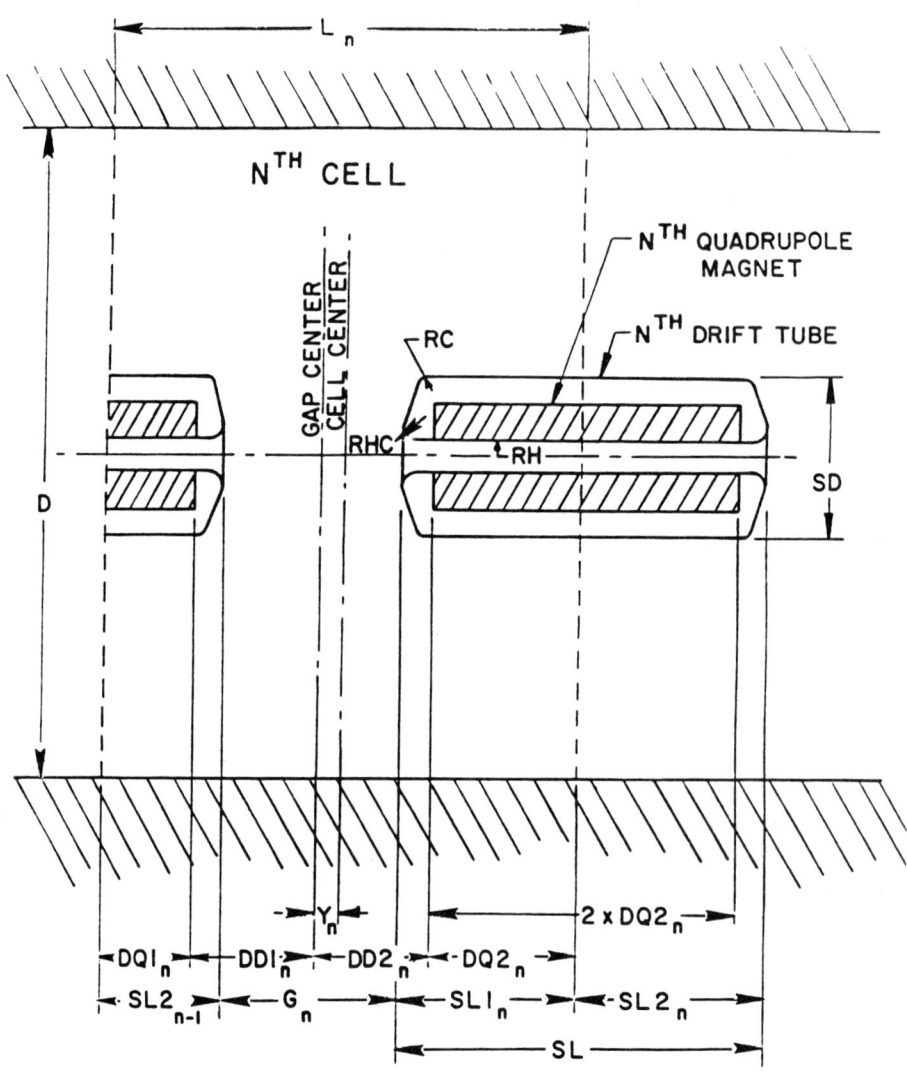

Fig. 1. Schematic of a typical DTL cell.

After generation, the user can change quad lengths and/or gradients, cell lengths, rf field amplitudes, etc. This is done according to directions PARMILA finds following CHANGE labels.

During the DTL generation, PARMILA also calculates and stores quantities that will be used later for beam-dynamic calculations. Prints of the geometrical and beam dynamical properties of the design can be requested. Figures 2 and 3 show portions of such printouts.

INPUT BEAM

As far as PARMILA is concerned, a particle is a set of six coordinates, X, X', Y, Y', φ and W. An input beam is a collection of such sets. PARMILA can read in an externally generated beam or it can generate one internally. PARMILA's internal generation capabilities include[14]:
- Five types of hyperspace
- Each hyperspace distribution can be
  ▷ Uniform,
  ▷ Gaussian to 3σ,
  ▷ Gaussian to 4σ,
  ▷ Conical, or
  ▷ Parabolic.

Other specialized types are allowed as well as beams made up of combinations of types.

LEBT AND HEBT LINE DYNAMICS

Transport lines are entered element by element using TRANS1 or TRANS2 labels in the PARMILA input file. TRANS1 labels define a LEBT and TRANS2 labels are used for a HEBT. The inclusion of LEBT line dynamics is triggered by a zero following the START label. HEBT dynamics are done if the STOP label indicates a cell number greater than the number of cells in the DTL. PARMILA moves the beam through the transport line element by element using a chain-matrix method. Extended elements, for example drifts or quadrupoles, include one or more space-charge calculations. Output can be requested at the end of any element. PARMILA currently recognizes the following transport line elements:

drift	quadrupole
bending magnet	buncher
circular aperture	rectangular aperture
thin lens	solenoid lens
steering magnet	special user defined element

linout subroutine no. 2   geometric parameters

tank no.   1   tank length   238.806 centimeters   40 cells

cell number	kinetic energy	cell length	sl1	sl	quad length	ezero mv/m	total length
	beta	gap length	sl2	y	quad gradient	phis	stem center
initial	2.0700			1.8820	2.3389		0.000
	0.0663		1.8820		19.5960		0.000
1	2.1264	4.7058	1.9095	3.8151	2.3705	1.9470	4.706
	0.0672	0.9143	1.9056	0.0138	-19.3615	-40.0000	4.704
2	2.1838	4.7691	1.9332	3.8623	2.4021	1.9470	9.475
	0.0681	0.9303	1.9291	0.0138	19.1320	-40.0000	9.473
3	2.2421	4.8325	1.9568	3.9095	2.4339	1.9470	14.307
	0.0690	0.9466	1.9527	0.0138	-18.9064	-40.0000	14.305
4	2.3013	4.8961	1.9804	3.9569	2.4657	1.9470	19.203
	0.0699	0.9629	1.9764	0.0139	18.6851	-40.0000	19.201
5	2.3615	4.9599	2.0042	4.0044	2.4976	1.9470	24.163
	0.0708	0.9792	2.0001	0.0139	-18.4688	-40.0000	24.161
6	2.4226	5.0238	2.0280	4.0518	2.5296	1.9470	29.187
	0.0717	0.9957	2.0238	0.0139	18.2575	-40.0000	29.185
7	2.4847	5.0879	2.0518	4.0992	2.5617	1.9470	34.275
	0.0726	1.0123	2.0475	0.0140	-18.0514	-40.0000	34.273
8	2.5477	5.1521	2.0755	4.1466	2.5938	1.9470	39.427
	0.0735	1.0292	2.0711	0.0140	17.8495	-40.0000	39.425
9	2.6116	5.2165	2.0992	4.1941	2.6260	1.9470	44.644
	0.0745	1.0461	2.0949	0.0140	-17.6517	-40.0000	44.642
10	2.6764	5.2810	2.1230	4.2416	2.6583	1.9470	49.925
	0.0754	1.0631	2.1186	0.0141	17.4579	-40.0000	49.923
11	2.7422	5.3456	2.1468	4.2892	2.6906	1.9470	55.270
	0.0763	1.0802	2.1424	0.0141	-17.2680	-40.0000	55.268
12	2.8089	5.4104	2.1706	4.3368	2.7230	1.9470	60.681
	0.0772	1.0974	2.1661	0.0141	17.0819	-40.0000	60.678
13	2.8766	5.4753	2.1945	4.3843	2.7555	1.9470	66.156
	0.0781	1.1147	2.1899	0.0142	-16.8993	-40.0000	66.154
14	2.9452	5.5403	2.2183	4.4319	2.7880	1.9470	71.696
	0.0790	1.1321	2.2136	0.0142	16.7210	-40.0000	71.694
15	3.0147	5.6053	2.2421	4.4794	2.8205	1.9470	77.302
	0.0800	1.1497	2.2373	0.0142	-16.5465	-40.0000	77.299
16	3.0851	5.6705	2.2658	4.5269	2.8531	1.9470	82.972
	0.0809	1.1673	2.2610	0.0142	16.3755	-40.0000	82.970
17	3.1565	5.7358	2.2896	4.5743	2.8858	1.9470	88.708

Fig. 2. Portion of geometrical parameter print. Definitions of sl1, sl2, sl, and y are in Fig. 1.

1 linout subroutine no. 1   dynamical parameters

tank no. 1   tank length 238.806 centimeters   40 cells   power= 0.11996 mw   frequency= 425. mhz

cell number	kinetic energy	beta	length	t	tp	s	sp	quad length	quad gradient	ezero mv/m	phis	total length
initial	2.0700000	0.06632										0.000
1	2.1264361	0.06721	4.7058	0.8041	0.0556	0.4565	0.0520	2.3389	19.5960	1.9470	-40.00	4.706
2	2.1837837	0.06811	4.7691	0.8062	0.0552	0.4547	0.0520	2.3705	-19.3615	1.9470	-40.00	9.475
3	2.2420726	0.06901	4.8325	0.8087	0.0545	0.4530	0.0525	2.4021	19.1320	1.9470	-40.00	14.307
4	2.3013168	0.06991	4.8961	0.8113	0.0538	0.4511	0.0530	2.4339	-18.9064	1.9470	-40.00	19.203
5	2.3615103	0.07081	4.9599	0.8137	0.0533	0.4492	0.0530	2.4657	18.6851	1.9470	-40.00	24.163
6	2.4226449	0.07172	4.9238	0.8159	0.0528	0.4475	0.0530	2.4976	-18.4688	1.9470	-40.00	29.187
7	2.4846988	0.07263	5.0879	0.8177	0.0523	0.4459	0.0530	2.5296	18.2575	1.9470	-40.00	34.275
8	2.5476773	0.07354	5.1521	0.8196	0.0518	0.4443	0.0530	2.5617	-18.0514	1.9470	-40.00	39.427
9	2.6115860	0.07446	5.2165	0.8214	0.0513	0.4428	0.0530	2.5938	17.8495	1.9470	-40.00	44.644
10	2.6764280	0.07537	5.2810	0.8232	0.0507	0.4412	0.0530	2.6260	-17.6517	1.9470	-40.00	49.925
11	2.7422069	0.07629	5.3456	0.8250	0.0502	0.4397	0.0530	2.6583	17.4579	1.9470	-40.00	55.270
12	2.8089278	0.07721	5.4104	0.8268	0.0497	0.4382	0.0530	2.6906	-17.2680	1.9470	-40.00	60.681
13	2.8765962	0.07813	5.4753	0.8286	0.0492	0.4366	0.0530	2.7230	17.0819	1.9470	-40.00	66.156
14	2.9451929	0.07905	5.5403	0.8301	0.0488	0.4353	0.0530	2.7555	-16.8993	1.9470	-40.00	71.696
15	3.0147049	0.07997	5.6053	0.8315	0.0486	0.4341	0.0530	2.7880	16.7210	1.9470	-40.00	77.302
16	3.0851361	0.08089	5.6705	0.8328	0.0483	0.4330	0.0530	2.8205	-16.5465	1.9470	-40.00	82.972
17	3.1564903	0.08182	5.7358	0.8341	0.0481	0.4318	0.0530	2.8531	16.3755	1.9470	-40.00	88.708
18	3.2287712	0.08275	5.8011	0.8354	0.0478	0.4306	0.0530	2.8858	-16.2077	1.9470	-40.00	94.509
19	3.3019826	0.08367	5.8665	0.8367	0.0475	0.4295	0.0530	2.9184	16.0431	1.9470	-40.00	100.375
20	3.3761284	0.08460	5.9320	0.8380	0.0473	0.4283	0.0530	2.9512	-15.8817	1.9470	-40.00	106.307
21	3.4512123	0.08553	5.9977	0.8394	0.0470	0.4271	0.0530	2.9839	15.7232	1.9470	-40.00	112.305
22	3.5272121	0.08647	6.0633	0.8404	0.0467	0.4262	0.0530	3.0168	-15.5677	1.9470	-40.00	118.368
23	3.6041282	0.08740	6.1291	0.8414	0.0464	0.4254	0.0530	3.0496	15.4156	1.9470	-40.00	124.498
24	3.6819636	0.08833	6.1949	0.8424	0.0460	0.4246	0.0530	3.0825	-15.2664	1.9470	-40.00	130.693
25	3.7607211	0.08927	6.2608	0.8434	0.0457	0.4237	0.0530	3.1154	15.1199	1.9470	-40.00	136.953
26	3.8404035	0.09020	6.3268	0.8444	0.0454	0.4229	0.0530	3.1484	-14.9761	1.9470	-40.00	143.280
27	3.9210138	0.09114	6.3928	0.8454	0.0450	0.4221	0.0530	3.1813	14.8348	1.9470	-40.00	149.673
28	4.0025381	0.09207	6.4588	0.8463	0.0448	0.4215	0.0532	3.2143	-14.6961	1.9470	-40.00	156.132
29	4.0849775	0.09301	6.5249	0.8471	0.0447	0.4211	0.0533	3.2474	14.5602	1.9470	-40.00	162.657
30	4.1683341	0.09395	6.5911	0.8479	0.0445	0.4206	0.0536	3.2804	-14.4267	1.9470	-40.00	169.248
31	4.2526104	0.09489	6.6573	0.8488	0.0444	0.4201	0.0538	3.3135	14.2955	1.9470	-40.00	175.905
32	4.3378085	0.09583	6.7235	0.8496	0.0442	0.4196	0.0538	3.3466	-14.1666	1.9470	-40.00	182.628
33	4.4239307	0.09677	6.7898	0.8504	0.0440	0.4191	0.0540	3.3797	14.0400	1.9470	-40.00	189.418
34	4.5109652	0.09771	6.8561	0.8511	0.0439	0.4186	0.0540	3.4129	-13.9155	1.9470	-40.00	196.274
35	4.5989113	0.09865	6.9225	0.8518	0.0438	0.4180	0.0540	3.4461	13.7934	1.9470	-40.00	203.197
36	4.6877706	0.09959	6.9889	0.8525	0.0436	0.4175	0.0540	3.4793	-13.6734	1.9470	-40.00	210.186
37	4.7775449	0.10053	7.0553	0.8531	0.0435	0.4170	0.0540	3.5125	13.5554	1.9470	-40.00	217.241
38	4.8682361	0.10147	7.1218	0.8538	0.0434	0.4164	0.0540	3.5457	-13.4395	1.9470	-40.00	224.363
39	4.9598459	0.10242	7.1883	0.8545	0.0432	0.4159	0.0540	3.5789	13.3254	1.9470	-40.00	231.551
40	5.0523761	0.10336	7.2548	0.8551	0.0431	0.4154	0.0540	3.6122	-13.2132	1.9470	-40.00	238.806
								3.6454	13.1029			

Fig. 3. Portion of PARMILA dynamical parameter printout.

BEAM DYNAMICS IN THE DTL

PARMILA proceeds through the DTL cell by cell. Each cell starts at the center of a quadrupole and ends at the center of the next downstream quadrupole. The PARMILA cell is shown in Fig. 1. The beam is moved through the cell in five steps.[4,5] The first step is a hard-edged quadrupole transformation through the distance $DQ1_n$, then a drift of length $DD1_n$ to the gap center position. At this point a gap transformation is applied. The cell transit is completed by a drift through the length $DD2_n$ followed by a hard-edge quadrupole transformation through the distance $DQ2_n$ to the end of the cell at the downstream quad center.

Appropriate transformations occur at the entrance and exit of each tank to move the beam to the start of the first cell and from the end of the last cell across the intertank spacing to the next tank. A space-charge calculation can be performed at the gap center position, the quadrupole center position, or at both positions. Output can be requested at the end of cells which correspond to the centers of the quadrupoles.

OUTPUT

The output processing represents a major difference between the AT–1 and AT–6 versions of PARMILA. The original PARMILA version dumped the entire set of beam coordinates to a disk file at each position where output was requested. A postprocessor, OUTPROC, was then used to obtain results. Because the AT–6 version was developed specifically to run enormous numbers of pseudoparticles, such dumps to disk would gobble a huge amount of disk space during the run, and the large file produced would be difficult to save after the run. To get around these problems, the vectorized version limits the number of full beam dumps. Instead it does "on-the-fly" calculations at the desired output positions and then dumps the results to various output files for printing, plotting, or further post-processing. The codes to do the plotting and postprocessing are not as fully developed as OUTPROC and are constantly being enlarged and improved.

Figures 4 through 8 show some examples of PARMILA output. Figure 4 shows phase-space scatter plots at cell 3 of a hypothetical DTL. Scales for the plots are obtained either from data following the OPTCON label or by a default procedure. Because the plotting of all points is time consuming, only the default number of points, 100, is plotted. The number of points can be changed using the OPTCON label. Transverse and longitudinal rms emittance plots are shown in Figs. 5 and 6. Figures 7 and 8 are profile plots in the transverse coordinates X and Y and in phase and energy. These plots are made by postprocessing the output TAPE26 file using the postprocessor code DTLPLOT.

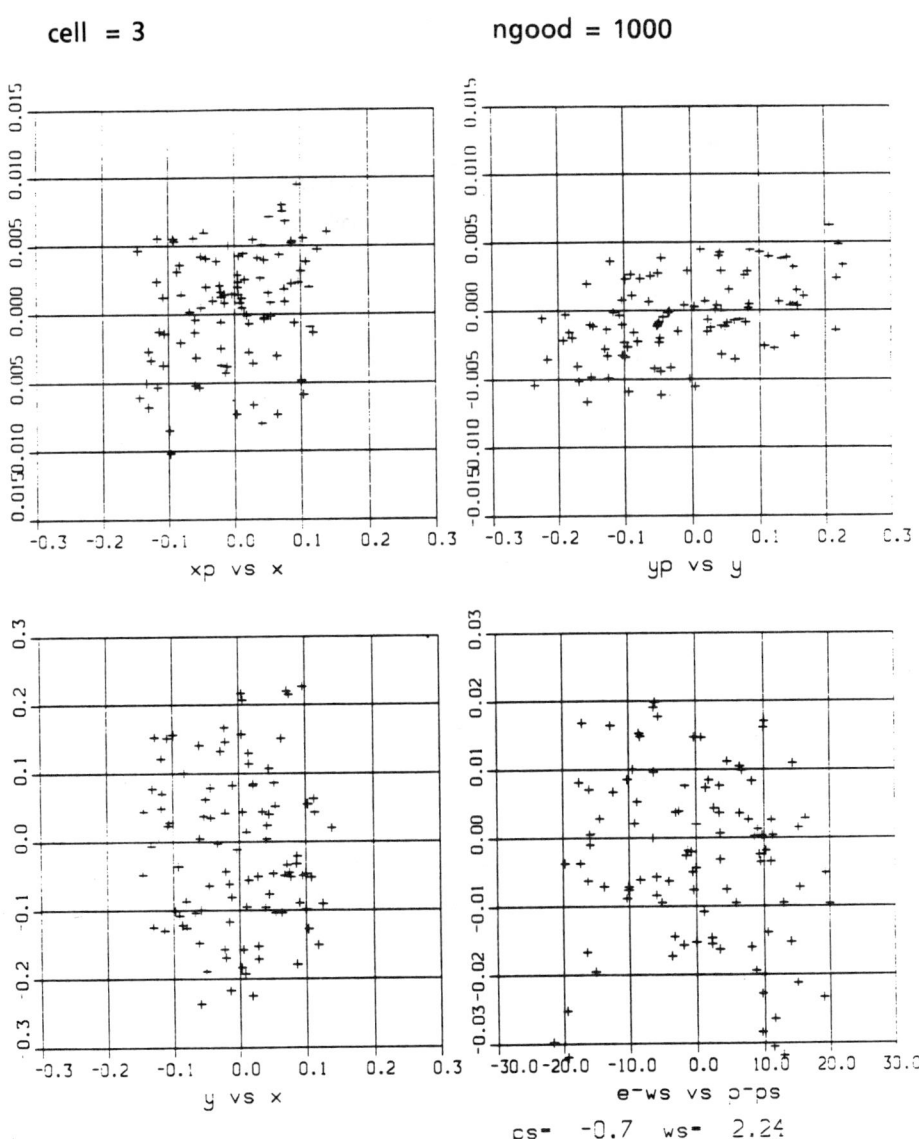

Fig. 4. Phase-space scatterplots.

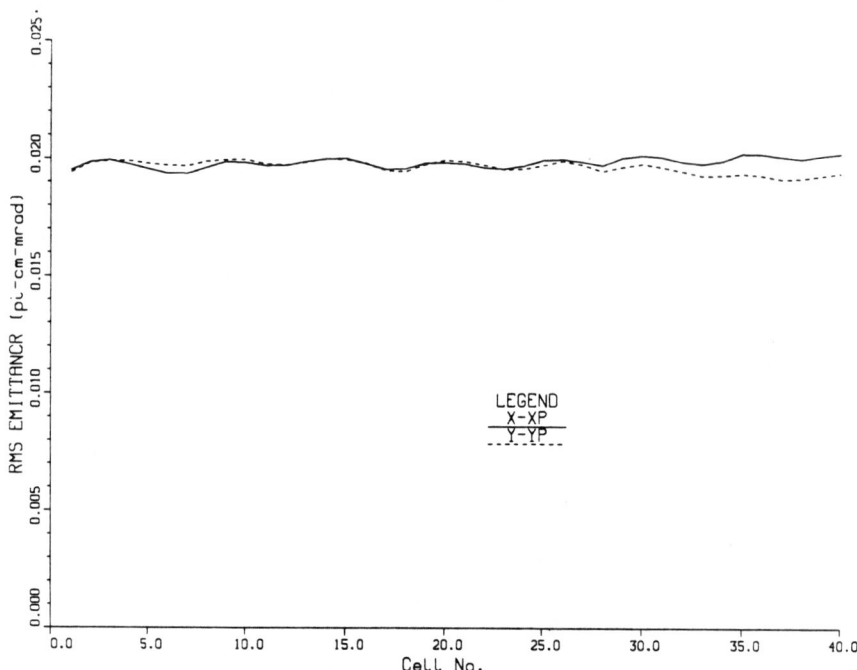

Fig. 5. Transverse rms emittance plots.

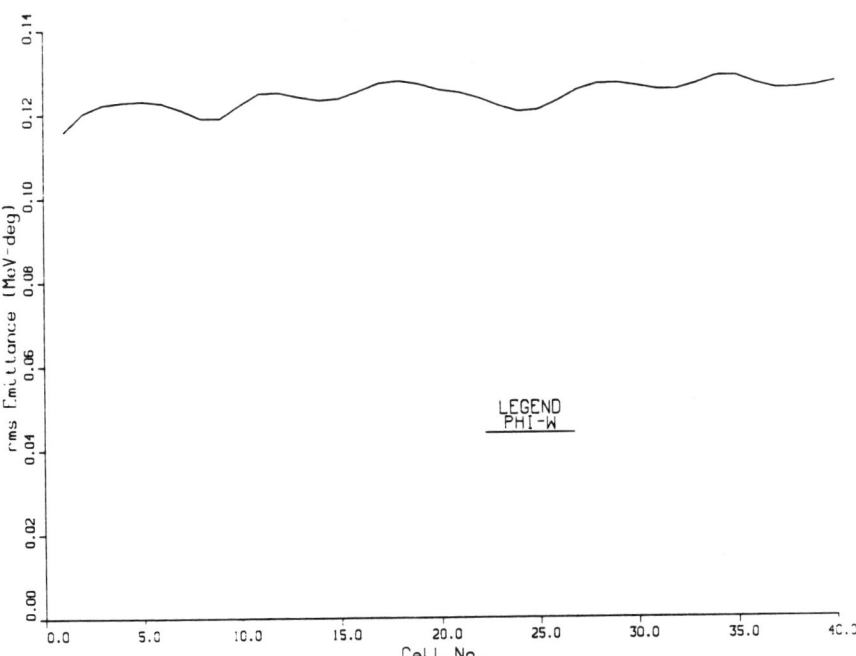

Fig. 6. Longitudinal rms emittance plot.

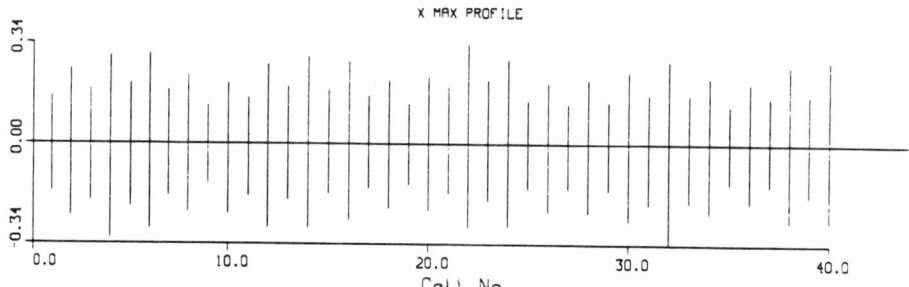

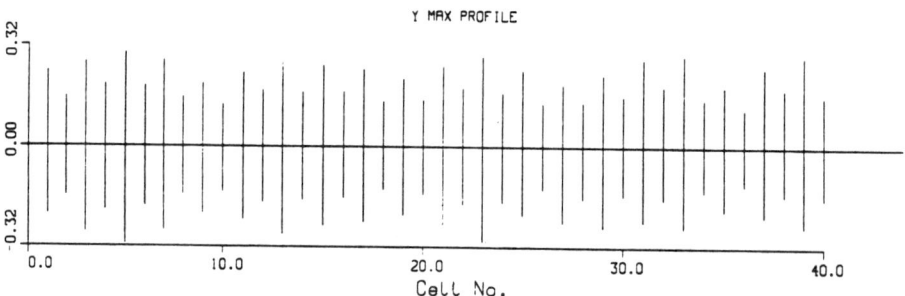

Fig. 7. Transverse profiles.

parmila short input file

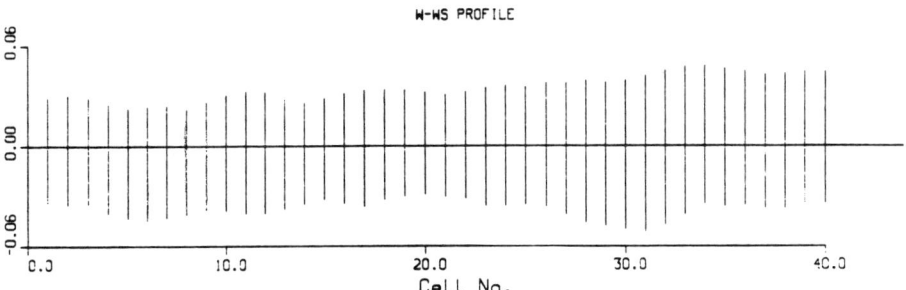

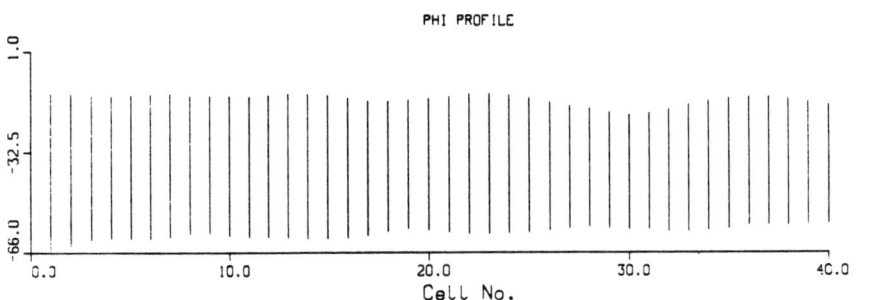

Fig. 8. Longitudinal profiles.

ERROR CALCULATIONS

The presence of errors in an accelerating system is inevitable. By errors, I mean deviations from the specified design operating conditions including misalignments and rf fluctuations. Accelerators must be designed so that the actual output beam properties are within specified limits and so that beam-spill losses are manageable. Part of the design procedure for a linac is an error study to predict the effects of errors on the output beam and beam spill and to make sure that the real operation of the linac will be within the required limits when the expected errors are present.

PARMILA's error handling capabilities allow the assessment of error produced effects of single error types and of combinations of error types. PARMILA handles the following types of errors

- quadrupole misalignments
  ▷horizontal and vertical displacements
  ▷pitch, yaw, and roll
- tank misalignments
  ▷horizontal and vertical displacements
  ▷pitch and yaw
- quadrupole exitation errors
- rf phase errors
- rf amplitude errors

Limits on the errors are input using ERROR, ERSPQD, and TNKERR labels. The code picks its applied errors randomly within the allowed limits. The error limits may be different for different tanks. Quadrupole errors applying to the first tank are also applied to quadrupoles in the LEBT if the sixth entry on the quadrupole's TRANS1 card is greater than 0. Similarly, quadrupoles in the HEBT can be given the same errors as the quadrupoles in the last tank. Fixed and specified rf phase and amplitude errors can be applied to individual tanks.

USER FRIENDLINESS

Since the advent of personal computers, the term "user friendly" has taken on a meaning of genie-like qualities. Some of us may remember when FORTRAN coding rather than machine-language coding made computers user friendly. Thus, user friendliness is in the eye of the user. In the case of PARMILA, an accelerator knowledgeable user will have less trouble using it than an accelerator novice. It is also true that, as the input forms become more familiar, the code seems more friendly. In some people's lexicon, user friendliness is catching input errors. In this sense, PARMILA is not user friendly.

Perhaps the best way to determine friendliness is to let the prospective users make up their own minds, so I will present a short example. In the following, I want to examine the short input file that was used to produce Figs. 4 through 8. The file is shown in Fig. 9.

ENTIRE INPUT FILE WITH SPACES SEPARATING FUNCTIONS

```
run 1
title
parmila short input file

linac 1 2.07 425. 1.000
tank 1 05.0 -40. .01947 0.0 .000 0 0 2.3100 +7.0000
      -.5 1 1   11 0 0.50 0.0 00 0
sfdata  .0659  .8030 .056   .458   .052  .19358  69.49 0. 0. 0.
        .0680  .8080 .055   .454   .052  .19541  70.68 0. 0. 0.
        .0690  .8110 .054   .452   .053  .19633  70.07 0. 0. 0.
        .0709  .8160 .053   .448   .053  .19791  71.52 0. 0. 0.
        .0744  .8230 .051   .442   .053  .20093  73.03 0. 0. 0.
        .0780  .8300 .049   .436   .053  .20390  73.69 0. 0. 0.
        .0851  .8400 .047   .427   .053  .20973  75.52 0. 0. 0.
        .0907  .8460 .045   .422   .053  .21433  76.98 0. 0. 0.
        .0964  .8510 .044   .419   .054  .21888  78.23 0. 0. 0.
        .1035  .8560 .043   .415   .054  .22451  79.55 0. 0. 0. -1

linout 2 1
output 2 1 1 13 0 0 1 40 1
output 5 1
output 1 1 1 13 0 1 40
optcon 1.8 .042 1.8 .042 1.8 1.8 1.8 .042 1.8 .042 0 0 0 0 0 0 00 00

input -6 1000 0.00 31.00 .0029 0.00 7.32 .0029 22. .041 0.
scheff 100. .500 .50 10 20 0 2 1

start 1
stop 40

begin
end
```

Fig. 9. PARMILA short input file.

The six functions performed by this file are separated by blank lines. (The presence of blank lines in an actual input file would cause difficulties.) I will go over the functions individually. The first three lines, shown in Fig. 10, serve as identification. The next three labels, Fig. 11, with their accompanying data define the drift-tube linac. The LINAC and TANK labels are followed by a set number of entries that define frequency, number of tanks, input energy, quadrupole configurations, etc. The SFDATA label is followed by two or more sets of data obtained from rf cavity code runs. As described above, this data, along with the data preceding it, allows PARMILA to calculate energy gain, cell lengths, gap lengths, and gap positions.

FUNCTION = IDENTIFICATION

```
run 1
title
parmila short input file
```

Fig. 10. The RUN and TITLE labels serve to enter a run number and a run title.

FUNCTION = DTL SPECIFICATION

```
linac 1 2.07 425.  1.000
tank  1 05.0 -40.  .01947 0.0  .000 0 0 2.3100 +7.0000
      -.5  1  1   11 0 0.50 0.0  00 0
sfdata  .0659  .8030  .056  .458  .052  .19358  69.49 0. 0. 0.
        .0680  .8080  .055  .454  .052  .19541  70.68 0. 0. 0.
        .0690  .8110  .054  .452  .053  .19633  70.07 0. 0. 0.
        .0709  .8160  .053  .448  .053  .19791  71.52 0. 0. 0.
        .0744  .8230  .051  .442  .053  .20093  73.03 0. 0. 0.
        .0780  .8300  .049  .436  .053  .20390  73.69 0. 0. 0.
        .0851  .8400  .047  .427  .053  .20973  75.52 0. 0. 0.
        .0907  .8460  .045  .422  .053  .21433  76.98 0. 0. 0.
        .0964  .8510  .044  .419  .054  .21888  78.23 0. 0. 0.
        .1035  .8560  .043  .415  .054  .22451  79.55 0. 0. 0. -1
```

Fig. 11. The LINAC, TANK, and SFDATA labels are followed by data that specifies the DTL.

The third set of labels, Fig. 12, is used to get output. The LINOUT label (in this case) triggers a call to a routine that prints the tables shown in Figs. 2 and 3. The OUTPUT labels are used to specify types and positions of output. The OPTCON label supplies scales for plotting.

FUNCTION = OUTPUT SPECIFICATION

```
linout  2 1
output  2 1 1 13 0 0 1 40 1
output  5 1
output  1 1 1 13 0 1 40
optcon 1.8  .042 1.8  .042 1.8 1.8 1.8  .042 1.8  .042 0 0 0 0 0 0 00 00
```

Fig. 12. The LINOUT, OUTPUT, and OPTCON labels specify types, positions, and forms of output.

The fourth set of labels, shown in Fig. 13, defines the input beam for the run. The INPUT label specifies the distribution type and the number of pseudoparticles to be used; it gives the geometrical limits inside which the particles are to lie. The SCHEFF label specifies the beam current, defines the space-charge mesh, and tells where and how the space-charge calculation is to be computed.

FUNCTION = BEAM SPECIFICATION

```
input -6 1000 0.00 31.00  .0029 0.00 7.32  .0029 22.  .041 0.
scheff 100.  .500  .50 10 20 0 2 1
```

Fig. 13. The INPUT and SCHEFF labels define the input beam.

The final two functions are simple but necessary. The two lines in Fig. 14 tell PARMILA to start the calculation at cell 1 and to end it at cell 40. In Fig. 15, the BEGIN label tells PARMILA that all the desired input has been read and that the computation should begin. After the calculation has been made, PARMILA encounters the final END label; being told that nothing further is wanted, it ends.

FUNCTION = SPECIFY START AND STOP POSITIONS

start 1
stop 40

Fig. 14. The START and STOP labels serve to tell PARMILA the positions where the calculation should start and stop.

FUNCTION = SPECIFY COMPUTATION START AND STOP

begin
end

Fig. 15. The BEGIN label causes PARMILA to do the calculation. On completion of this task, PARMILA encounters the END label and quits.

I do not want to leave the impression that this input file is typical. The file from which it was made (Fig. 16) was more than twice as long. The extra lines serve to refine the linac design and to add LEBT and HEBT lines. Multiple runs may also increase the input file length.

The present user manual situation **does not help** PARMILA's user friendliness. Swenson and Stovall[4] started to write a manual for PARMILA about 1979. This incomplete manual exists, but it is out of date. A manual is in preparation for the AT-6 version, but progress on it has been slow for two reasons. First, we are doing a certain amount of code rewriting, both to add new features and to simplify the input; this cuts into the manual-writing effort. Second, there is no funding for the manual so the writing is being done as a labor of love. We expect to have something available in about a year.

```
run 2
title
parmila example input file
linac 1 2.07 425. 1.000
tank 1 05.0 -40. .01947 0.0 .000 0 0 3.000 -3.000
    -.5 1 1   11 0 0.50 0.0 00 0
sfdata  .0659  .8030 .056  .458   .052  .19358  69.49 0. 0. 0.
        .0680  .8080 .055  .454   .052  .19541  70.68 0. 0. 0.
        .0690  .8110 .054  .452   .053  .19633  70.07 0. 0. 0.
        .0709  .8160 .053  .448   .053  .19791  71.52 0. 0. 0.
        .0744  .8230 .051  .442   .053  .20093  73.03 0. 0. 0.
        .0780  .8300 .049  .436   .053  .20390  73.69 0. 0. 0.
        .0851  .8400 .047  .427   .053  .20973  75.52 0. 0. 0.
        .0907  .8460 .045  .422   .053  .21433  76.98 0. 0. 0.
        .0964  .8510 .044  .419   .054  .21888  78.23 0. 0. 0.
        .1035  .8560 .043  .415   .054  .22451  79.55 0. 0. 0. -1
change 10 19178.5  1  1 +1
change 10 19210.2  1  1
change 10 19235.6  2  2
change 10 18983.6  3  3
change 10 19162.3  4  4
change 10 19182.1  5  5
change 10 18846.0  6  6
change 10 19255.3  7  7
change 10 19230.5  8  8
change 10 19323.3  9  9
change 10 19453.8 10 10
change 10 19466.2 11 11
change 10 19360.2 12 12
change 10 19371.2 13 13
change 10 19381.2 14 14
change 10 19350.7 15 15
change 10 19320.7 16 16
change 10 19211.0 17 17
change 10 19296.3 18 18
change 10 19343.1 19 19
change 10 19309.6 20 20
change 10 19277.0 21 21
change 10 19086.0 22 22
change 10 19170.3 23 23
change 10 19136.1 24 24
change 10 19218.5 25 25
change 10 19222.7 26 26
change 10 19187.9 27 27
change 10 18955.8 28 28
change 10 19195.1 29 29
change 10 19159.3 30 30
change 10 19083.5 31 31
change 10 19125.4 32 32
change 10 18971.4 33 33
change 10 18698.0 34 34
change 10 19133.0 35 35
change 10 19056.1 36 36
change 10 19058.2 37 37
change 10 18903.3 38 38
change 10 18905.1 39 39
change 10 18957.5 40 40
change 11 2.54 1 40 1
trans1 1  1  1.72  1.     .5 0 1 00 13
trans1 2  3 +19236.8 2.54 .5 1 1 00 13 quad
trans1 3  1  .883945 1 .5 00.0 1 00 13 drift
trans1 4  2 0.0000 -90 1 .5 1 00 13 buncher
trans1 5  1  .883945 1 .5 0 1 00 13
trans1 6  3 -18407.2 2.54 .5 1 1 00 13
trans1 7  1 2.14 1 .5 0 1 00 13 drift
trans1 8  3 +19178.5 1.27 .5 1 1 00 13 12 half quad
trans2 1  3 18957.5 1.27 .5 0 1 00 13
trans2 2  1 29.6146 3 2.54 00.0 1 00 13
linout 2
start 0
stop 40
elimit 40.
error 0 0 0 0 0 0 0 0 0 1 0 0
output 2 1 12 13 00 00 1 40 1
output 1 1 12 13 12 00 40
optcon 1.8 .042 1.8 .042 1.8 1.8 1.8 .042 1.8 .042 0 0 0 0 0 0 00 00
scheff 00.00 .500 .50 10 20 0 2 1
adjust -10
begin
end
```

Fig. 16. A more typical PARMILA input file.

ACCURACY

The determination of the accuracy of a code is exceedingly difficult. This fact is not generally understood. Most people assume that benchmarking a code is a simple matter of comparing experimental results with code results. Nothing could be further from the truth. The problem is not with the computer codes. The computations are easy. Anything calculated by the code is accessible. A little extra work may be required, but it can be obtained. Results are always repeatable. The DTL and transport lines are exactly defined, operating voltage and phase are exactly set, and the properties of the input beam can be determined exactly.

The problem is in defining and determining experimental quantities accurately. Without them an accurate comparison is impossible. Usually the code user assumes that the experimental conditions are the design conditions, which is a poor assumption. Because there are differences between the design and experimental conditions, disagreements between code results and experimental results do not prove the code is wrong. Conversely, agreement between the code and the experiment does not benchmark the code.

However, the experimentalist should not be blamed. His lot is not a happy one. Most things going on in an accelerator are inaccessible to measurement. The accelerator and associated transport lines are never as designed. At best, the deviations from design are known only to within tolerances. At worst, a slip of a draftsman's pencil or a machining mistake may occur and be included in the final assembly unknown to anyone. Errors of the latter kind, if egregious enough, cause a search for the problem, and they are uncovered. That is why we know they happen. But if such mistakes are present and unknown, they can lead to differences between the code and the experimental results, and the code will be blamed.

There are other problems inherent in the accelerator experiment. Many measurements require the collection of data over a period of time. During the measurement period operating conditions can change. The actual rf amplitude is known to only about $\pm 1\%$, and it varies with time. The rf phase is known to only about $\pm 1°$ and varies with time. The beam ion-source voltage drops during the course of a single macropulse. The currents between bunches can vary by more than 5%. There are time variations in the input match.

Given that there are difficulties, what can be said about PARMILA's accuracy? One of the better recommendations for PARMILA is that DTLs and lines designed using it, work. But there are some less general observations that can be made. PARMILA was used successfully to optimize the longitudinal tuning of LAMPF.[13] In a study by Jameson, Mills, and Sander,[11,12] is the following observation:

"...by combining certain amounts of mismatch and mis-steering, one could ...produce emittance growths in the simulation model which match the observed experimental behavior."

(In the case they studied, measurements of the match and misalignments were not available.) They concluded that PARMILA accurately predicts beam behavior for at least the central 90–95% of the beam.

DESIRABLE OPTIONS AND IMPROVEMENTS

Because this is a code workshop, it seems appropriate to describe some options and improvements that would be nice to have in the PARMILA code. There are a number of accelerator component properties that are suspected or known to have effects on the beam quality. Those properties that are known to be important or are easily modeled are already in the code. However, there are a number of properties that may or may not cause beam problems depending on the level of the property or on its interaction with the properties of other components.

One function of a beam dynamics code is to determine which causal properties produce which effects and, also, to determine at what level the causal properties become important to the beam quality. The inclusion of these causes in a code allows them to be checked either singly or in combination with other causes. Some of the causal properties it would be useful to have in PARMILA include

- overlapping fringe-field soft-edged quads with and without field harmonics
- image charge in drift tubes and beam pipes
- fully 3–D space charge at more locations
- electric field misalignments in gaps

There is another item that would be useful to have; this is a time-dependent version of PARMILA. There are two versions of the RFQ code PARMTEQ, a space-dependent and a time-dependent version. Each code acts as a check on the other. The time-dependent version seems to be able to handle certain non-design operating conditions better than the space-dependent version. This could also be the case with PARMILA.

ACKNOWLEDGMENTS

I would like to thank R. K. Cooper and J. L. Merson for their helpful advice and comments and C. L. Beckmann and L. H. Schilling for their typing and editing help.

REFERENCES

1. D. A. Swenson, "Application of Calculated Fields to the Study of Particle Dynamics," 1964 Linear Accelerator Conference, July 20–24, 1964, MURA report No. 714, p. 328.
2. W. K. H. Panofsky, "Linear Accelerator Beam Dynamics," UCRL–1216, February 1951.
3. P. L. Morton, " Particle Dynamics in Linear Accelerators," MURA report No. 679, December 13, 1963.
4. B. Austin, T. W. Edwards, J. E. O'Meara, M. L. Palmer, D. A. Swenson, and D. E. Young, "The Design of Proton Linear Accelerators for Energies Up to 200 MeV," MURA report No. 713, July 1, 1965.
5. M. T. Menzel and H. K. Stokes, "Users Guide for the POISSON/SUPERFISH Group of Codes," Los Alamos National Laboratory report LA–UR–87–115, January 1987.
6. "Reference Manual for the POISSON/SUPERFISH Group of Codes," Los Alamos National Laboratory report LA–UR–87–126, January 1987.
7. D. A. Swenson and J. Stovall, "PARMILA," Los Alamos Scientific Laboratory internal report MP–3–19, January 1968.
8. D. A. Swenson, "Generation of Geometrical Dimensions for Drift Tube Linacs," Los Alamos National Laboratory internal report MP–3/DAS–1, June 29, 1967.
9. L. Smith and R. L. Gluckstern, "Focusing in Linear Accelerators," Rev. Sci. Instrum. **26** (2), 220 February 1955.
10. J. E. Stovall, "Selection of Quadrupole Strengths for Drift Tube Linac," Los Alamos Scientific Laboratory internal report MP–3–38, July, 3, 1968.
11. R. A. Jameson, R. S. Mills, and O. R. Sander, "Report on Foreign Travel," Los Alamos Scientific Laboratory internal report AT–DO:351(U) MP–9, Dec. 28, 1978.
12. R. A. Jameson, "Emittance Growth in the New CERN Linac–Transverse Plane, Comparison Between Experimental Results and Computer Simulation," Los Alamos Scientific Laboratory private communication, AT–DO–377(U), January 15, 1979.
13. R. A. Jameson, W. E. Jule, R. S. Mills, E. D. Bush, Jr., and R. L. Gluckstern, "Longitudinal Tuning of the LAMPF 201.25–MHz Linac Without Space Charge." Los Alamos Scientific Laboratory report LA–6863, March 1978.
14. R. A. Jameson, "PARMILA Input Subroutine and Particle Distribution Fitting Techniques," Los Alamos Scientific Laboratory internal memo MP–9, June 22, 1978.

PARMTEQ — A BEAM-DYNAMICS CODE FOR THE RFQ LINEAR ACCELERATOR*

Kenneth R. Crandall
AccSys Technology, Inc., P.O. Box 5247, Pleasanton, CA 94566

and

Thomas P. Wangler
Los Alamos National Laboratory, MS-H817, Los Alamos, NM 87545

ABSTRACT

The PARMTEQ code is used for generating the complete cell design of a radio-frequency quadrupole linear accelerator and for multiparticle simulation of the beam dynamics. We present a review of the code, with an emphasis on the physics used to describe the particle motion and the cell generation.

INTRODUCTION

The radio-frequency quadrupole (RFQ) linear accelerator[1] has undergone a remarkable expansion in development and application during the past decade[2] because of its unequaled ability to focus, bunch, and accelerate low-velocity ion beams with high current and low emittance. During this time, the beam-dynamics code PARMTEQ has been used for nearly all the RFQ design work at accelerator laboratories throughout the world.

PARMTEQ (Phase And Radial Motion in Transverse Electric Quadrupole linacs) is a multiparticle code, adapted from the well-known linac beam-dynamics code PARMILA for application to the RFQ. It was written originally at Los Alamos by K. R. Crandall for use in (1) generating the complete cell-by-cell linac design and (2) predicting the beam performance by multiparticle simulation of the dynamics, including space-charge effects. Figure 1 shows the geometry of the unit cell specified by the cell length L_c, the minimum radial aperture a, and the maximum radial aperture ma, where m is the vane modulation parameter. These parameters vary throughout the RFQ as the beam energy and the bunching requirements change.

---

*This work was supported by Los Alamos National Laboratory Program Development funds, under the auspices of the United States Department of Energy.

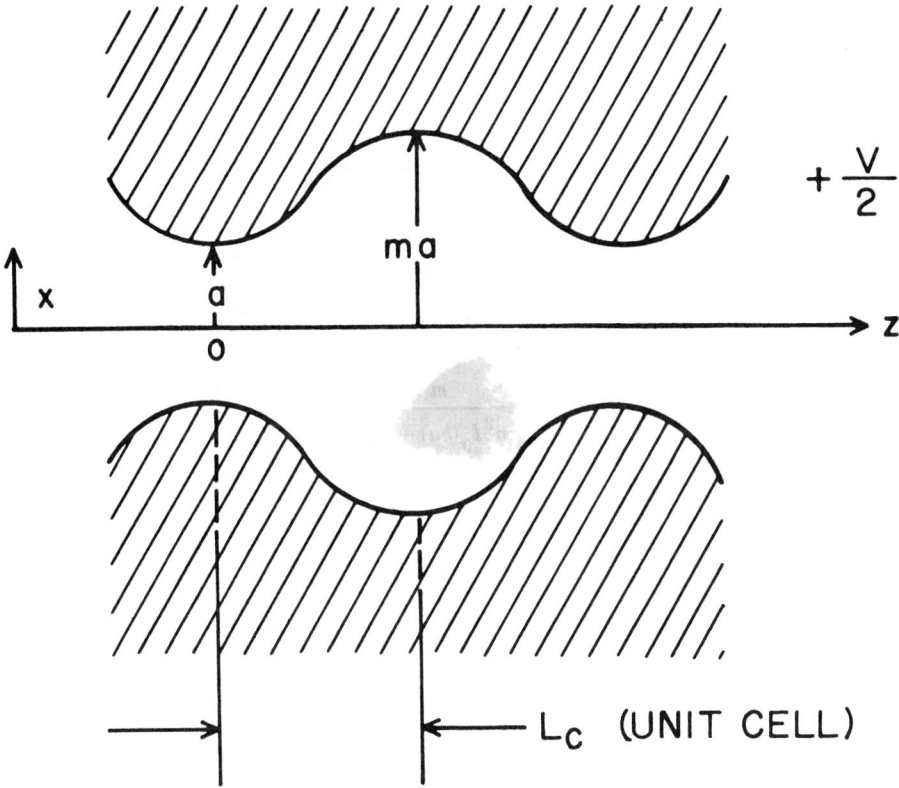

Fig. 1. Geometry of the RFQ unit cell in the x-z plane, where z is the beam direction. The cell length is L_C, the minimum aperture is a, the vane modulation parameter is m, and V is the intervane voltage.

The radio-frequency electric fields in the RFQ are used for focusing, bunching, and acceleration. The fields are described analytically by the two lowest-order terms of a general multipole expansion that is a solution of Laplace's equation.[1,3] The 3-D electric fields used in PARMTEQ for each cell are given in cartesian coordinates x, y, and z as

$$E_x = -\left[\frac{XV}{a^2} + \frac{kAV}{2}\frac{I_1(kr)}{r}\cos kz\right] x \sin \phi, \quad (1)$$

$$E_y = -\left[-\frac{XV}{a^2} + \frac{kAV}{2}\frac{I_1(kr)}{r}\cos kz\right] y \sin\phi, \quad (2)$$

and

$$E_z = \left[\frac{kAV}{2} I_0(kr)\sin kz - \frac{d}{dz}\left(\frac{XV}{a^2}\right)\frac{(x^2-y^2)}{2}\right] \sin\phi, \quad (3)$$

where $r = \sqrt{x^2 + y^2}$, V is the intervane voltage, and X and A are the quadrupole focusing and acceleration efficiency parameters given as

$$A = \frac{m^2 - 1}{m^2 I_0(ka) + I_0(mka)} \quad (4)$$

and

$$X = 1 - A I_0(ka). \quad (5)$$

The origin is assumed to be on-axis at the beginning of the cell. The quadrupole gradient XV/a^2 and the axial voltage AV are assumed to vary linearly over each cell. The wave number k for each cell is defined in terms of the cell length L_c, as $k = 2\pi/L_c$.

The quantity ϕ is the phase of the field as a given particle travels through the cell; ϕ depends on the particle position z and velocity β (relative to light speed) as

$$\phi = \phi_i + \frac{2\pi}{\lambda}\int_0^z \frac{dz}{\beta(z)}, \quad (6)$$

where λ is the rf wavelength, and ϕ_i is given as the phase when the particle is at the entrance to the cell. The phase convention used in PARMTEQ is not the usual linac convention. It is defined so that $\phi = 0$ corresponds to zero crossing of the rising accelerating field, and $\phi = \pi/2$ corresponds to the peak accelerating field. A synchronous phase parameter, ϕ_s, is specified for each cell and corresponds to the phase when the synchronous (or design) particle is at the beginning of that cell.

There are two versions of PARMTEQ, one with longitudinal position z as the independent variable (written by Crandall at Los Alamos) and one with time as the independent variable (written by Chidley at Chalk River). The quadrupole focusing and space charge are treated more easily when choosing time as independent, but the cell-dependent acceleration effects are easier with position as independent. In practice, both versions usually are in good agreement, and in this report, we will describe the version that uses z as the independent variable.

Because the RFQ accelerates low-velocity ions, large fractional velocity changes can occur within a given cell. Consequently, it is important to treat the beam dynamics using a method that accurately describes the variation of the particle velocity throughout the cell. The approach used in PARMTEQ is to numerically integrate the trajectories using a "drift/thin-lens/drift" approximation.

CELL GENERATION

The input for cell generation must be supplied by the designer of the RFQ and is usually obtained by using separate preprocessor design codes (see the discussion in "Summary for the Working Group on Linac Codes," these proceedings), which select the parameters to optimize the bunching and focusing of the beam. The main PARMTEQ input parameters are charge, mass, rf frequency, and the initial and final energies. Local input parameters must be given as a function of distance along the linac. These include a normalized quadrupole gradient, called B (or the aperture a as an alternative), synchronous phase ϕ_s, vane modulation m, and intervane voltage V.

The linac is then generated one cell at a time by interpolating on these local input parameters. The cell generation is an iterative procedure in which each cell length is determined so that the synchronous particle, whose input phase is given, passes through the cell in the correct time corresponding to a total phase shift of $\Delta\phi_s = \pi + \delta\phi_S$. The quantity $\delta\phi_s$ is the change in the synchronous phase parameter over the cell length, as determined from the local input data. For a constant value of ϕ_s in the input data, the design phase shift reduces to $\Delta\phi_s = \pi$.

The method is iterative because the local accelerating field depends on the cell length, which is not known initially. Likewise, the cell length depends on the phase shift for the synchronous particle, which is not known until the synchronous particle velocity profile within the cell is obtained by integration through the local fields. The procedure in subroutine RFQGEN begins with an initial guess for the cell length $L_c = \beta_i\lambda/2$, where β_i is the known synchronous particle velocity at the beginning of the cell. Then the analytic expressions for the fields can be evaluated. The cell is divided into N segments (typically $N = 10$) for the numerical integration. The synchronous energy is incremented at the end of each segment by multiplying the E_z field at the center of the segment by the segment length. After stepping through all segments in the cell, the phase shift $\Delta\phi_s$ for the synchronous particle is compared with the design value $\pi + \delta\phi_s$. If these quantities are not equal, the cell length is adjusted, and a new iteration step begins.

The procedure converges within five steps, and after each cell is generated, the cell output parameters are stored in a two-dimensional array CELL(I, NC), where I is the parameter and NC is the cell number. The output cell parameters, evaluated at the end of the cell, include synchronous energy, velocity and phase, intervane voltage, aperture a, modulation m, normalized quadrupole gradient, acceleration efficiency, average axial electric field, cell length, and total accumulated length. The value n is subtracted from the output synchronous phase before it is stored. The output values for a given cell determine the input conditions for the next cell to be generated. Cells are usually generated until the final linac energy is reached; typically, the total number of cells can be a few hundred.

BEAM–DYNAMICS SIMULATION

For the beam-dynamics simulation, a variety of randomly generated initial particle distributions can be generated at the RFQ entrance using the subroutine INPUTT. These include 4-D transverse phase-space distributions for a continuous beam with uniform phase and zero energy spread and 6-D phase-space distributions for a bunched input beam. Simulation runs with several thousand particles are generally adequate for calculating the average properties of the beam. The simulation models the dynamics of a single bunch within a phase extent of 2π.

The beam-dynamics calculations are made one cell at a time using the subroutine RFQDYN. Each cell is again divided into N segments for numerical integration (typically $N = 10$). Initial coordinates x, $x' = dx/dz$, y, $y' = dy/dz$, ϕ, and W, where W is the particle energy, are transformed to final values through each segment using a "drift/thin-lens/drift" approximation with the fields given by Eqs. 1-3. In this procedure, the particle coordinates x, y, and ϕ are drifted to the center of the segment using the initial values of x', y', and W. An energy gain impulse and transverse momentum impulses are delivered to each particle using fields evaluated at the center of the segment. New values of x', y', and W are then calculated and used to drift the coordinates x, y, and ϕ to the end of the segment. When the particles arrive at the center of each cell, all coordinates are transformed to values at the time when the accelerating field is zero and rising ($\phi = 0$). A single space-charge impulse per cell is then delivered to all particles, depending on their transformed coordinates. Then all coordinates are transformed back to the center of the cell, and the integration is continued through the second half of the segments in that cell. The Jacobian for a segment step can be shown to equal unity in each of the three degrees of freedom, corresponding to conservation of each normalized phase-space area (without space charge).

The space-charge calculation is done in subroutine SCHEFF and can be described as a particle-in-cell (PIC) method, which uses a Green's function approach for obtaining the actual space-charge fields from the particle distribution. Cylindrical symmetry is assumed, and a 2-D r-z mesh (typically 10×20) is superimposed on the bunch at the time corresponding to $\phi = 0$, when the matched beam cross section is circular. The rectangular mesh defines a set of source rings, from which a table of space-charge field components E_r and E_z for a given unit charge has been precalculated and stored for each mesh point. Each particle charge is distributed on the mesh using an area weighting, and the total charge per source ring is obtained by summing over all particles. The total space-charge field at each particle is obtained by summing over the contributions of each source ring and interpolating on the four nearest mesh points. The space-charge impulses are calculated in x', y', and W for each particle, treating these space-charge fields as the average value over the whole cell. Forces from adjacent bunches are included, assuming a periodic string of identical bunches. Because the adjacent bunch contribution depends on the bunch spacing, the table with space-charge field components must be regularly updated. Image forces are not accounted for in the present version. For the RFQ radial matching section, a different space-charge calculation[4] is used to handle the space-charge forces for a continuous, converging beam.

In general, the space-charge mesh does not extend over an entire 360° of phase. A small fraction of particles that are not captured into a stable bucket can drift behind the main bunch and fall outside the mesh. The space-charge force on these particles is obtained by replacing the particle distribution by a single point located at the centroid, with a charge equal to the total value for the bunch. When the uncaptured particle has drifted behind the synchronous particle by more than half an rf cycle, its phase value is reduced by 2π. It then appears at the head of the bunch, to approximate a particle from an upstream bucket drifting through the bunch that is modeled in the simulation. The uncaptured particles will fall further and further behind the synchronous energy; eventually the space-charge approximations, which are used when the particles are outside the mesh, may lead to large accumulated errors. Rather than carry these particles along throughout the entire simulation, they can be removed when their energy falls behind the synchronous energy by an amount equal to a variable ELIMIT, which can be set by the user. This procedure is used because it is believed that the small fraction of off-energy particles are not modeled correctly.

Most of the simulation output is written to tape and is processed by an output processor code called OUTPROC. Output can be obtained at the beginning and end of the RFQ, and within each cell at $\phi = 0$. Tabulated output includes the number of transmitted particles, the distribution moments, rms ellipse parameters, and rms emittances in each plane. Plotted output includes phase-space plots and projections, beam cross-section plots, and profile plots versus cell number along the RFQ.

Conclusions

The PARMTEQ code has probably been the most important tool for RFQ beam-dynamics design, and has contributed greatly to the remarkable development of the RFQ. Comparison of the PARMTEQ results with experimental measurements have been made for several operating RFQs.[5,6,7] The results for the Los Alamos proof-of-principle RFQ[5] showed agreement between measurement and predictions of an unexpected double peak in longitudinal phase-space.

Desirable improvements for PARMTEQ might include (1) better treatment of entrance and exit fields,[8] (2) a treatment of image-charge fields, (3) higher-order multipoles to allow the study of errors in vane positioning, and (4) fields resulting from trapped ions[9] produced by collisional ionization between the beam and residual gas.

REFERENCES

1. I. M. Kapchinskiy and V. A. Teplyakov, "Linear Ion Accelerator With Spatially Homogeneous Strong Focusing," Prib. Tekh. Eksp. $\underline{119}$, No. 2, p. 19 (1970).

2. Richard H. Stokes and Thomas P. Wangler, "Radio-Frequency Quadrupole Accelerators and Their Applications," to be published in *Annual Rev. Nucl. Sci.*, Vol. 38 (1988).

3. K. R. Crandall, R. H. Stokes, and T. P. Wangler, "RF Quadrupole Beam Dynamics Design Studies," Proc. 10th Linear Accelerator Conf., Montauk, New York, September 10-14, 1979, Brookhaven National Lab report BNL-51134, p. 205 (1980).

4. K. R. Crandall, "A Space-Charge Calculation Appropriate for Radial Matching Sections of RFQ Linacs," Los Alamos Group AT-1 memorandum AT-1:83-251, August 26, 1983.

5. J. E. Stovall, K. R. Crandall, and R. W. Hamm, "Performance Characteristics of a 425 MHz RFQ Linac," IEEE Trans. Nucl. Sci. NS-28, p. 1508 (1980).

6. E. Boltezar, H. Haseroth, Ch. Hill, F. James, W. Pirkl, G. Rossat, P. Tetu, and M. Weiss, "Performance of the CERN RFQ," Proc. 1984 Linac Conf., GSI Darmstadt report GSI-84-11, p. 56 (1984).

7. G. P. Boicourt, O. R. Sander, and T. P. Wangler, "Comparison of Simulation With Experiment in an RFQ," IEEE Trans. Nucl. Sci. NS-32, p. 2562 (1985).

8. K. R. Crandall, "RFQ Radial Matching Sections and Fringe Fields," Proc. 1984 Linac Conf., GSI Darmstadt report GSI-84-11, p. 109 (1984).

9. M. S. de Jong, "Background Ion Trapping in RFQs," Proc. 1984 Linac Conf., GSI Darmstadt report GSI-84-11, p. 88 (1984).

TRACE, AN INTERACTIVE BEAM DYNAMICS PROGRAM

K.R. Crandall

AccSys Technology, Inc., Pleasanton, CA 94566

ABSTRACT

The purpose and applicability of both the three-dimensional and the newer two-dimensional versions of TRACE are discussed, along with the basic method used for the dynamics calculations. The transport system elements and the interactive commands are listed. The matching and optimization capabilities are mentioned briefly, along with the method for finding the solutions. Typical examples are presented.

INTRODUCTION

TRACE is an interactive, first order, beam-dynamics program that calculates beam envelopes through a user defined transport system. Linear space-charge forces are included. The beam envelopes are graphically displayed as they are calculated. This feature helps the user to understand the optics in the transport system. A bad situation is immediately obvious. Because the program is interactive, adjustments to the transport system and/or to the initial beam characteristics can be made quickly. A variety of matching options and transport system optimizations are possible.

The original version[1] of TRACE was written in 1972 and was developed for use on the controls computer at the Los Alamos Medium-Energy Physics Facility (LAMPF). The objective was to provide an aid in tuning the Low-Energy Beam Transport (LEBT) line. The LEBT included emittance measuring hardware at three locations. The tuning technique consisted of measuring the emittances, calculating the dynamics, adjusting the LEBT parameters, and iterating until satisfied. TRACE was found to be extremely useful in this tuning procedure. A modified version[2] was written in 1977 for the linac controls computer at the European Organization for Nuclear Research (CERN) in Geneva, Switzerland. In 1979, the CERN version was adapted to the Pion Generator for Medical Irradiations (PIGMI) controls system in Los Alamos.

The computer operating systems, under which TRACE was run, were multi-tasking systems. More than one program was resident in memory at the same time. Consequently, TRACE had to be as short as possible and had to use as little computing time as possible. The dynamics calculations were done for a continuous beam having no energy spread, and the motions in the x and y directions were uncoupled, except for the space-charge forces.

The interactive graphics mode of operation also made TRACE useful as a tool for designing beam transport systems. TRACE was expanded extensively and adapted for use on the CDC-7600 computer at the central computer facility and on the AT Division VAX 11/750 at Los Alamos.[3] Throughout these modifications, TRACE retained its two-dimensional dynamics. When the need arose for a similar program for bunched beams, TRACE 3-D was written.[4] The dynamics portion of the program was not only modified to handle bunched beams with coupling between x, y, and z, but completely restructured to make modifications and expansions much easier. A new two-dimensional version, TRACE 2-D, retains the features of TRACE 3-D, except that the beam is continuous instead of bunched. Both TRACE 3-D and TRACE 2-D run on personal computes as well as on large computers.

DYNAMIC CALCULATIONS

The basic assumption of TRACE is that all forces are linear or can be linearized. If the coordinates of a particle are known at some location s_1 along the transport system, then at s_2 the coordinates can be calculated by a single matrix multiplication. That is,

$$\vec{u}(s_2) = R\,\vec{u}(s_1) \;, \tag{1}$$

where $\vec{u}(s)$ is the column vector of the coordinates at location s, and R is the transfer matrix between locations s_1 and s_2. The elements of the R-matrix depend on the transport elements between s_1 and s_2 and on the size of the beam (for computing space-charge forces) in this interval.

The beam is represented by a beam matrix σ. The elements of the σ-matrix are proportional to the second moments of the beam coordinates. If the transfer matrix between s_1 and s_2 is known, and if the beam matrix at s_1 is known, then the beam matrix at s_2 is calculated by

$$\sigma(s_2) = R\,\sigma(s_1)\,R^T \;, \tag{2}$$

where R^T denotes the transpose of R. The dynamics calculations in TRACE are done by a sequence of the above matrix transformations. Starting with an initial σ-matrix, a transfer matrix is constructed from the external forces for a small transport interval and a new σ-matrix is calculated. The size of the beam is obtained from elements of the σ-matrix, and linear space-charge forces are calculated. An R-matrix is constructed for the space-charge impulse, and a new σ-matrix is calculated. This process is repeated until the beam has been followed through the specified transport elements.

For the space-charge forces to be linear, one must postulate a beam having a uniform charge distribution in real space. Although real life is rarely so accommodating, it has been shown, for distributions having ellipsoidal symmetry, that the evolution of the rms beam envelope depends almost exclusively on the linearized part of the self-forces.[5] Consequently, for calculational purposes the "real beam" may be replaced by an "equivalent uniform beam" having identical rms properties. The total emittance of the equivalent uniform beam in each phase plane is $(n+2)$ times the rms emittance in that plane, where n is 2 for TRACE 2-D and 3 for TRACE 3-D. The beam envelopes displayed by TRACE are $\sqrt{n+2}$ times their respective rms values. (Real beams have ill-defined boundaries. In general, one can expect a few percent (<10%) of the particles in a real beam to be outside the boundaries displayed by TRACE.)

In TRACE 3-D, the internal coordinates are x, x', y, y', z, and $\Delta p/p$, where x, y, and z are horizontal, vertical and longitudinal displacements from the center of the beam, x' and y' are p_x/p and p_y/p, where p_x and p_y are the horizontal and vertical momentum components and p is the longitudinal momentum of the beam center, and Δp is $p_z - p$, the difference in the longitudinal momentum from that of the beam center. For input and output, z and $\Delta p/p$ are usually replaced by $\Delta\phi$ and ΔW, the phase and energy displacements from the beam center. The transfer matrices are dimensioned six by six. In TRACE 2-D the beam is longitudinally continuous, so there is no z coordinate and the R-matrices are dimensioned five by five.

TRANSPORT SYSTEM ELEMENTS

The transport elements recognized by TRACE 3-D are: (1) drift; (2) thin lens; (3) quadrupole; (4) permanent-magnetic quadrupole (PMQ); (5) solenoid; (6) symmetric doublet; (7) symmetric triplet; (8) bending magnet; (9) edge-angle on bending magnet; (10)

rf gap; (11) radio-frequency quadrupole (RFQ) cell; (12) rf cavity; (13) coupled-cavity tank; (14) special (user defined); (15) axis rotation; and (16) identical (with another element). The field in a PMQ extends beyond its own boundaries into neighboring drifts, other PMQs, and RFQ cells, where the fields are superimposed. The fields in all other elements do not extend beyond their own boundaries.

TRACE 2-D recognizes the same elements, with the following exceptions; (9) accelerating column; (11) current change; and (12) ramped solenoid. The fields is an accelerating column also extend beyond its own boundaries.

MATCHING AND OPTIMIZATION

TRACE is very useful in matching and optimizing beam lines. It can be used for finding the beam-ellipse parameters that are matched to a specified periodic transport system, and for finding values of transport system parameters that will produce a specified set of ellipse parameters at a specified location. TRACE 3-D also has the capability of adjusting transport parameters to obtained specified values for specified elements of the σ-matrix or of the R-matrix.

The matching or optimization is accomplished by finding the solution to N nonlinear, simultaneous equations, where N is the number of conditions to be satisfied. The method used is that of *regula falsi*, which reduces to the secant method for one equation. Starting with an initial "guess" at the solution, an additional N guesses are generated using pseudo random numbers. From then on, the next guess is calculated from the previous "best" $N+1$ guesses until a solution is found or until the number of tries has exceeded a specified limit (defaulted to 10 tries). If a solution is not found, the variables are set to the values that produced the best results that were found. At this point, the user usually asks TRACE to try again, starting closer to the solution than before.

During the search for a solution, the convergence factor and the values of the variables are displayed at each iteration. If no progress is made toward a solution (after the first N random guesses) as evidenced by the convergence factor, there is usually something wrong with the data. However, it is also possible that a solution does not exist, or that the initial guess was too far from the solution.

PROGRAM INTERACTION

The interaction with TRACE is via a keyboard and a graphics display. Upon running the program, the user is asked whether the initial input data is on TAPE 30 (an ASCII file in the NAMELIST format) or on TAPE10 (an unformatted file). After reading the data and initializing, the program waits for the user to issue a command via the keyboard. The command is a single letter which tells the program to take a certain action. A list of the valid commands for TRACE, along with the action to be taken, is given in Table I. After any action in which any parameter is changed, the TAPE10 file is rewritten, keeping it continually up to date and allowing for easy restarts.

EXAMPLES

An example in which TRACE 3-D is extremely useful is the problem of matching the beam from the exit of an RFQ into the entrance of a drift-tube linac (DTL). This matching problem has three stages: (1) determine the matched beam characteristics at the exit of the RFQ; (2) determine the matched beam characteristics at the entrance to the DTL; and (3) design a transport system between the RFQ and DTL to transform the matched RFQ beam into the matched DTL beam.

Table I. Commands Recognized by TRACE.

Command	Action
a (add)	Add (insert) elements in transport system
b (beam)	Print beam parameters
d (delete)	Delete elements from transport system
e (end)	Terminate the program
f (phase)	Calculate and print phase advances
g (go)	Draw graphics background and follow beam through transport system
i (input)	Enter new parameters
j (projections)	Plot initial and final beam projections on the x–y, x–z, and x–$\Delta p/p$ planes
k^1 (periodic)	Generate periodic focusing system (RFQ or DTL)
l (ellipse)	Determine emittance ellipses from three profile measurements
m (match)	Perform matching
o (mismatch)	Calculate and print mismatch factors
p (print)	Print parameters for beam, control, graphics, and transport
q^2 (RFQ input)	Find matched input for RFQ
r (R-matrix)	Print R-matrix from latest run
s (save)	Save ellipse parameters and σ-matrix
u (update)	Replace ellipse parameters and σ– matrix by their stored values
x	Toggle printed output to printer
w (φ-W)	Print phase and energy information
z (σ-matrix)	Print modified σ-matrix

[1]TRACE 3-D only.
[2]TRACE 2-D only.

The first stage is accomplished by creating a transport system consisting of two RFQ cells (one transverse focusing period), specifying the beam current, energy, and the emittances in the three phase planes, and telling TRACE to find the match by issuing the "m" command. The results are shown in Figure 1, a graphics display produced by the personal computer version of TRACE 3-D. The input phase-space ellipses are shown at the upper left, x–x' and y–y' together on the top and the phase-energy ellipse below. The output ellipses, shown at the upper right, are the same as the input ellipses, showing that the beam is matched. A schematic of the transport system, which includes two DTL cells as well as two RFQ cells, is shown in the lower portion of the display, along with the x, y and phase envelopes in the RFQ. The beam current, the input and output energy of the beam (acceleration was suppressed), and the initial and final emittances are shown at the top center of the display. Below this are shown the results of issuing the "f" command, which gives the phase advances in each plane along with the Twiss parameters, all of which are calculated from the R-matrix. The results of the "w" command are shown at the bottom, giving additional information about the longitudinal properties of the beam.

The second stage is accomplished by telling TRACE to find the matched conditions at the entrance to the DTL, which consist of PMQs and rf gaps. The results are shown in Figure 2.

The final stage is accomplished by inserting two PMQ's and a buncher (rf gap) between the RFQ and DTL, and asking TRACE to find the values of the quadrupole gradients, the buncher voltage, and the length of one drift space that transform the matched beam from the RFQ into the matched beam for the DTL. The resultant values are shown at the center of the display in Figure 3, and the mismatch factors for the three phase planes (produced by the "o" command) are written at the bottom of the display. Ideally, the mismatch factors should all be zero, but in this example there were six conditions to be satisfied and only four variables were used. A good match was obtainable under these circumstances because the focusing strengths per unit length in the RFQ and in the DTL were approximately equal, as evidenced by the phase advances shown in Figures 1 and 2.

An example of the usefulness of TRACE 2-D is that of finding the matched-beam characteristics into an RFQ and then designing a LEBT system to produce the matched beam. The first step is to create the radial matching section and the first two cells of the RFQ, and find the matched beam parameters (at zero phase) at the beginning of the first two cells through the "m" command. The match at the beginning of the radial matching section is obtained using the "q" command. TRACE automatically runs the beam backward through the radial matching section starting from the beginning of each of the two RFQ cells and averages the resultant Twiss parameters to get the matched input to the radial matching section. The results are shown in Figure 4. The ellipses at the beginning of the first cell are shown at the upper right, and the ellipses at the entrance to the radial matching section are shown at the upper left. There are actually three superimposed ellipses in each phase plane displayed at the upper left: the two ellipses produced by following beam backward and the averaged ellipse. The mismatch factors between these ellipses are written at the center of the display.

The results of using a LEBT consisting of two solenoids to match an axisymmetric beam to the RFQ is shown in Figure 5.

REFERENCES

1. K.R. Crandall, "TRACE: An Interactive Beam-Transport Program," Los Alamos Scientific Laboratory report LA-5332, October 1973.
2. K.R. Crandall, "TRACE: An Interactive Beam-Transport Program for Unbunched Beams," and "Addendum to Program TRACE," CERN/PS/LIN/Note 77-3, Geneva, Switzerland, February 1977.
3. K.R. Crandall and D.P. Rusthoi, "Documentation for TRACE: An Interactive Beam-Transport Code," Los Alamos National Laboratory Report LA-10235-MS, January 1985.
4. K.R. Crandall, "TRACE 3-D Documentation," Los Alamos National Laboratory report LA-11054-MS, August 1987.
5. F. Sacherer, "RMS Envelope Equation with Space Charge," CERN/SI/Internal Report 70-12, Geneva, Switzerland 1970.

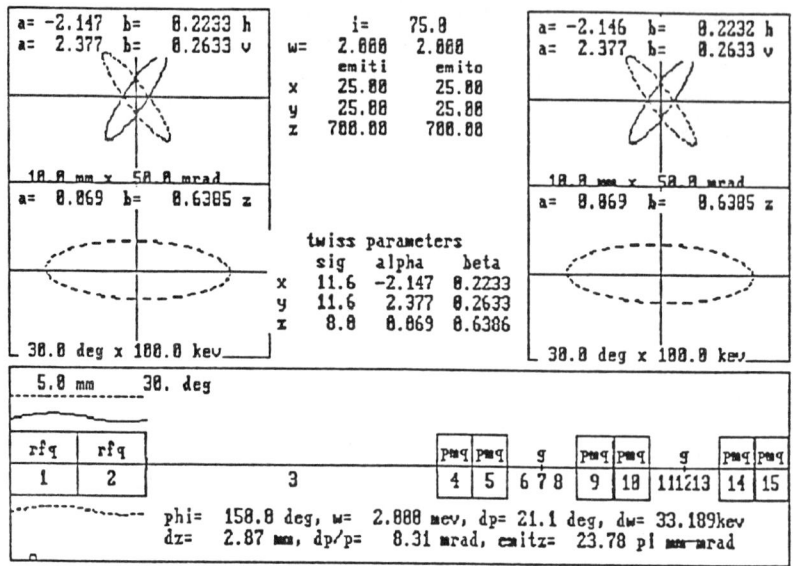

Fig. 1. Matched beam at exit of RFQ. (TRACE 3-D)

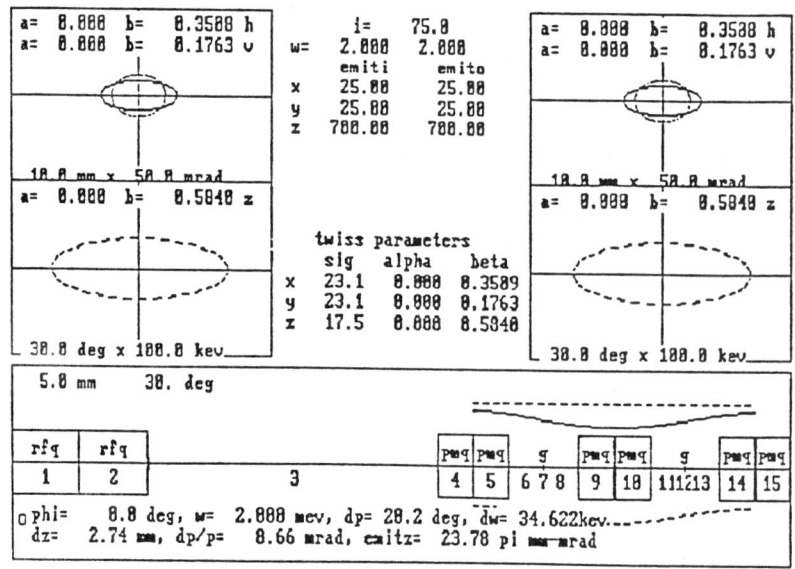

Fig. 2. Matched beam at entrace to DTL. (TRACE 3-D)

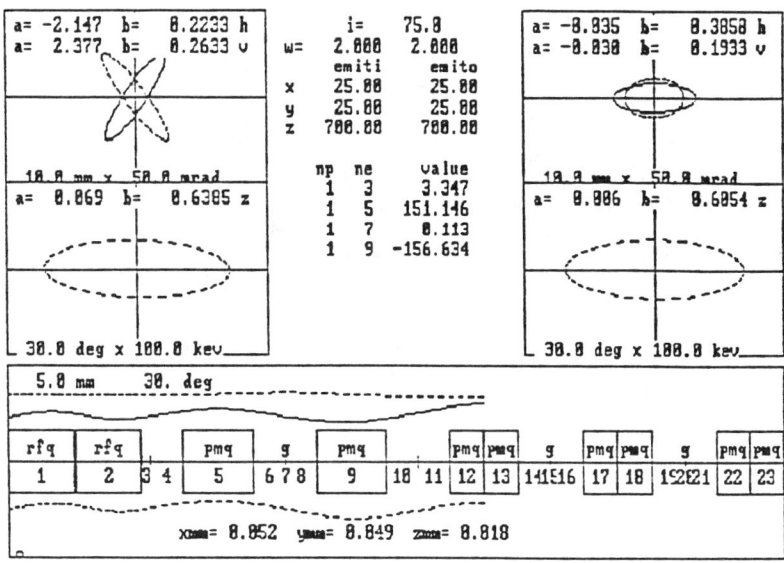

Fig. 3. Results of matching between an RFQ and a DTL. (TRACE 3-D)

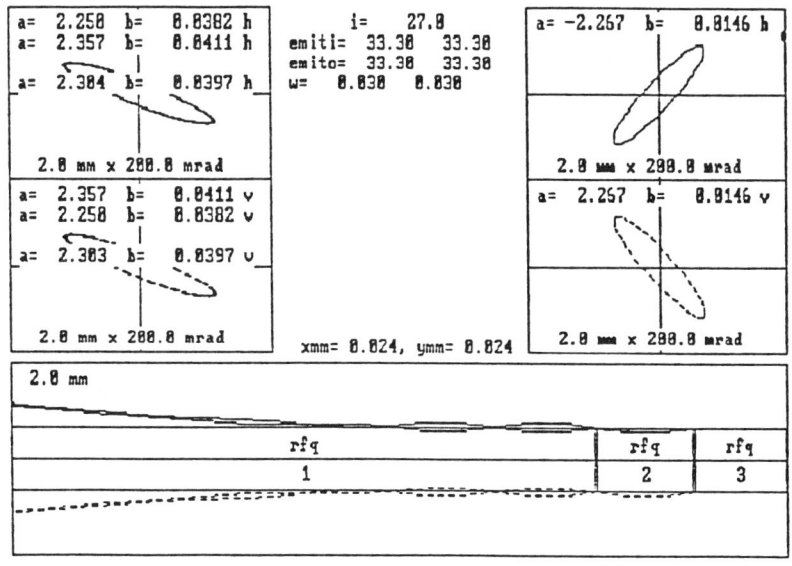

Fig. 4. Matched input to an RFQ. (TRACE 2-D)

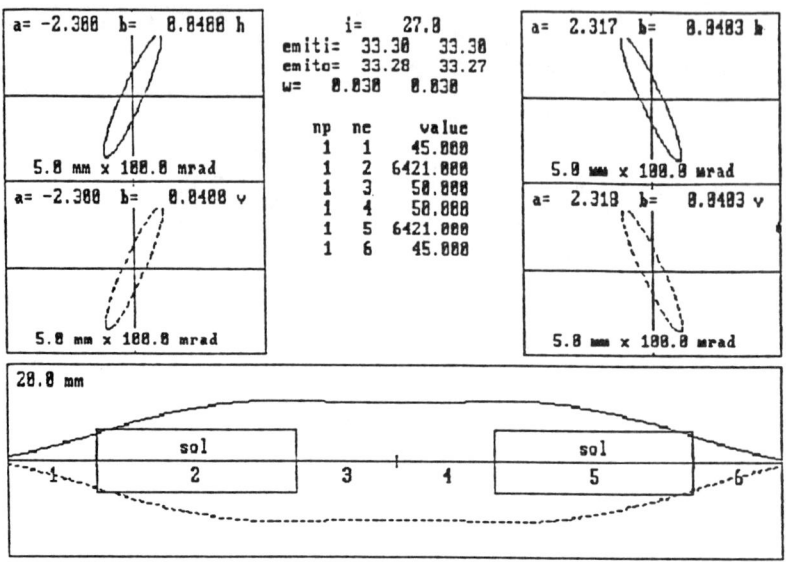

Fig. 5. LEBT for matching into an RFQ. (TRACE 2-D)

LTRACK — BEAM-TRANSPORT CALCULATION INCLUDING WAKEFIELD EFFECTS [*]

K. C. D. Chan and R. K. Cooper
Los Alamos National Laboratory, Los Alamos, NM 87545, U.S.A.

ABSTRACT

LTRACK is a first-order beam-transport code that includes wakefield effects up to quadrupole modes. This paper will introduce the readers to this computer code by describing the history, the method of calculations, and a brief summary of the input/output information. Future plans for the code will also be described.

INTRODUCTION

LTRACK is a simple first-order beam-transport code that has the special feature of including wakefield[1] effects of a beam bunch. When a beam bunch traverses beamline elements, it interacts with the elements and generates electromagnetic fields called wakefields. The wakefields induced by the front of the bunch will provide additional forces for the latter part of the bunch, and different parts of the bunch will be transported differently. At the end of the beam transport, there will be variations of beam parameters along the beam bunch. When the beam parameters of the whole bunch are considered, these variations produce a growth in energy spread and transverse emittance. When the beam current is increased, this deterioration of beam qualities increases to become significant.

Beam-transport codes, including wakefield effects, began to appear in the last few years for two reasons. First, the introduction of computer codes like TBCI[2] has allowed the calculation of wakefield effects in the time domain, including wakefield effects in beam-transport codes has become a straightforward task. Second, the design of accelerators with high peak currents and good beam qualities, requires codes that treat wakefield effects properly. There are two important examples of this type of design. The first example is the injector for a free-electron laser. The light conversion efficiency of a wiggler increases with the peak beam current and with improved beam qualities. It is imperative that wakefield effects be considered when the injector is designed. The second example is the TeV collider, where high current and good beam qualities are necessary to produce high luminosity for the high-energy physics experiments. At the Workshop of New Developments in Particle Acceleration Techniques held at Orsay, France, this past summer, a high-frequency rf linac was proposed as a way to build tomorrow's e^+e^- colliders.[3] With wakefield effects scaling with the

---

[*] Work supported by Los Alamos National Laboratory Institutional Supporting Research, under the auspices of United States Department of Energy.

powers of rf frequency, these colliders have to be designed with beam-transport codes including wakefield effects.

HISTORICAL BACKGROUND

LTRACK is a by-product of the design effort of the Stanford Linear Collider (SLC). In 1983, Chao and Cooper wrote LTRACK to investigate the transverse quadrupole wakefield effects in the SLC.[4] Around 1984, Bane adapted LTRACK for use in investigating the effectiveness of Landau damping for curing single bunch beam breakup caused by the dipole wakefield in the SLC.[5] Currently, we have incorporated LTRACK as a part of the wakefield code set being developed in AT-6 of the Los Alamos National Laboratory. This effort is supported with Institutional Supporting Research and Development (ISRD) funds whose goal is to develop a set of beam-transport codes to be used in the design of high-energy electron linacs with high bunch-charge density, where wakefield effects are important. At the present time, this set of codes includes: LTRACK, a first-order code used for quick preliminary designs; and PARMELA,[6] a particle tracking code used for final designs with all the higher-order beam-optics included.

METHOD OF CALCULATIONS

At the present time, LTRACK is used in conjunction with TBCI, which produces wakefield information. Figure 1 shows the running sequence of these codes and the associated files. The wakefield information is produced by TBCI in the form of wake functions and are stored in the output file TBCIWAK of TBCI. Using a small program called WORK1, these wake functions are processed and output to a file called LTRAKWAK, in a form useful to LTRACK. If analytical forms of the wake functions are available, LTRAKWAK can also be replaced by a user-defined subroutine in LTRACK that contains these analytical forms. The rest of information required by LTRACK, i.e., beamline arrangement, beam parameters, etc., are specified in LTRAKIN, the principal input file for LTRACK.

During a calculation using LTRACK, a beam bunch is divided into a number of slices in the longitudinal direction. Each slice is characterized with the following parameters:

1. number of particles in the slice;
2. longitudinal position of the slice relative to the bunch center, i.e., z;
3. energy of the slice;
4. parameters describing the location and divergence of the centroid of the slice, i.e., (x, x', y, y'); and
5. parameters describing the transverse phase-space distribution of particles in the slice, i.e., a 4 x 4 Σ-matrix.

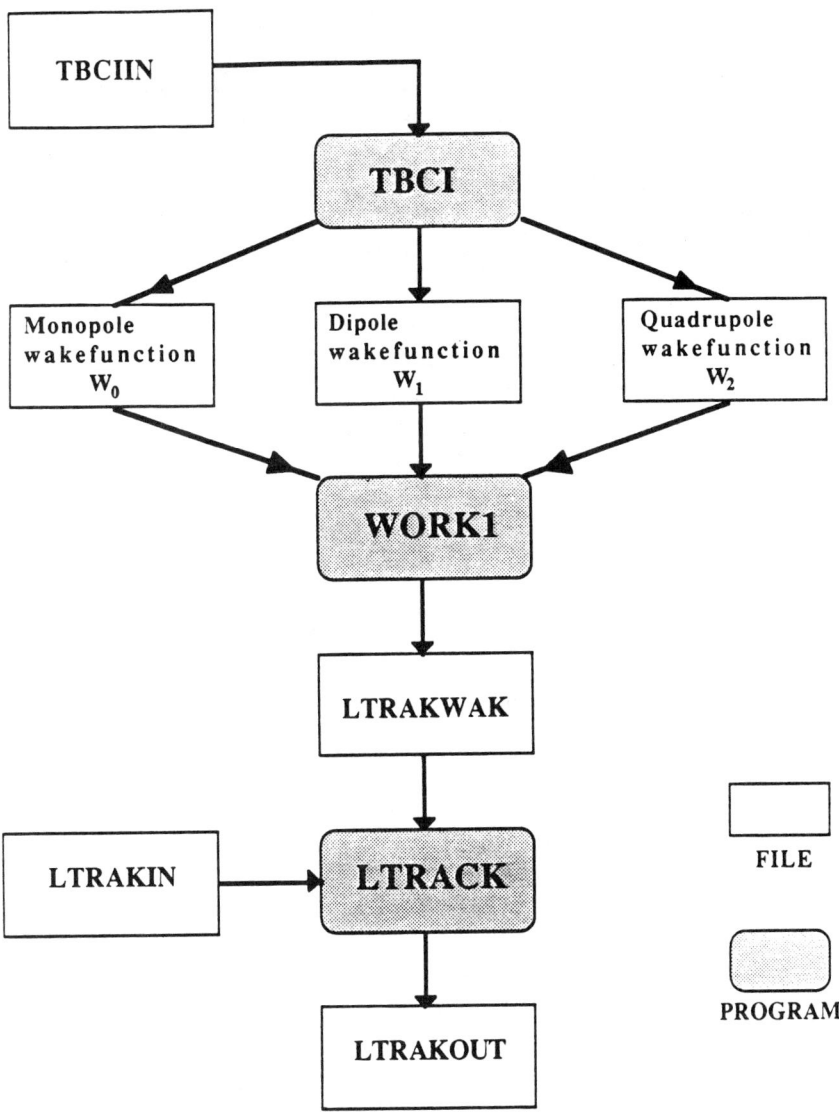

Fig. 1. Operation sequence of LTRACK and the associated files.

LTRACK has been designed for particles that are travelling at the speed of light; therefore, the numbers of particles in the slice and the relative positions of the slices (parameters in 1 and 2) are assumed to be unchanged during the transport. The other parameters (parameters in 3 to 5) are changed when the slices are transported through the beamline under two influences: the beamline elements and the wakefield forces.

The treatment of the beamline elements is similar to that of the computer code TRANSPORT.[7] The elements are represented by transfer matrices **T**. The centroid of a slice is tracked according to

$$\begin{bmatrix} x \\ x' \\ y \\ y' \end{bmatrix}_{out} = \mathbf{T} \begin{bmatrix} x \\ x' \\ y \\ y' \end{bmatrix}_{in}, \qquad (1)$$

and the transverse distribution of particles in the slice according to:

$$\Sigma_{out} = \mathbf{T} \Sigma_{in} \tilde{\mathbf{T}}, \qquad (2)$$

where Σ is the 4 x 4 matrix describing the transverse phase space and the tilde indicates the transpose of a matrix. The transfer matrix **T** is allowed to depend on the energy of the slice being tracked. Types of beamline elements currently available are:

1. drift,
2. bending magnet with edges,
3. quadrupole,
4. accelerating section, and
5. solenoid.

The treatment of monopole, dipole, and quadrupole wakefields, respectively, is explained as follows. Using the impulse approximation, wakefield effects are applied at regular intervals during beam transport.

The monopole wake has the effect of changing the energy of the slices. The kth slice suffers an energy change of ΔE_k, which is given by

$$\Delta E_k = -L \sum_{i=1}^{k-1} N_i W_0(z_i - z_k), \qquad (3)$$

where

L is the length over which wake effect is to be considered,

N_i is the number of particles in the ith slice,

$W_0(s)$ is the monopole wake function per unit length, and

$\sum_{i=l}^{k-1}$ is sum over all slices ahead of the kth slice.

The dipole wakefield causes energy loss and deflections of particles. The energy loss is not taken into account in LTRACK because it is usually small compared to that caused by the monopole wake effect. The dipole transverse forces for all the particles in a slice are the same; therefore, they will only introduce a deflection of the centroid of the slice without affecting the Σ-matrices of slices. These deflections are represented by changes of divergences of x' and y' for the centroid of the kth slice:

$$\begin{bmatrix} \Delta x'_k \\ \Delta y'_k \end{bmatrix} = \frac{r_e L}{\gamma_k} \sum_{i=l}^{k-1} N_i W_1(z_i - z_k) \begin{bmatrix} x_i \\ y_i \end{bmatrix} , \qquad (4)$$

where $W_1(s)$: the dipole wakefunction per unit length,

r_e : classical electron radius, and

γ_k : energy of the kth slice in units of the electron rest mass.

The quadrupole wakefield changes the energy, the centroid parameters, and the Σ-matrices of the slices. As in the case of the dipole wakefield, the energy changes are not taken into account in LTRACK. The quadrupole transverse wake force is in a form very similar to that of a quadrupole. The force, induced by the ith slice at distance z behind it, is given by

$$e^2 W_2(z)[Q_{1i}(x\hat{x} - y\hat{y}) + Q_{2i}(y\hat{x} - x\hat{y})] , \qquad (5)$$

where

$Q_{1i} = \Sigma_{11i} - \Sigma_{33i} + x_i^2 - y_i^2 ,$

$Q_{2i} = 2\Sigma_{13i} + 2x_i y_i$, and

$W_2(s)$ is to the quadrupole wake function per unit length.

The terms Q_{1i} and Q_{2i} are the regular and skewed quadrupole moments, respectively. These moments contain contributions from both centroid displacements and beam shapes. A transfer matrix $T_{\text{quad wake}}$ similar to those for a beamline element can be constructed as

$$\mathbf{T}_{\text{quad wake}} = \begin{bmatrix} 1 & 0 & 0 & 0 \\ q_1 & 1 & q_2 & 0 \\ 0 & 0 & 1 & 0 \\ q_2 & 0 & -q_1 & 1 \end{bmatrix}, \tag{6}$$

where

$$q_{1,2} = \frac{r_e L}{\gamma k} \sum_{i=1}^{k-1} N_i W_2(z_i - z_k) Q_{1,2i} \ .$$

The quadrupole transverse wake effects on the kth slice are treated by operating with this matrix on the kth slice as in Eqs. (1) and (2).

INPUT/OUTPUT INFORMATION

Figure 2 shows a summary of the input to LTRACK contained in the input file LTRAKIN. It is shown in a form as echoed in the LTRACK output file, called LTRAKOUT and is very much self-explanatory. The input information can be classified into three types. First, there are initial beam parameters. Second, there are parameters specifying the options, which include the choice of positrons or electrons, including alignment errors for the accelerator, the beam-position monitor (BPM), and the quadrupoles and the simulation of the BPM and the corrector pair as in automatic beam-position controls. Third, there is information on the beamline lattice, which takes the same form as in the computer code TRANSPORT (not shown in Figure 2). In LTRAKOUT, after echoing the information from LTRAKIN, the wake function information as supplied via LTRAKWAK will be echoed. After reading in all the input information, LTRACK starts transporting the bunch and printing out the bunch parameters averaged over all the slices at each of the beamline elements. These parameters include the centroid and first moment of the transverse bunch positions and divergence, the transverse emittance growth, the bunch energy, and the energy spread. LTRACK also prints out the parameters of individual slices after the last beamline element.

FUTURE PLAN

There are three major areas in which LTRACK will be improved in the near future. First, a graphic package convenient to the user will be added. Second, the capability of using more than one set of wake functions corresponding to different beamline elements will be implemented. At the present time, LTRACK only accepts one set of wake functions, which are used at accelerator sections. Third, a more complete User's Guide will be written.

```
==================
=input parameters=
================================================================
        wakefield existence flag, monopole  : lwak0   =              t
                                    dipole  : lwak1   =              f
                                quadrupole  : lwak2   =              f
        acc sect. slice length / m (if wake) : aclenm =     5.7500e-02
        initial ztot (no scraping for ztot<0): ztot0  =     0.0000e+00
        flag for scattering at each element  : lthrua =              f
        flag for scraping at raper= 1cm      : lscrap =              f

        bunch,  total number of particles    : qntot  =     1.0000e+11
                rms length / m               : sigz   =     3.8200e-03
                trunc. num. of sigz in front: nsgfro  =              3
                trunc. num. of sigz in back : nsgbak  =              3
                number of transverse slices  : nsli   =             37
                flag for positrons           : lposi  =              f
                flag for electrons           : lelec  =              t
                slice retention limit / m    : rout   =     1.0000e+03

        orbit correction, flag x             : lcorx  =              f
                          flag y             : lcory  =              f
                          scale factor       : rcor   =     1.0000e+00
                          scat.-for-corr.flag: lcosca =              t
                          mon.s/block forward: nmofor =             20
                          mon.s/block bakward: nmobak =              4

        errors, accelerator, generation flag : lacgen =              f
                             rms error x / cm: sigacx =     0.0000e+00
                             rms error y / cm: sigacy =     0.0000e+00
                monitor,     generation flag : lmogen =              f
                             rms error x / cm: sigmox =     1.0000e-02
                             rms error y / cm: sigmoy =     1.0000e-02
                quadrupole,  generation flag : lqugen =              f
                             rms error x / mr: sigqux =     1.0000e-02
                             rms error y / mr: sigquy =     1.0000e-02
                             generation seed  : iseedg =         678345

        card enabling flags, corrector kicks : lcorrt =              f
                             accelerator errs: lacert =              f
                             monitor     errs: lmoert =              f
                             quadrupole  errs: lquert =              f

        bunch,  initial energy      / gev    : e0      =    1.0000e-02
                initial energy spread / gev  : sige0   =    0.0000e+00
                initial x offset    / m      : cen0(1) =    1.0000e-04
                initial x prime     / mr     : cen0(2) =    0.0000e+00
                initial y offset    / m      : cen0(3) =    1.0000e-04
                initial y prime     / mr     : cen0(4) =    0.0000e+00

        bunch,  initial sigma matrix for e+  :  1.5329e-06  1.4452e-06  0.0000e+00  0.0000e+00
                                                1.4452e-06  2.0440e-06  0.0000e+00  0.0000e+00
                                                0.0000e+00  0.0000e+00  1.5329e-06 -1.4452e-06
                                                0.0000e+00  0.0000e+00 -1.4452e-06  2.0440e-06

        random number generator seed for scat: iseed   =          13579
        number of points to scatter          : nsca    =           2000
        flag for echoing the input           : lecho   =              t
        index for printing at each element   : itabl   =              2
        indicator for printing at the end    : iprint  =              6
        plotting, flag                       : lplot   =              t
                  plot selection string      : splot   = 111m11111111
        debugging, flag                      : sdebug  =    0011001000

        initial emittances:
        gam         =   1.9569e+01
        gam2*eps_xy =   4.0005e-10
        gam*eps_x   =   2.0001e-05
        gam*eps_y   =   2.0001e-05
```

Fig. 2. Input information to LTRACK as echoed by LTRAKOUT.

REFERENCES

1. Reviews of wakefield effects can be found in the following two references:
   A. W. Chao, Coherent Instabilities of a Relativistic Bunch Beam, SLAC-PUB-2946 (1982), also Physics of High Energy Accelerators (SLAC Summer School 1982), AIP Conf. Proc. 105.
   P. B. Wilson, High Energy Electron Linacs: Applications to Storage Ring RF Systems and Linear Colliders, SLAC-PUB- 2884 (1982); also Physics of High Energy Particle Accelerators (Fermilab Summer School, 1981), AIP Conf. Proc. 87.
2. T. Weiland, Transverse Beam Cavity Interaction, Part I: Short Range Forces, DESY 82-015, March 1982.
3. G. Coignet and S. Turner, Foreword to the Proceedings of the Workshop of the New Developments in Particle Acceleration Techniques, CERN 87-11, October (1987) pp. vii.
4. A. W. Chao and R. K. Cooper, Transverse Quadrupole Wake Field Effects in High Intensity Linacs, Particle Accelerators, Vol. 13, pp. 1-12 (1983).
5. K. L. F. Bane, Landau Damping in the SLAC Linac, IEEE Trans. Nucl. Sci. **32** (5), 2389 (1985).
6. PARMELA, particle tracking code for electrons written by K. R. Crandall. For further information consult Computer Codes used in Particle Accelerator Design, Los Alamos National Laboratory report LA-UR-86-3320, (1987).
7. D. C. Carey, TRANSPORT Manual, Stanford Linear Accelerator Center report SLAC-91, 1973 revised.

ELECTRON RAY TRACING PROGRAMS FOR GUN DESIGN AND BEAM TRANSPORT*

W. B. HERRMANNSFELDT

Stanford Linear Accelerator Center,
Stanford University, Stanford, CA 94309

ABSTRACT

Computer simulation of electron and ion sources is made by using a class of computer codes known as gun design programs. In this paper, we shall first list most of the necessary and some optional capabilities of such programs. Then we will briefly note specific codes and/or authors of codes with attention to specialized applications if any. There may be many more such programs in use than are treated here; we are only trying to cover a range of examples, not perform a comprehensive survey.

CAPABILITIES AND LIMITATIONS OF GUN PROGRAMS

At the most basic level, an electron gun program should be able to accurately define the boundaries and the electric fields that result from imposing voltages on these boundaries. Ease of entering the boundary data is important but should not be so limited as to compromise the necessary versatility. Such details as isolated grid wires, dielectrics and shadow grids require significant amounts of information to be defined. Generally, there are two ways in which the needed detail can be expressed:

1. Through the use of a regular square or rectangular mesh with interpolated fractional meshes to define boundaries, and
2. Through the use of a deformable, triangular mesh.

Most of the paper will concentrate on specific methods used in the program EGUN.[1] We will discuss the mathematical algorithms and discuss their implementation with respect to an example of a Pierce diode.

A gun program, as distinguished from other beam transport programs, must be able to treat longitudinal space-charge effects and, in particular, must be able to calculate the space-charge limited current of an emitter.

Any charged particle transport program must be able to accurately integrate individual particle motion through the defined fields. A gun program must do this from very low velocities, typified by the emission of ions, through

---

*Work supported by the Department of Energy, contract DE-AC03-76SF00515.

fully relativistic velocities. Not only must the equations of motion be completely defined, with no approximations, but also the mathematical methods must not break down, especially at very low velocities.

Beam transport calculations should include the effects of self-fields, both space-charge and self-magnetic fields. There are essentially two ways in which this is typically done:

1. by calculating the forces on each individual particle, or ray, from every other ray, or

2. by solving for Poisson's equation with space-charge and separately calculating the self-magnetic field term.

The first of these has the advantage that, for relativistic beams, the self-fields nearly cancel each other, to $(1 - v_z^2/c^2)$, and thus the method is intrinsically more accurate. This method, however, lacks the flexibility required to perform many of the functions of a general purpose program. This is because it is necessary to keep particles in step with each other, which is not always possible as, for example, in the case of a depressed collector. A third approach, which effectively combines the above two, is to reduce the space-charge by $(1-v_z^2/c^2)$ when there are no significant longitudinal fields, e.g., in the transport of an intense beam through a drift pipe. This makes it possible to use the second method near the cathode, and since particles are not usually highly relativisitic there, this approach is generally quite accurate.

Externally imposed magnetic fields play a critical role in many electron guns and transport systems. In some ways this subject is larger than the subject of electron guns themselves. A recipe for the design of a magnetically focused device is as follows:

1. Make a preliminary test without a magnetic field.

2. Perform the actual design using an ideal magnetic field that can perform the needed transport function. This field can be expressed in any form that is compatible with Maxwell's equations. For example, if the field on the axis is chosen, it can be expressed in terms of ideal point coils, or ideal solenoids, thus avoiding problems inherent with off-axis expansions.

3. Convert the ideal field to a form that can be built using realistic magnetic elements, aided by a magnet design program.

4. Take output from the magnet design program and use it to check the effects of the realistic fields on the beam transport problem.

From the above, it follows that the ray tracing program should have a variety of options for specifying magnetic fields including a simple uniform field, ideal coils, and by accepting results from a magnet design program.

Optional capabilities are usually designed to fulfill objectives of a specific application. Some of these include:

1. treating plasma discharge ionization sources including finding the emission surface of a plasma,

2. including neutralization, second species or secondary emission,

3. operating in other coordinate systems besides the usual two-dimensional cylindrical or rectangular coordinates. This includes full three-dimensional as well as mixed round beams in rectangular coordinates or round beams offset in cylindrical coordinates.

4. Optional capabilities also include other applications of Laplace's Equation such as finding peak surface fields, calculating capacitance and accounting for dielectrics.

We should also note some of the limitations of electron gun programs of this type (we are not here dealing with fully electromagnetic particle-in-cell programs):

1. Two-dimensional calculations of electric fields: either cylindrical or rectangular symmetry. In cylindrical coordinates, a cylindrically symmetric beam is propagated along the axis. In rectangular coordinates, both the electrodes and the beam extend infinitely far in the directions normal to the "plane of the paper" on which the problem is shown. In both symmetries, the nominal direction of propagation of the beam lies in the "plane of the paper" but transverse motion is allowed. Thus, for example, the spiral motion of a beam in an axial magnetic field can be simulated.

2. Time independence: these are dc calculations after the beam has reached "steady state." A common characteristic of such programs is that if steady state cannot be achieved, for example, if an attempt is made to propagate a beam beyond the space-charge limit, then the programs will not converge to a satisfactory steady solution. Under such conditions, one is not justified in claiming any physical reality for the results.

3. Idealized computer models: the nature of modeling programs is to ignore various real complications. Such things as tolerances out of cylindrical symmetry, stray electrons or ions, partially poisoned cathodes, etc., may play large parts in any real device but are usually ignored in models. Other aspects of models; finite elements, numbers of trajectories, discrete iteration steps, etc., may also affect the accuracy of the results. One should not expect a computer code to yield exactly correct predictions of operating parameters. One should expect that the effects of varying input parameters, particularly for small perturbations, should be reliably reflected in changes in the real device. Of course, some predictions are

better than others; for example, the EGUN[1] program typically predicts gun perveance correctly to within a few percent but has a somewhat harder time in predicting beam diameter for a high intensity beam.

It is useful to examine the limitation on time independence. If an electron gun is suddenly turned on (for example, with a laser photocathode), the initial burst of current can substantially exceed the space-charge limited current, and the kinetic energy of the beam can exceed the product eV, where V is the diode voltage. The electron optics of the front of the pulse closely resembles the electrostatic solution without space-charge. Later on in the pulse, the conditions will converge to the space-charge limited solution, assuming that the cathode can emit adequate current. The time constant for this transition is typically the transit time of the space-charge limited particles across the gap of the diode, at most a fraction of a nanosecond for electrons. Accurate simulation of pulses shorter than the transit time must use the particle-in-cell approach, while pulses longer than the transit time can use the more economic electrostatic programs.

CHARACTERISTICS OF DIFFERENT PROGRAMS

Since there are several authors of gun design programs at this conference, it will not be necessary to include much detail about other programs. The program SNOW, written by Jack Boers[2] of Varian Associates, is predominantly a program for the design of ion sources. Boers is reporting on the development of a three-dimensional version of SNOW at this meeting.

The Darwin model of solving for electromagnetic fields has been used by John Boyd[3] of LLNL to develop a program called DPC, for Darwin Particle Code. This program bridges a gap between electrostatic programs, of the type discussed here, and the fully electromagnetic PIC codes. The Darwin method relaxes the so-called "Courant Condition" for electromagnetic programs and is especially useful for applications involving pulses of a few tens of nanoseconds in length.

The best example of the triangular mesh approach is in a program by Richard True[4] of Litton Industries. This is a general purpose program which is particularly useful for the design of high area-convergence guns, such as for TWT's, because of the feature allowing the mesh to be concentrated near the axis.

The program EBQ, written by Art Paul[5] of LLNL, uses the direct cancellation of self-magnetic and space-charge forces and is thus especially suited for intense beam transport.

A pair of general purpose ion and electron source programs from GSI have been reported by Spädke.[6] They are AXCEL-GSI and KOBRA3, the latter being a three-dimensional code.

EGUN

We will now briefly describe the organization of the program EGUN[1] and also describe in some detail some of the important algorithms. In operation, the program starts by reading and checking the input boundary data. The program then solves Laplace's equation, i.e., Poisson's equation without space-charge. The result of this calculation, together with all the boundary information is then printed.

The Poisson solver in EGUN is a column matrix inversion routine that compares favorably in speed to conventional point-by-point over relaxation schemes. Boundary interpolation, both for the Poisson solver and for partial differentiation of potentials near boundaries, is based on

$$V_w = V_p + \frac{V_b - V_p}{\Delta x} \quad ,$$

where V_w = potential of a mesh point behind the boundary,

V_p = potential of a mesh point nearest to the boundary,

V_b = potential assigned to the boundary,

Δx = vector distance to the boundary from the point at 'p'.

Next, the first iteration of electron trajectories is started. These are initiated by one of four schemes:

1. "GENERAL" cathode in which electrons are started assuming Child's law holds near a surface designated as the cathode. This surface can be of any arbitrary shape and may include holes and shadow grids.

2. "SPHERE" for a spherical cathode (cylindrical in rectangular coordinates) in which the electrons are assumed to be emitted at right angles to the surface defined by a radius of curvature and a radial limit. Child's law for space-charge limited current is again used.

3. "CARDS" in which the specific starting conditions for each ray are specified in an 80-column card format.

4. "GENCARD" which combines the versatility of "CARDS" with the assumptions of Child's law from "GENERAL." This is especially useful for cases involving very nonuniform current emission.

The three methods for initiating space-charge limited flow all include a Busch's Law calculation to account for magnetic flux through the cathode.

On the first iteration cycle, space-charge forces are calculated from the assumption of paraxial flow. As the rays are traced through the program, space-charge is computed and stored in a separate array. After all the electron trajectories have been calculated, the program begins the second cycle by solving Poisson's equation with the space-charge from the first cycle. For problems meeting the paraxial assumptions, especially if relativistic electron beams are involved, this one cycle may be sufficient to solve the entire problem. For other problems in which space-charge is negligible, e.g., spectrometers and phototubes, a single cycle is usually adequate.

Subsequent iteration cycles (as many as are requested) follow the above pattern. The Child's law calculations for the starting conditions are remade by averaging the perveance used for the previous cycle with the perveance calculated directly from the solution of Poisson's equation.

An additional starting option is "LAPLACE" intended for any application of Laplace's equation not involving electron ray tracing. In this case the number of cycles is used simply to improve the accuracy of the solution of Laplace's equation. The "LAPLACE" option includes a provision for inputting arbitrary data in the "space-charge" array. The output from LAPLACE includes a list of the fields on the entire boundary. This can be used to find local peak field strengths and to calculate the capacity of part or all of some configuration.

The program always operates in two dimensions; either R and Z in cylindrical coordinates or Y and X in rectangular coordinates. The rectangular coordinate output retains the R and Z labels, however. Electron orbits are calculated through azimuthal changes (labeled "PHI") referenced to the Z axis. In rectangular coordinates, PHI is actually the third Cartesian coordinate.

EGUN uses a four step Runge-Kutta method of solving the relativistic differential equations given below. Suitable substitutions are used to reduce the three second-order equations to six first-order differential equations.

The independent variable is time but the time interval is calculated from the allowed iteration step and the velocity. It is necessary to use fairly short steps because of the auxiliary calculations that must be made at each mesh unit. Thus it is generally not helpful to use any self-checking "corrector" solving routine. If some unusual application requires shorter iteration steps, the results usually show this by their internal inconsistency.

The relativistic differential equations are

$$\ddot{Z} = \alpha(1-\beta^2)^{1/2}\left[-E_z(1-\dot{Z}^2) + \dot{Z}\dot{R}E_r + \dot{Z}\dot{A}E_\phi - c\dot{R}B_\phi + c\dot{A}B_r\right]$$

$$\ddot{R} = \alpha(1-\beta^2)^{1/2}\left[-E_r(1-\dot{R}^2) + \dot{Z}\dot{R}E_z + \dot{R}\dot{A}E_\phi + c\dot{Z}B_\phi - c\dot{A}B_z\right] + \frac{\dot{A}^2}{R}$$

$$\ddot{A} = \alpha(1-\beta^2)^{1/2}\left[-E_\phi(1-\dot{A}^2) + \dot{Z}\dot{A}E_z + \dot{R}\dot{A}E_r - c\dot{Z}B_r - c\dot{R}B_\phi\right] - \frac{\dot{R}\dot{A}}{R}$$

where
$$\beta^2 = \dot{Z}^2 + \dot{R}^2 + \dot{A}^2 \quad \text{and} \quad \beta = v/c \quad .$$

The constant $\alpha = e\lambda/m_o c^2$ where e is the charge on the particle, $m_o c^2$ is the rest energy of the particle, and λ is the constant of proportionality between the real coordinates and the dimensionless coordinates. Thus,

$$z = \lambda Z, \quad r = \lambda R, \quad a = \lambda A \quad \text{and} \quad ct = \lambda T \quad .$$

By an arbitrary choice, $\lambda = 5.11 \times 10^5$ mesh units so that $\alpha = 1.0$ mesh unit per volt. Inspection of the differential equations shows that they are dimensionally correct if the electric fields are specified in volts per mesh unit.

Dimensionally, $E = vB$, so that in mksa units E is in volts per meter, v is in meters per second and B is in webers per meter.^{2} Then, cB has units of volts per meter. To convert to program fields of volts per mesh unit, magnetic fields are multiplied by the value UNIT in meters per mesh unit. Magnetic field input to the program is in gauss, which is the common engineering unit, and is internally converted to webers/meter^2.

The azimuthal magnetic field B_ϕ comes from the current in the electron beam and is called the self-magnetic field of the beam. The magnetic field created by an axial current is

$$B_\phi = \frac{\mu_o}{2\pi} \frac{I}{r} \text{ webers/meter}^2 \quad .$$

The field is assumed to be due to an infinite conductor which is a good approximation in the area in which the field is significant. After multiplying B_ϕ by the scale factor and expressing r in meters which requires multiplying r by the scale factor also, the scale factor cancels as might be expected. Thus, the scale factor only enters for external magnetic fields. The current I is the summation of the current in the trajectories at lower radii than the trajectory being calculated, but including the one being calculated.

Two field components are neglected. The azimuthal electric field is neglected because of the axial symmetry assumed. The axial magnetic field can have a contribution from the beam due to azimuthal velocity of the beam. The magnitude has been shown to be less than one gauss in most practical cases and so is neglected.

The space-charge is calculated to supply the right side of Poisson's equation which is

$$\nabla^2 V = \frac{\rho}{\epsilon_o} = \frac{J}{v\epsilon_o} .$$

The element of area for J is $r \times 1.0$ square mesh units where r is the particle radius. The velocity is only the Z-component since the space-charge is being spread between adjacent points on the same column. The one mesh unit space between adjacent points accounts for the 1.0 in the area expression above.

In the finite difference form, the right-hand side becomes

$$RO = \frac{36\pi \times 10^9 I(K) \times 10^{-6}}{ABS(ZDOT) \times 3 \times 10^8} = \frac{(3.77 \times 10^{-4})I(K)}{ABS(ZDOT)} ,$$

where RO is to be spread between two adjacent mesh points in inverse ratio to the distance from the ray to each point, $I(K)$ is the current in the one radian segment of the ray (in microamperes) and $ZDOT$ is the velocity in units of c. If the angle of inclination, dR/dZ, exceeds 45°, the calculation is made for $RDOT$. The absolute value of $ZDOT$ is used to allow a negative $ZDOT$. The explicit value of R is canceled by the R which would convert the current to current density, thus avoiding special problems as $R \to 0$.

In practice, however, there are still some space-charge problems near the axis. In rectangular coordinates, if the axis is a plane of symmetry, then any trajectory between $R = 0$ and $R = 1$ has a mirror image between $R = 0$ and $R = -1$. (A reminder...when in rectangular coordinates, the axis still retain their cylindrical labels.) To account for all the space-charge on the axis, the calculated charge is doubled. In cylindrical coordinates, the algorithm for distributing the space charge proportionately to the distance between the adjacent points is not a very accurate solution within one mesh unit of the axis. Good smooth laminar flow near the axis results by simply making the space-charge on the axis equal to that found for the first row.

Magnetic fields, except for the self-magnetic field of a beam, are input directly in one of three ways:

1. by specifying the field along the Z-axis,
2. by specifying a set of coils (giving position, radius and current), or
3. by using the vector potential output from a magnet program such as Poisson. It is interesting to note that Colman[7] has converted several accelerator physics programs including Poisson to run on the IBM-AT.

In cylindrical coordinates, the field is interpreted as an axial magnetic field with radial terms as required by Maxwell's equations. The off-axis fields can be made by either a sixth-order expansion from the axial fields or, for the case of a set of coils, by directly using the appropriate elliptic functions. When the vector potential input has been used, local interpolation is used in place of the expansion.

In rectangular coordinates the magnetic field can be defined to be principally in any one of the three Cartesian directions. Off-median plane expansions are made in the direction of the field on the median plane. If the median plane is the R-Z plane, then the field is in the PHI direction and the field extends to infinity in the R-direction. This fits the configuration of the pole face of a dipole magnet. (Remember that R, Z and PHI are here taken to be orthogonal Cartesian coordinates.) If the median plane lies normal to the plane of the problem, through the Z-axis, then the field extends to infinity in the PHI direction. In this case, the direction of the field on the median plane can be either in the Z-direction or in the R-direction, depending on the symmetry of the coils that produce the field. The off-median-plane expansions in rectangular coordinates satisfy Maxwell's equations to second order.

Self-magnetic fields are calculated for both coordinate systems from the current in the rays on the present cycle. A built-in sort routine insures that the rays are sequentially numbered from the axis outwards. The self-magnetic field calculation assumes all the current from the previous rays lies on the axis in an infinitely long conductor. If the ray being calculated crosses the last preceding ray, then the current from that ray is dropped. If the ray continues to cross other rays, then the current from those rays is only dropped if the ray goes below the minimum radius of a previous ray. Note that if the self-magnetic field is very significant, then almost by definition, one is dealing with an intense relativistic beam. This problem is generally better suited to the paraxial ray approach, as solved in the first cycle, in which the space-charge is offset by the self-magnetic field directly, rather than by the offsetting effects of two large terms. Best results can be obtained for such problems if the electron gun region can be separated from the drift and focusing regions in which the self fields are so important. For cases where the beam is already relativistic in the gun, a new option allows the user to define a kinetic energy above which the direct cancellation of space-charge by the self magnetic field is used, as described earlier, instead of the normal separate terms. This permits the Child's Law calculation to be used near the cathode and the paraxial calculation to be used when the beam is at higher energy.

In rectangular coordinates, the self-magnetic field assumes symmetry about the $Y = 0$, $(R = 0)$ plane. If this is not correct, or if for other reasons it is desired to turn off the self-magnetic field, then an external field of strength zero can be specified. In any case, in rectangular coordinates, the self-magnetic field functions only if there is no external field.

A single variable controls plotting. Normally, at least the last cycle is plotted. The first cycle may also be plotted or one may even plot every cycle. Ray tracing plots may include equipotential plots, either separate or overlaid with the trajectory plots. Figure 1 is an example of the graphic output showing a Pierce diode with equipotential lines and trajectory paths. If there is an external magnetic field, then this field is also plotted, overlaid on the trajectory plots. A feature that is especially useful if a magnetic field is present, is an option that allows one to choose a single trajectory for which the azimuthal position PHI, is plotted as a function of Z. Finally, there are a pair of simple plots; current density versus radius and Alpha versus radius, where Alpha = arctan dR/dZ. The latter plot is equivalent to a phase-space plot.

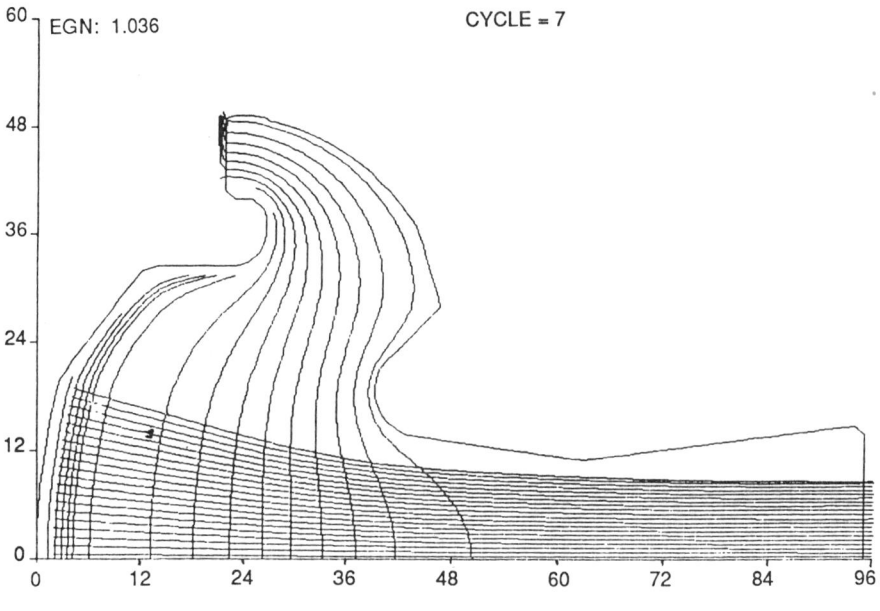

Figure 1. The computer drawn simulation of a Pierce diode, as described in the text. This particular sample was plotted on a dot matrix printer using data sent directly from the PC, not using a screen dump routine.

PERFORMANCE

EGUN has been available for many years in various Fortran versions and is now available from the author as EGN87c, written in C and including an integrated plotting package for the IBM class of PC's. The problem shown in Fig. 1 required 22 seconds to run on the IBM 3080 and 23 minutes on an AT-equivalent machine with 8 MHz clock. The PC had the Intel[8] 80287 (R) Math co-processor installed and 1.1 Mbyte storage, which permits writing the program output directly to virtual disk space. This factor of roughly 60 between the PC and the IBM 3080 seems to hold for a variety of configurations for this program. Faster AT clones and PS-2 machines, both based on the 80286 processor, reduce this time by about half while 80386 based machines with the 32-bit architecture reduce the time a further factor or two to three. IBM XT-type PC's based on the 8086 processor are only about 50% slower.

As usual with a space-charge-dominated configuration such as this, the problem is run for several cycles in order to achieve convergence. In this case, the problem iterated for seven cycles which has long been the default standard for EGUN. There are several convergence processes going on simultaneously in the program. First there is the solution to Poisson's equation, which is the only one that is even remotely guaranteed mathematically to converge. Then there is the convergence of the Child's Law problem on the cathode. This can be divided into two parts, local and global. The global convergence of perveance is achieved by constantly averaging the perveance found by solving Poisson's equation, with the perveance used in the previous cycle. The perveances found on successive cycles for the problem in Fig. 1 illustrate this process. The numbers (all in units of microperveance, defined as $IV^{-3/2} \times 10^6$) are: 1.06, 0.84, 0.71, 0.70, 0.76, 0.76, 0.75. The final value of 0.75 agrees to within the expected tolerance of the observed value for this tube of 0.70. The perveance value before averaging is obtained by defining a starting surface at least two mesh units in front of the cathode and calculating the current that would flow to that surface given the voltage found on the surface by solving Poisson's equation. For the first cycle, this initial value of the perveance is then divided by two. The user can specify an initial value of perveance, and can also request that this value be maintained for a specified number of cycles. This is one way in which temperature limited emission can be modeled. Sometimes knowing the desired perveance helps the program to converge faster, but the method described above, dividing the no-space-charge value by two, frequently converges faster than the alternatives.

Usually the local perveance convergence, that is, convergence of local current density along the cathode surface from cycle to cycle, follows the global calculation without showing hot spots. Exceptions may occur for particularly "bad" designs, or for cases when the starting surface is too close to the cathode. It is necessary to define a starting surface that is at least two, and preferably three mesh units from the cathode.

Another convergence process concerns the final beam configuration. If there are enough rays, i.e., at least one per mesh unit on the starting surface, then usually the seven cycles are enough to give radial convergence to the particle distribution. More cycles may be needed if the problem concerns intense relativistic beams for the reasons discussed earlier. Guns requiring high area convergence ratios may also require more program cycles to find a self-consistent solution to the beam diameter. Strong magnetic fields somewhat aid in the convergence, just as they add stability to the beam in practice. The normal criterion for stability is to examine the output to see whether there is much change from cycle to cycle. The program always reduces the iterative step by a factor of two on the last cycle and also reduces the error criterion for the Poisson solver by a factor of ten. These two measures test for robustness of the solution to calculational limits. In the example of the problem in Fig. 1, the beam radius after seven cycles was 8.7 mesh units. After six it was 8.8 mesh units, a small enough difference to consider the problem converged.

DIAGNOSTICS

The program contains three classes of diagnostics:

1. Input data diagnostics, particularly for the NAMELIST entries, which must correspond to the list of expected elements, and for the boundary data that must fulfill a variety of criteria to be used to determine difference equations for the Poisson solver. Boundary diagnostics include definite errors that would cause the problem to fail, and warnings about data points that do not fit some predetermined criteria of what boundary data should look like, but which may actually be correct. The user should of course be certain that he fixes or clearly understands why each warning is given. There are also fairly extensive diagnostics provided for magnetic field input data, including a plot of the axial field profile that can be plotted at any chosen radius (but usually on the Z-axis), and a table of the on-axis and off-axis magnetic fields.

2. Program operation diagnostics, including messages about the progress of EGN sent to the PC terminal, tabulation of the time spent in each subroutine and the time required for each program cycle, as well as the total time. Convergence data for the Poisson solver is printed on-line and in the listing.

3. Physics diagnostics including tables of potentials and space-charge, and initial and final data for each trajectory, with an option of printing each iteration step from up to six trajectories. The final trajectory information is used to calculate the emittance of the beam.

Even though, by its very nature, EGUN is not a statistical program, aberrations due to nonlinear fields and nonuniform space-charge distributions cause the beam to fill an area in phase space from which emittance can be derived. For the problem illustrated in Fig. 1, the calculated normalized emittance area is about $10\,\pi$ mm-mrad. This is an especially low value for such a gun, which shows that this is an especially good design. For a uniform beam, which is the ideal of an electron gun, the normalized emittance is

$$\epsilon_n = 4\beta\gamma(<x^2><x'^2> - <x \times x'>^2)^{1/2} \quad ,$$

where $x' = dx/dz$, the $<>$ brackets signify weighted averages for all the particles, and $\beta\gamma$ is the relativistic velocity times the relativistic mass. It is the $\beta\gamma$ product that makes this the "invariant" or "normalized" emittance, as it is usually called. It is invariant because subsequent acceleration does not cause the emittance area to change.

ACKNOWLEDGEMENTS

We wish to acknowledge the continued support and encouragement of Dr. Roger Miller (SLAC). The origin of EGUN is based on the work of Vladimir Hamza[8] at Stanford University. Finally, our thanks go to all of the users of EGUN, many of whom have made helpful suggestions and provided important benchmarks to test the program.

REFERENCES

1. W. B. Herrmannsfeldt, SLAC Electron Trajectory Program, SLAC-226, November 1979. Also available as a revised version in draft form, January 1988.
2. Jack Boers, SNOW and SNOW 3-D, papers at this conference.
3. John Boyd and Dennis Hewett, Bull. Am. Phys. Soc. **29**, 1436, (1984).
4. Richard True, IDEM Technical Digest, pp. 257-60, December 1985.
5. A. C. Paul and V. K. Neil, IEEE Trans. on Nucl. Sci. **NS-26**, No. 3, June 1979.
6. Peter Spädke, Computer Simulation of High-Current DC Ion Beams, Proceedings of the 1984 Linac Conference, Seeheim.
7. Judith Colman, The Use of the IBM PC Computer in Accelerator Design Calculations, Proceedings of the 1986 Linear Accelerator Conference, Stanford Linear Accelerator Center, Stanford, CA 94309, June 1986. Ms. Colman's address is Neutral Beam Division, Brookhaven National Laboratory, Upton, NY 11973.
8. Intel is a registered trademark of the Intel Corporation.
9. Vladimir Hamza et al., NASA TN-1323, TN-1665, TN-1711, Lewis Research Center, Cleveland, Ohio (1962).

STATUS OF THE "PATH" MAGNETIC OPTICS DESIGN CODE*

R. J. Kashuba, R. J. Schmitt,
and P. F. Meads, Jr.[†]
McDonnell Douglas Astronautics Company, St. Louis, MO 63166

ABSTRACT

The PATH codes are used to design magnetic optics subsystems for neutral particle beam systems. This paper describes several checks and modifications that have been made to PATH. The third order hard-edge quadrupole model in PATH has been checked against direct numerical integration (raytracing) and found to be accurate to better than 0.5 microradian. The original third order linear ramp quadrupole fringe field model in PATH has been checked against raytracing and against an alternative linear ramp fringe field model and found to be accurate to better than 0.1 microradian. An alternative third order nonlinear ramp quadrupole fringe field model has been added to PATH. This model is a close approximation to the fringe field of a Halbach ring magnet and has been checked by raytracing and has been found to be accurate to better than 0.1 microradian. PATH has been modified to model combined quadrupole/octupole elements through third order. The paper describes this model and presents results obtained by optimizing quadrupole and octupole elements to produce lowest beam divergence performance for a neutral particle beam expander.

INTRODUCTION AND HISTORY

The PATH (PARMILA and TURTLE Hybrid) code has been developed from several codes which have been in existence for over 20 years (see Figure 1). Unlike many other codes, PATH has, from its inception, been developed with the purpose of designing neutral particle beam (NPB) relevant optics systems. PATH's linear accelerator modeling code, LINAC, is a variation of the standard ion linac code PARMILA (Phase and Radial Motion in Ion Linear Accelerators). Farrell and Rusthoi (LANL) adapted PARMILA[1] for use in PATH by incorporating a common input/output file structure into PARMILA for inter-communication purposes. To model generalized magnetic optic beamlines, PATH uses models for magnetic elements (dipoles, quadrupoles, sextupoles, and octupoles) obtained from the standard beamline design codes TRANSPORT[2] and TURTLE.[3] These models are collected into PATH's optical beamline modeling code called TRAVEL. Farrell and Rusthoi also added a 2-1/2 D nonlinear space charge model

---

* Supported by McDonnell Douglas Independent Research and Development Program.
† Consultant

(PARMILA's subroutine SCHEFF) and a third order quadrupole linear ramp fringe field model to TRAVEL. We added nonlinear fringe field models in 1987. During the same period most of the magnetic element subroutines were re-analyzed to eliminate possible bugs. These models are particularly useful for designing high performance NPB beam expander optical systems.

As an aid in checking out the TRAVEL code, parts of the MIT RAYTRACE[4] code were added to TRAVEL. TRAVEL and RAYTRACE now use the same set of initial rays and can calculate the same first, second, and third order beamline matrix components for comparison purposes. RAYTRACE is a standard magnetic optics design code used for high precision systems such as particle spectrometers.

We succeeded in adding an unconstrained optimizer (minimizer) to the TRAVEL code. This optimizer (called M-O-M) was developed at LANL by C. T. Mottershead and K. Overly. Using M-O-M, magnetic optics designs problems can now be solved automatically (e.g. quadrupole and octupole fields can be determined iteratively in a single computer run to minimize beam spot size on a distant target). Design productivity improvements by a factor of at least ten have been realized using this optimizer.

SECTION I. PATH STRUCTURE

Below we present the basic codes which comprise PATH (see Figure 2).

(1) BEAMGEN - A beam generator program that uses a random number generator to create a beam file of up to 5,000 6-D particle coordinates. Uniform, gaussian, Kapchinskij-Vladimirskij (K-V), and binominal distributions are available.

(2) TRAVEL - The beam optics program TRAVEL, a modified version of TURTLE, transports the particle beam through a series of optical elements. There are the usual elements such as drifts, quadrupoles, sextupoles, octupoles, and bending magnets, and several extra options such as space charge (2-1/2 D), buncher cavity, accelerator column, and both DC and RF linac accelerating gaps. Some elements are modeled to third order (e.g. quadrupoles, octupoles). TRAVEL also has models for Halbach-type quadrupole and octupole magnets and can handle overlapping fringe field regions.

(3) PLOT - This program can generate a selection of histograms, contour, and scatter plots at the desired locations along the beam line. It can produce graphics with the output from BEAMGEN, TRAVEL, or LINAC.

(4) LINAC - This linac program is a modified version of PARMILA that generates output for the PLOT program. In addition, LINAC generates a linear accelerator beamline

comprised of quadrupoles, drifts, and accelerating gaps
that can be used as beamline elements in TRAVEL. The
LINAC program simulates an Alvarez linear accelerator.

Both TRAVEL and LINAC can use the same beam data file, either that
generated by BEAMGEN or the beam data files generated as output by
the two programs themselves. A listing of the input and output
files for each program is as follows:

Program	Input Files Required	Output Files Generated
BEAMGEN	none	beam.dat
TRAVEL	beam.dat travel.in	rays.dat (for PLOT) travel.out (summary) beamt.dat
LINAC	beam.dat linac.in	rays.dat (for PLOT) linac.out (summary) line.dat (for TRAVEL) beaml.dat
PLOT	rays.dat	plot.out (summary)

The beam files (beam.dat, beamt.dat, and beaml.dat) are the same
in format and may be used as the input beam in either LINAC or
TRAVEL. The PLOT program will also accept a beam file in lieu of
rays.dat. The "beam" files and the "rays" files are binary.

SECTION II. CODE ELEMENTS

In this section, we describe three of the basic models
necessary for the design of a beam expander. We discuss the space
charge model (because of its importance), the nonlinear quadrupole
fringe field model and integration technique, and, lastly, the
optimizer model. Models which are not discussed explicitly are of
the standard form.

SPACE CHARGE − The PATH code is basically a simulation code
(i.e., a lumped-element raytrace) for the transport of a particle
beam through a number of optical elements. Each particle in the
PATH code is representative of some rather large number of
particles ($\sim 10^{15}$) in an actual beam for the currents of
interest for a neutral particle beam (0.1 to 1.0 ampere).

PATH uses a 2-1/2 D nonlinear space charge model (called
SCHEFF) to simulate both continuous (DC) particle beams and
bunched beams produced by linear accelerators. SCHEFF establishes
a mesh in cylindrical coordinates (r, ϕ, z) which covers a
particle bunch or a portion of a DC beam. The ϕ-dependence is
integrated out during the space charge calculation (i.e., the
model is 2-1/2 D). Nonlinear space charge forces are calculated
between rings of charge using elliptic integrals.

The importance of the space charge forces for beam expander design is illustrated in Figure 3. The RMS emittance is plotted versus drift length for a 100 mA beam with ΔP/P=0.026 percent. The transverse emittance continues to grow linearly with beam expander length. The nature of the emittance increase can be understood by examining Figure 4. The results are plotted with the ellipses rotated to lie horizontally. We can see from both the x-x' and y-y' plots the nonlinear distortions produced by the space charge forces. This should be contrasted with the case of a general beam dilution where the x-x' and y-y' plots would remain more or less elliptical. In the case illustrated here, most of this emittance growth can be compensated by using octupole fields in the objective lens. This is discussed in more detail in Section III.

FRINGE FIELDS - PATH contains both linear ramp and nonlinear fringe field models for quadrupole magnets. The linear ramp model was included in the original version of PATH by Farrell and Rusthoi. We added the nonlinear fringe field models in 1987.

In order to illustrate the methodology used in PATH to calculate higher order terms, we consider an quadrupole magnet with a nonlinear fringe field. The model for the magnet and fringe field is illustrated in Figure 5. For the fringe field, a fifth-order polynomial is used to model the magnetic strength function in the ramp region which we take to be equal to twice the bore radius, a. Thus, the fringe field region extends to a distance "a" both outside and inside the physical magnet. The form of the polynomial is chosen such that both the first and second derivatives of k(z) are zero at the end of the ramp region and, in general, are non-zero inside the ramp region. These form part of the contribution to the third-order aberration terms. This polynomial was investigated because of its analytic simplicity and its general agreement with the Halbach model as illustrated in Figure 6.

For purposes of calculation, the ramp region is divided into a number of steps. During each step, the particle position can be calculated from the integral equation

$$x(\zeta) = x_0 + x_0' \zeta + \int k(\zeta') x(\zeta')(\zeta-\zeta') d\zeta' \qquad (1)$$

which is just the solution of Newton's equation of motion for a first order quadrupole, i.e.

$$\frac{d^2x}{dz^2} = k(z) x(z). \qquad (2)$$

A sufficiently accurate solution can be obtained by a double iteration of Eq. (1). These solutions are given by:

$$x_1(\zeta) = x_0 \{ 1 \mp (\frac{k\zeta^2}{2} + \frac{k'\zeta^3}{6} + \frac{k''\zeta^4}{24})$$

$$+ \frac{k^2\zeta^4}{24} + \frac{kk'\zeta^5}{30} + (\frac{7kk''}{4}+k'^2) \frac{\zeta^6}{180} + \frac{k'k''}{840}\zeta^7 + \frac{k''^2\zeta^8}{2688} \}$$

$$+ x'_0 \{ \zeta \mp (\frac{k\zeta^3}{6} + \frac{k'\zeta^4}{12} + \frac{k''\zeta^5}{40})$$

$$+ \frac{k^2\zeta^5}{120} + \frac{kk'\zeta^6}{120} + (\frac{13kk''}{10}+k'^2) \frac{\zeta^7}{504} + \frac{k'k''\zeta^8}{840} + \frac{k''^2\zeta^8}{5760} \}$$

(3)

$$x'_1(\zeta) \left[\equiv \frac{dx(\zeta)}{d\zeta} \right]$$

$$= x_0 \{ \mp (k\zeta + \frac{k'\zeta^2}{2} + \frac{k''\zeta^3}{6})$$

$$+ \frac{k^2\zeta^3}{6} + \frac{kk'\zeta^4}{6} + (\frac{7kk''}{4}+k'^2) \frac{\zeta^5}{30} + \frac{k'k''\zeta^6}{48} + \frac{k''^2\zeta^7}{336} \}$$

$$+ x'_0 \{ 1 \mp (\frac{k\zeta^2}{2}+\frac{k'\zeta^3}{3}+\frac{k''\zeta^4}{8})$$

$$+ \frac{k^2\zeta^4}{24} + \frac{kk'\zeta^5}{20} + (\frac{13kk''}{10}+k'^2) \frac{\zeta^6}{72} + \frac{k'k''\zeta^7}{105} + \frac{k''^2\zeta^8}{640} \}$$

(4)

The $y_1(\zeta)$ and $y'_1(\zeta)$ solutions are obtained by selecting the lower sign in the "$\mp$" symbol.

The second and third order terms can then be calculated using straightforward perturbation techniques. The second and third order terms are given by:

$$\Delta x_3 = k \, (\frac{\Delta P}{P_0}) \, x_1 \, (1-\frac{\Delta P}{P_0})$$

$$- \frac{3}{2}kx_1x'_1{}^2 - \frac{1}{2}kx_1y'_1{}^2 + (kx'_1+k'x_1) \, y_1y'_1$$

$$+ k''x_1 \, (\frac{1}{3}x_1{}^2 + y_1{}^2) \, / \, 4$$

(5)

$$\Delta y_3 = -k \left(\frac{\Delta P}{P_0}\right) y_1 \left(1-\frac{\Delta P}{P_0}\right)$$

$$+ \frac{3}{2} k y_1 {y'}_1^2 + \frac{1}{2} k y_1 {x'}_1^2 - (k{y'}_1 + k'y_1) x_1 x'_1$$

$$- k'' y_1 \left(\frac{1}{3} y_1^2 + x_1^2\right) / 4 \tag{6}$$

Note that the values of k, k', and k" are evaluated at $(\zeta + h)$.

The solutions through third order are given by the summations,

$$\begin{aligned} x_3 &= x_1 + h^2 \Delta x_3 \\ x'_3 &= x'_1 + h \Delta x_3 \\ y_3 &= y_1 + h^2 \Delta y_3 \\ y'_3 &= y'_1 + h \Delta y_3 \end{aligned} \tag{7}$$

Basically, we obtain a solution for a single interval and then step through the quadrupole magnet. In order to obtain solutions valid through third order over a wider interval such as the complete fringe region, more sophisticated techniques would have to be employed. This problem is quite difficult and, in fact, a number of errors exist in the results reported in the literature. However, the simple summing technique given here is quite accurate as shown in the following table, which compares the results to those obtained by numerically solving the differential form of Newton's equation of motion using Runge-Kutta techniques.

NONLINEAR FRINGE FIELD (NFF) MODEL VS.
DIRECT NUMERICAL INTEGRATION (DNI)

Z(m)		X(m)	X'(Radian)	Y(m)	Y'(Radian)
0.4583	DNI	5.221070E-02	4.30111E-03	5.237268E-02	5.69982E-03
	NFF	5.221076E-02	4.29999E-03	5.237262E-02	5.70094E-03
0.4889	DNI	5.2340350E-02	4.18396E-03	5.2548632E-02	5.81738E-03
	NFF	5.2340353E-02	4.18303E-03	5.2548630E-02	5.81831E-03
0.5194	DNI	5.246637E-02	4.06503E-03	5.272821E-02	5.93684E-03
	NFF	5.246634E-02	4.06447E-03	5.272824E-02	5.93740E-03
0.5500	DNI	5.258876E-02	3.94602E-03	5.291144E-02	6.056517E-03
	NFF	5.258873E-02	3.94603E-03	5.291148E-02	6.056501E-03

The additional effects due to space-charge forces can then be added at some or all of the steps. Since the quadrupole must be divided into a number of steps in order to calculate the space charge effects, there is no point in developing techniques which give the solution for a single particle through the complete fringe field rather than using a summing solution which is valid over a smaller segment of the magnet (i.e. Eq. (7)).

M-O-M OPTIMIZER - This optimizer is a special-purpose unconstrained minimization routine which finds the minimum of a nonlinear function $f(p_i)$, i=1, 40,

$$f = \sum_{i=1}^{40} [(p_+ - p_*)_i / \delta_i]^2 \qquad (8)$$

where $(p_+)_i$ = present value of the ith beam parameter,
       $(p_*)_i$ = desired value of the ith beam parameter,
       δ_i = tolerance allowed in $(p_+ - p_*)_i = 0 \pm \delta_i$

For example, p_i could be the RMS beam radius and specifying $(p_*)_i = 0$ would cause the M-O-M optimizer to minimize this variable by adjusting quadrupole and octupole fields. Each beam parameter is a function of up to 10 beamline independent variables b_j,

$$p_i = p_i(b_1, b_2, \ldots b_{10}).$$

The variable b_j has the form

$$b_j = b_{oj} + \delta b_j$$

where b_{oj} is the initial value of the jth independent beamline variable and δb_j is the change in the jth variable recommended by the M-O-M optimizer on a particular iteration. The independent variables b_j typically are quadrupole and octupole magnetic field strengths and drift lengths.

The values of p_j are calculated via a raytrace using PATH with typically 100-2000 rays. PATH uses the M-O-M optimizer as an oracle, supplying it with current values of b_j and $f(p_j)$ and then receiving advice in terms of a better vector variable $(\delta b_j)_+$ in return. The M-O-M optimizer is designed to reduce the number of calls to PATH for calculating the function $f(p_j)$ in exchange for more linear algebra overhead. This is appropriate because the typical PATH raytrace calculation with space charge and third order effects takes from 2 to 10 minutes of MicroVAX computer time. M-O-M employs a quasi-Newton optimization algorithm with convergence to a local minimum ensured by making use of a trust region globalization strategy.[5]

In an attempt to make best use of each function call to PATH, M-O-M does not calculate gradients using finite differences. Instead, given n independent variables, b_i, M-O-M keeps track of the previous best $(n+1)(n+2)/2$ points to which it fits and minimizes a quadratic form. This corresponds to modeling the function f with a second order Taylor series, using a secant approximation to the gradient and to the second derivative Hessian matrix. M-O-M then solves a linear system to find the new values of b_i at the minimum of the quadratic form. This process is repeated until no further progress is realized in minimizing the function $f(p_i)$.

SECTION III. CODE APPLICATION

In this section we describe a beam expander design problem using the PATH code. In addition, we show the differences between several alternate models of the fringe fields.

The beam expander considered here consists of a quadruplet eyepiece and doublet objective. In addition, preceding the eyepiece is a two-gap momentum compactor which yields $\Delta P/P=0.026$ percent. The beam current is 100 mA. The design is carried out through third order using a 2-1/2 D space charge model and the nonlinear fringe model described above.

For this beam expander with a 30 cm aperture, we have the results shown in Figure 7. The design allowed for approximately 4 percent of the particles to be scraped off. Figures showing x-x' and y-y' have the typical shapes indicating the nonlinear aberration terms. Most of the aberration is produced in the fringe field region. With the addition of 40 cm aperture octupoles, we have the results illustrated in Figure 8. Note the reduction in scale of the transverse divergence compared to Figure 7. This is due to the fact that the octupoles reduce the effects of cubic aberration terms.

Because of the difficulty of correctly modeling the fringe field, we examined the difference between various fringe field models. The results of this calculation are illustrated in Figure 9. Here we have plotted the fraction of the beam propagating inside a normalized divergence angle, Θ_N. The models considered are: A, the nonlinear fringe field model; B, a linearized version of the same model; and C, the original linear model in the Farrell-Rusthoi version of the code. We see that the difference in the results of the two linear models are very small and not significant. With the addition of the nonlinear terms, the beam quality decreases due to the aberration effects proportional to k". The interesting aspect is that the nonlinear fringe field model does not produce a large effect for the particular beam expander considered here.

Finally, we consider the beam expander design using the M-O-M optimizer discussed above. A case was run in which the optimizer found a solution after 24 iterations. The convergence criterion versus number of iterations is plotted in Figure 10. The quadrupoles and octupoles were adjusted to yield a double waist at

the beam expander output plane and a minimum spot size on target.
A comparison of the beam expander design obtained by the optimizer
with the one which was found using hand optimization is shown in
Figure 11. Curve A corresponds to the same curve in Figure 9 and
curve A' is the design performance of the M-O-M optimized beam
expander. The results are nearly identical with the optimizer
yielding a slightly better design with greatly reduced labor
expenditure.

SECTION IV. CONCLUSIONS

The PATH code is an excellent tool for magnetic optics
design. Its modular aspect and standardized subroutine
construction allow for easy use and modification. Currently, it
can accurately model standard quadrupole/octupole magnetic systems
with a nonlinear fringe field through third order, including the
effects of nonlinear space charge. The code also features an
optimizer which can improve productivity by a factor of ten. Of
course, further improvements can be carried out.* These include
the incorporation of a full 3-D nonlinear space charge model and
an improved optimizer. Another useful feature would include
better interfaces to other analysis codes such as NASTRAN. This
would allow better modeling of beamline imperfections such as
misalignment.

REFERENCES

1. G. P. Boicourt, "PARMILA - An Introduction," LA-UR 88-106, 19 January 1988.
2. K. L. Brown, F. Rothacker, D. C. Carey, Ch. Iselin, "TRANSPORT - A Computer Program for Designing Charged Particle Beam Transport Systems," SLAC-91, May 1977.
3. D. C. Carey, "TURTLE - A Computer Program for Simulating Charged Particle Beam Transport Systems," NAL-64, December 1971.
4. S. Kowalski, H. A. Enge, "RAYTRACE," Laboratory for Nuclear Science and Department of Physics, MIT, 1 February 1986.
5. J. E. Dennis, Jr., and R. B. Schnabel, "Numerical Methods for Unconstrained Optimization and Nonlinear Equations," Prentice-Hall, Inc. Englewood Cliffs, NJ, 1983.

---

*Note: Recent additions to TRAVEL include off-axis quadrupoles and RF deflectors for beam funnel designs and the capability to model beamlines with overlapped quadrupole and octupole fringe fields.

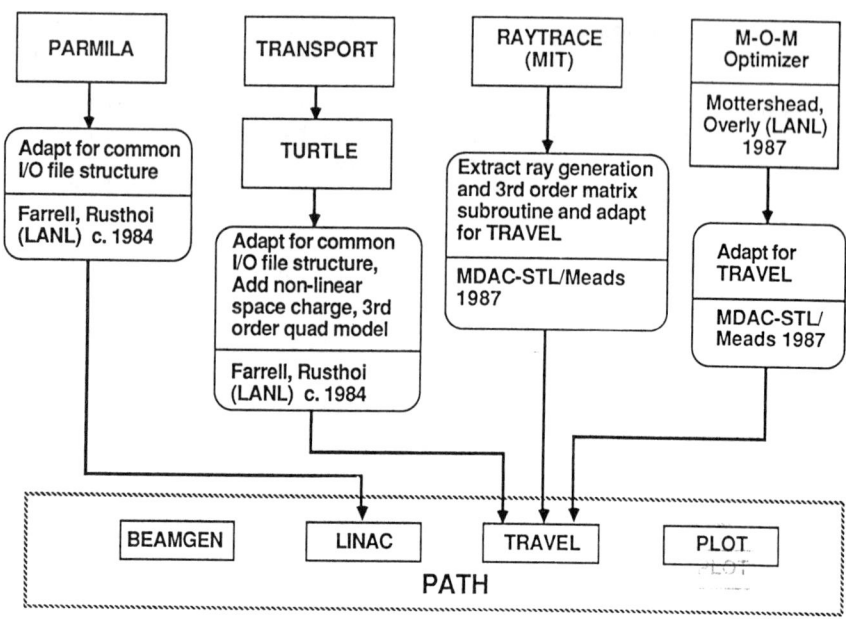

Figure 1. PATH History

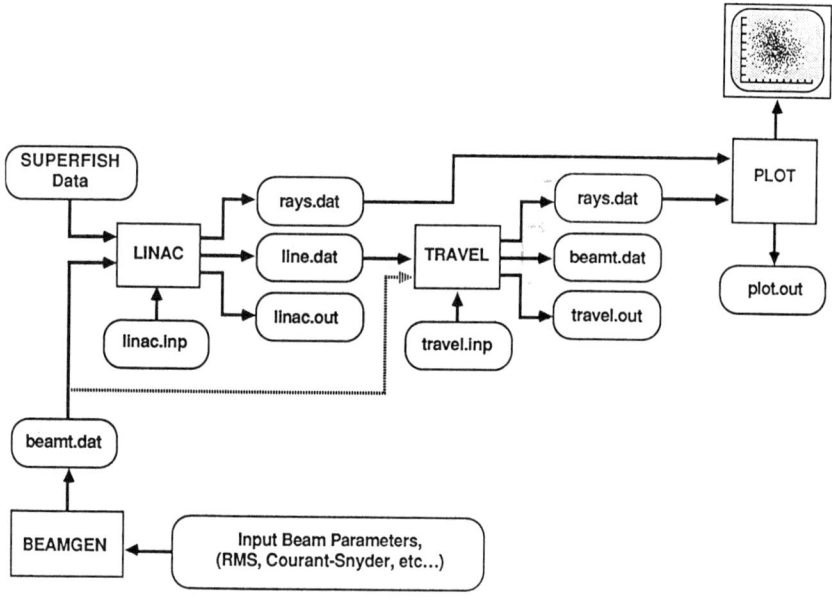

Figure 2. PATH Structure

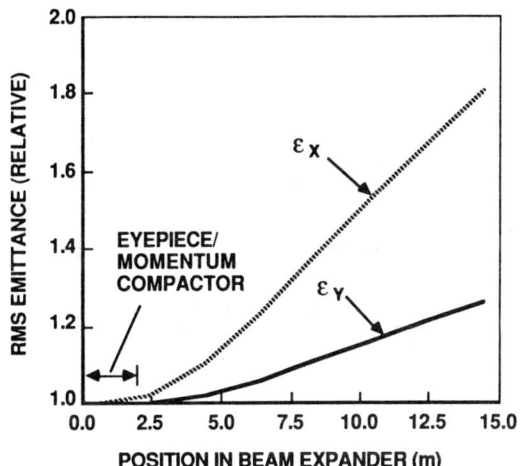

Figure 3. Space charge induced transverse emittance growth in a beam expanding telescope. The (x-x') and (y-y') particle distributions are gaussian. Beam current is 100 mA (H⁻) and the SCHEFF space charge model is used.

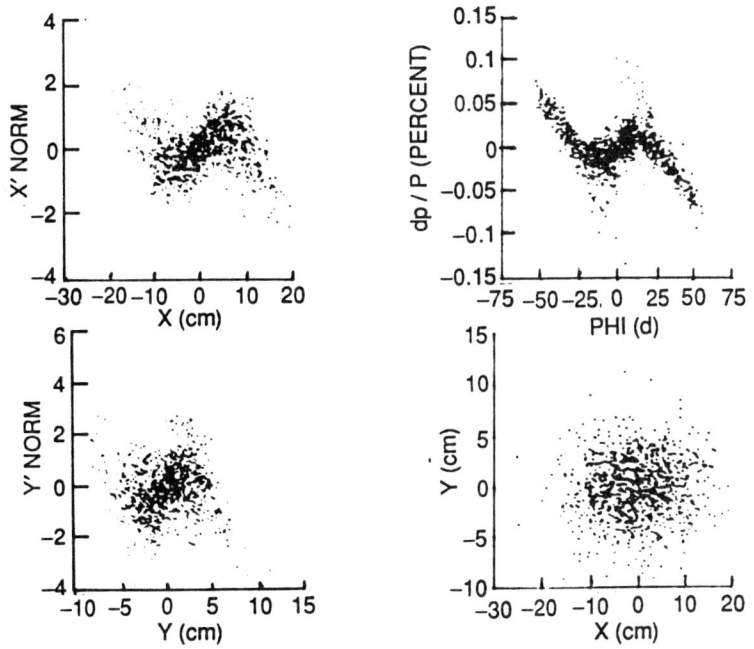

Figure 4. Beam phase space plots at the beam expander objective magnet input plane. The linear correlation in (x-x') and (y-y') has been removed (i.e., the ellipses have been rotated nearly horizontal) to reveal the distortion due to nonlinear space charge effects.

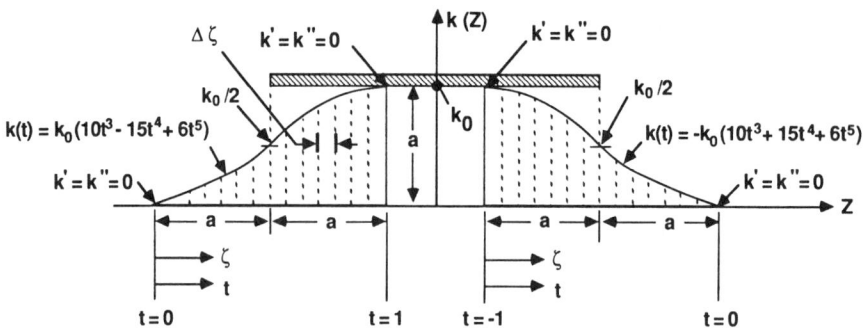

Figure 5. The new quadrupole nonlinear fringe field model added to PATH.

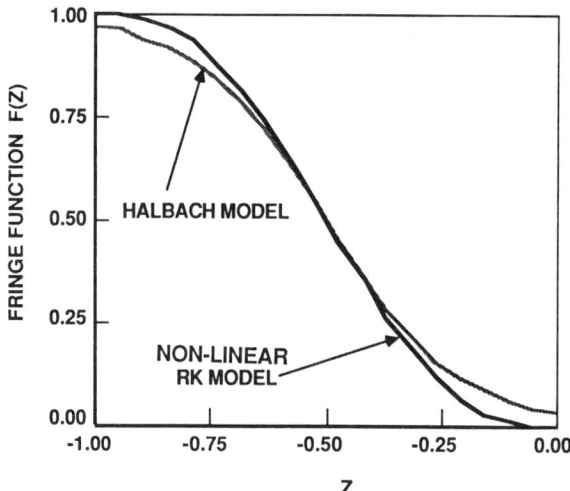

Figure 6. Comparison of quadrupole fringe field models. Both of these models are available in the present version of PATH. The nonlinear model usually executes slightly faster than the Halbach model.

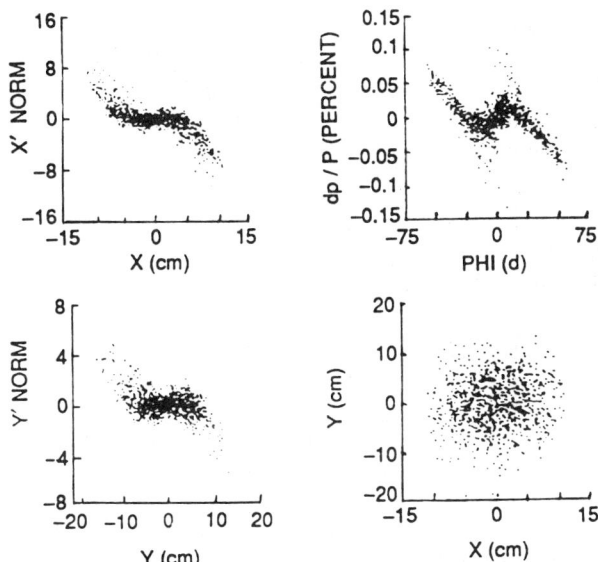

Figure 7. Scatter plots for a beam expander doublet objective without octupole correction. The (x-x') and (y-y') scatter plots show the characteristic aberration signature due to third order quadrupole effects in the fringe field regions and to nonlinear space charge.

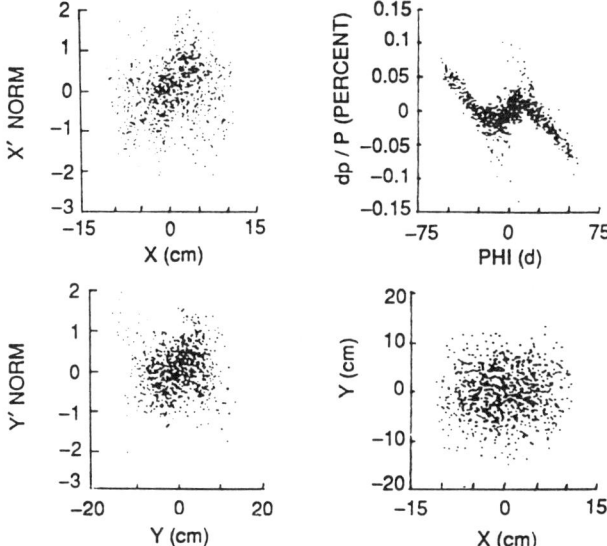

Figure 8. Scatter plots for a beam expander doublet objective with compensation octupoles added. Note the change in scale for x' and y' compared to Fig. 7. The octupoles increase current on a distant target by approximately 50 percent by correcting the aberrations of Fig. 7.

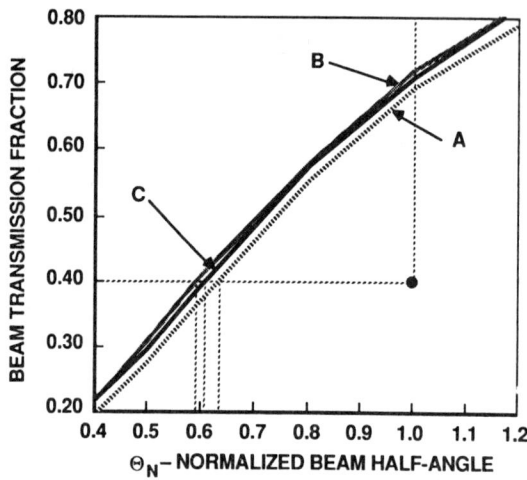

Figure 9. Comparison of fringe field models used in the beam expander doublet objective design.
Curve A: Nonlinear fringe field model of Fig. 5.
Curve B: Linear fringe field model based on analysis of Fig. 5 model.
Curve C: Linear fringe field model in the original PATH code.

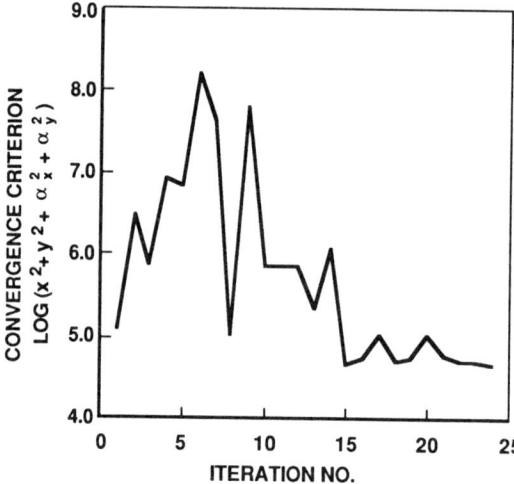

Figure 10. M-O-M optimizer performance obtained in adjusting the quadrupoles and octupoles of the doublet objective of Fig. 8. Note: x and y are RMS beam radii; α_x and α_y are Courant-Snyder parameters.

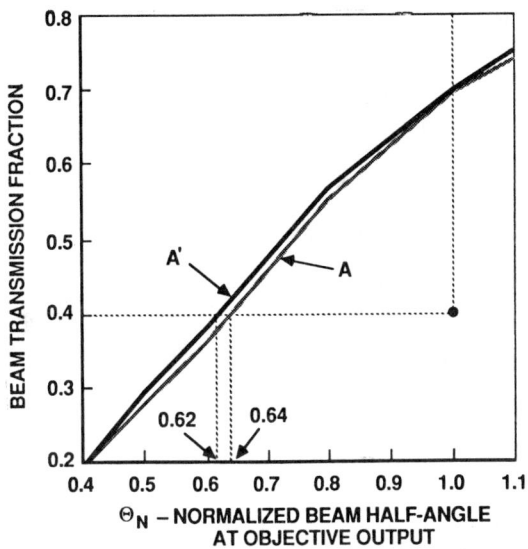

Figure 11. Beam expander optimization. Curve A' is the performance obtained via the M-O-M optimizer. Curve A is the hand-optimized design. The M-O-M optimizer solution required approximately 1 hour of computer time (batch mode) while hand optimization required about 12 hours of interactive computer work.

PRINCIPLES OF GIOS AND COSY

H. Wollnik, B. Hartmann and M. Berz

2. Physikalisches Institut, Universität Giessen, 6300 Giessen, W. Germany

ABSTRACT

The underlying principles of the transfer–matrix programs GIOS and COSY are outlined. Especial emphasis is given to the calculation of particle trajectories and of image aberrations. Input and output of a second–order calculation of GIOS is shown.

1. INTRODUCTION

Over the years we have developed the programs GIOS and COSY in order to facilitate the design of particle spectrometers and of beam guidance systems. GIOS[1,2] has been usable since about 1976 and COSY[3] since about 1986. Both programs are being improved constantly so all particle–optical problems can be solved as they arise.

2. METHODS OF CALCULATION

GIOS and COSY are programs which describe optical systems by describing deviations of arbitrary particle trajectories from particles moving along the optic axis, the z–coordinate of a curvilinear coordinate system[4]. These deviations are:

$$x, a, \delta_t, \delta_K, \delta_m, y, b .$$

Here x and y are coordinates perpendicular to the optic axis along which a reference particle of charge q_0, energy K_0 and rest mass m_{00} can move. Furthermore, an arbitrary particle with energy-to-charge and rest mass-to-charge ratios

$$\frac{K}{q} = \frac{K_0}{q_0}(1 + \delta_K) \qquad \frac{m_0}{q} = \frac{m_{00}}{q_0}(1 + \delta_m)$$

is assumed to have velocity components v_x, v_y, v_z and corresponding momentum components p_x, p_y, p_z with p_0 being the momentum of a reference particle. These quantities define the inclinations of the corresponding particle trajectory as:

$$a = p_x/p_0 \qquad b = p_y/p_0 .$$

In detail, an optical system is characterized by transfer matrices that relate the final deviations $x_n, a_n, \delta_{tn}, \delta_{Kn}, \delta_{mn}, y_n, b_n$ with the initial deviations $x_0, a_0, \delta_{t0}, \delta_{K0}, \delta_{m0}, y_0, b_0$. Here the quantity δ_t is defined by

$$T = T_0(1 + \delta_t).$$

where T_0 is the time a reference particle needs to move along the optic axis while T is the time an arbitrary particle needs to move along its trajectory.

In general, GIOS assumes midplane symmetry for the total particle optical system and uses third–order transfer matrices. These transfer matrices describe all relations of Gaussian optics of narrow beams, i.e., linear deviations in $x, a, \delta_t, \delta_K, \delta_m, y, b$, as well as all aberrations of second and third order, i.e., quadratic and cubic deviations in $x, a, \delta_t, \delta_K, \delta_m, y, b$.

The first order transfer matrices of GIOS read:

$$\begin{pmatrix} (x,x) & (x,a) & (x,\delta_t) & (x,\delta_K) & (x,\delta_m) \\ (a,x) & (a,a) & (a,\delta_t) & (a,\delta_K) & (a,\delta_m) \\ (\delta_t,x) & (\delta_t,a) & 1 & (\delta_t,\delta_K) & (\delta_t,\delta_m) \\ 0 & 0 & 0 & 1 & 0 \\ 0 & 0 & 0 & 0 & 1 \end{pmatrix}$$

$$\begin{pmatrix} (y,y) & (y,b) \\ (b,y) & (b,b) \end{pmatrix}.$$

GIOS allows the different optical elements to be rotated with respect to each other only for the problem of calculating intensity distributions by tracing a number of numerically defined particles through the in principle finalized optical system. COSY allows such rotations for all calculations and for this reason uses larger transfer matrices which contain skew terms and in addition to GIOS transfer matrices include aberrations of fourth and fifth order.

In both programs, these transfer matrices are concatinated by a matrix multiplication subroutine. In principle, the transfer matrices are of order N^2 where N is 49+34 in GIOS and 461 in COSY. However, there are only six independent variables characterizing an arbitrary particle and thus 45+32 dependent ones in GIOS and 455 in COSY. To save computer storage, in GIOS only the first 4+2 lines and in COSY only the first 6 lines of the corresponding transfer matrices are stored so that in GIOS transfer matrices of $4 \times 49 + 2 \times 34 = 264$ elements are sufficient and in COSY transfer matrices of $6 \times 461 = 2766$ elements. The only disadvantage of the usage of these drastically smaller transfer matrices is that the concatination routine is not a simple matrix multiplication.

Note here that between the different matrix elements interrelations exist[4,5] because the calculated particle trajectories must fulfill the canonical equations of motion. This so-called symplectic condition was tested and shown to be fulfilled for all elements of GIOS and of COSY transfer matrices.

2.1 PARTICLE MOTION IN THE MAIN FIELDS

The motion of charged particles can be described algebraically in separated or superimposed magnetic or electrostatic quadrupoles, hexapoles, and octopoles as well as in rotationally symmetric inhomogeneous sector fields in which the optic axis is a circle of radius ρ_0 in the xz-plane. In all these cases the field distribution does not vary with the z-coordinate measured along the optic axis, so that the electrostatic and the scalar magnetic potentials can be described by

$$V(x,y) = \sum_{i,j=0}^{\infty} a_{ij} \frac{x^i y^j}{i!\, j!} . \tag{1}$$

In order for the final fields to fulfill Maxwell's equations, this potential must fulfill Laplace's equation which reads in cylindrical coordinates $\Delta V = (\rho_0 + x)^{-1} \partial V/\partial x + \partial^2 V/\partial x^2 + \partial^2 V/\partial y^2 = 0$. From this relation one finds generally[4,6]

$$a_{i,j+2} = -a_{i+2,j} - i a_{i-1,j} - (i+1) a_{i+1,j} .$$

Using this relation the electrostatic and scalar magnetic potentials can be determined with the side condition that because of symmetry the a_{ij} must vanish for odd j in the electrostatic and for even j in the magnetic case, yielding:

$$V_E(x,y) = a_{10} x + a_{20} \frac{x^2}{2} - (a_{10} + a_{20}) \frac{y^2}{2} + a_{30} \frac{x^3}{6}$$
$$+ (a_{10} - a_{20} - a_{30}) \frac{xy^2}{2} + \ldots \tag{2a}$$

$$V_B(x,y) = a_{01} y + a_{11} xy + a_{21} \frac{x^2 y}{2} - (a_{11} + a_{21}) \frac{y^3}{6} + \ldots \tag{2b}$$

Thus in both cases it is only necessary to know the components $E_x(x,0) = -(\partial V_E/\partial x)_{y=0}$ of the electrostatic field and $B_y(x,0) = -(\partial V_B/\partial y)_{y=0}$ of the magnetic flux density in the plane $y = 0$, i.e., the coefficients a_{i0} or a_{i1}, respectively, to the desired order κ with $i \leq \kappa$ in the electrostatic case or $i \leq \kappa - 1$ in the magnetic case.

From the potential distributions of Eqs. (2) the distributions of the electrostatic field $\vec{E}(x,y)$ or the magnetic flux density $\vec{B}(x,y)$ are obtained as $-\mathrm{grad}\, V$. Using these $\vec{E}(x,y)$ and $\vec{B}(x,y)$, the differential equations of motion of charged particles can be determined, the solutions of which have been obtained manually for GIOS, and for COSY by HAMILTON[7], an especially designed formula manipulator program. In both cases the resultant formulae are evaluated in FORTRAN-subroutines which determine the elements of the corresponding transfer matrices.

2.2 PARTICLE MOTION IN THE FRINGING FIELDS

In GIOS as well as in COSY, great care is taken that the particle motion is described everywhere as precisely as possible. This is especially important in the fringing-field regions[4,8,9,10]. To illustrate the occurring difficulties, one may observe the motion of charged particles in the fringing field of a homogeneous magnetic sector or similarly in an electrostatic sector. As shown in Figure 1, a charged reference particle moves in the initially field-free region along a straight trajectory the curvature of which then increases more and more until finally the trajectory becomes a circle in the main-field region. One may now extrapolate the initially straight trajectory in a forward direction and the finally circular trajectory in a backward direction and observe that the inclinations of the two extrapolated trajectories have different values for all but one line $\zeta = \zeta_{eff}$ parallel to the pole face boundary. This line is called the effective field boundary (see Fig. 1) since if the main field were extended to this line, the particle would be deflected the same as if it were moving in the really existing field[4,8,9,10].

To describe a bundle of charged particles, one now must describe how all corresponding trajectories deviate from the above defined optic axis. In detail this amounts to assuming that all particle trajectories are straight before and circular after the effective field boundary and that at the effective field boundary these particle trajectories experience bends and shifts the magnitudes of which depend on how much each individual trajectory deviates from the optic axis.

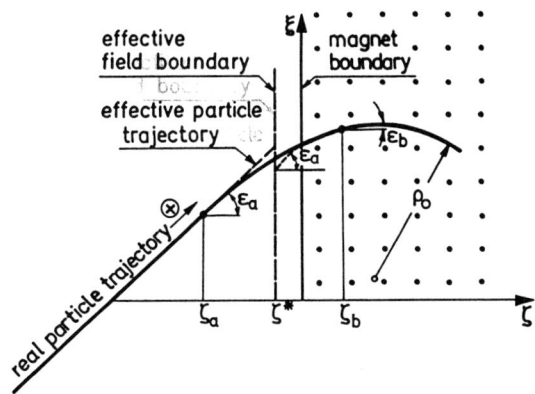

Figure 1. The optic axis, i.e., the trajectory of a reference particle is shown in the fringing field of a homogeneous sector magnet. Note that the curvature of this trajectory increases monotonically from left to right. Note also that outside of the fringing field as well as deep inside the main-field the real particle trajectory is identical to an "ideal" one that is straight up to the effective field boundary where it is slightly shifted and continues as a circle of radius ρ_0.

2.2.1 USE OF THE EFFECTIVE FIELD BOUNDARY ONLY

The simplest procedure to describe particle trajectories is to assume that the field of an optical element really starts and stops abruptly at its effective field boundaries ζ_{eff}. In this case the main first–order action of the fringing field is taken into account but all the bends and shifts of the individual particle trajectories at the effective field boundaries are neglected. Since such trajectory calculations assume a field distribution that does not fulfill Maxwell's equations, no real physical situation is described. Thus, the theoretically predicted particle trajectories still differ considerably from the finally observed ones, at least as far as the second–, third– and higher–order approximations are concerned, while moderate, though by no means negligible, effects are also present in the first–order approximation.

2.2.2 THE USE OF FRINGING–FIELD INTEGRALS

To include fringing–field effects in more detail one must know the fringing–field distribution in space using theoretical calculations or detailed field measurements and then somehow solve the equations of motion of given charged particles. Analogously to Eq. (1), one can define the electrostatic and the scalar magnetic potential in the ξ, η, ζ–coordinate system of Figure 1 as :

$$V(\xi,\eta,\zeta) = \sum_{i,j=0}^{\infty} a_{ij}(\zeta) \frac{\xi^i \eta^j}{i! \, j!} . \qquad (3)$$

Here the Laplace equation reads $\Delta V = \partial^2 V/\partial\xi^2 + \partial^2 V/\partial\eta^2 + \partial^2 V/\partial\zeta^2 = 0$ so that one finds from Eq. (3):

$$a_{i,j+2}(\zeta) = -a_{i+2,j}(\zeta) - \frac{\partial^2 a_{ij}(\zeta)}{\partial \zeta^2} .$$

Using this relation the electrostatic and the scalar magnetic potentials can be determined with the side condition that because of symmetry the $a_{ij}(\zeta)$ must vanish for odd j in the electrostatic and for even j in the magnetic case, yielding

$$V_E(\xi,\eta,\zeta) = a_{10}\xi + a_{20}\frac{\xi^2 - \eta^2}{2} + a_{30}\frac{\xi^3}{6} - (a_{30} + a_{10}'')\frac{\xi\eta^2}{2} + \ldots \qquad (4a)$$

$$V_B(\xi,\eta,\zeta) = a_{01}\eta + a_{11}\xi\eta + a_{21}\frac{\xi^2\eta}{2} - (a_{21} + a_{01}'')\frac{\eta^3}{6} + \ldots \qquad (4b)$$

where all $a_{ij}(\zeta)$ are abbreviated as a_{ij} and where $a_{ij}(\zeta)'' = \partial^2 a_{ij}(\zeta)/\partial\zeta^2$. In both cases it is sufficient to know the components $E_\xi(\xi,0,\zeta) = -[\partial V_E/\partial\xi]_{\eta=0}$ of the electrostatic field strength and $B_\eta(\xi,0,\zeta) = -[\partial V_B/\partial\eta]_{\eta=0}$ of the magnetic flux density in the plane $\eta = 0$, i.e., the coefficients $\partial a_{i0}^\mu/\partial \zeta^\mu$ or $\partial a_{i1}^\mu/\partial\zeta^\mu$ with $\mu = 0, 1, 2\ldots$ to the desired order κ with $i + \mu \leq \kappa$ in the electrostatic or

$i + \mu \leq \kappa - 1$ in the magnetic cases, respectively.

In GIOS the equations of motions are solved algebraically[8,9,10] in arbitrary fringing–field distributions, where these field distributions are expanded in power series around $\vec{E}(0,0,\zeta)$ or $\vec{B}(0,0,\zeta)$. Since this procedure requires complex manipulations with fringing–field integrals, no fourth– and higher–order effects have been treated. To illustrate this method assume a magnet that has a straight field boundary, i.e., the case where all a_{ij} in Eq. (4b) vanish for $i \neq 0$. With the side condition of $\eta = 0$ being a plane of symmetry one finds in this case

$$B_\xi(\xi,\eta,\zeta) = 0$$
$$B_\eta(\xi,\eta,\zeta) = a_{01}(\zeta) - \frac{\zeta^2}{2}\left(\frac{\partial^2 a_{01}(\zeta)}{\partial \zeta^2}\right) + \cdots$$
$$B_\zeta(\xi,\eta,\zeta) = \eta\left(\frac{\partial a_{01}(\zeta)}{\partial \zeta}\right) + \cdots .$$

where at a position ζ_a in the field free region $B_\xi(0,0,\zeta_a) = B_\eta(0,0,\zeta_a) = B_\zeta(0,0,\zeta_a) = 0$ and at a position ζ_b in the main field $B_\xi(0,0,\zeta_b) = B_\zeta(0,0,\zeta_b) = 0$ as well as $B_\eta(0,0,\zeta_b) = a_{01}(\zeta_b) = B_0$ (see Fig. 1). Determining the forces in ξ- and η-directions for a particle of charge q and mass m moving with a velocity $v = \sqrt{\dot{\xi}^2 + \dot{\eta}^2 + \dot{\zeta}^2}$, one finds

$$m\ddot{\xi} = \dot{\eta}q\left(\eta\frac{\partial a_{01}(\zeta)}{\partial \zeta} + \cdots\right) - \dot{\zeta}q\left(a_{01}(\zeta) - \frac{\zeta^2}{2}\frac{\partial^2 a_{01}(\zeta)}{\partial \zeta^2} + \cdots\right) \quad (5a)$$
$$m\ddot{\eta} = -\dot{\xi}q\left(\eta\frac{\partial a_{01}(\zeta)}{\partial \zeta} + \cdots\right). \quad (5b)$$

Integrating these relations over time one obtains $\dot{\xi} = \int \ddot{\xi}dt$ and $\dot{\eta} = \int \ddot{\eta}dt$ which involve several integrals that must be evaluated from ζ_a to ζ_b. One of these integrals, $\int a_{01}(\zeta)d\zeta/B_0$, determines the position of the effective field boundary[13] while two others,

$$\int \frac{\partial a_{01}(\zeta)}{\partial \zeta}\frac{d\zeta}{B_0} = 1 \quad \text{and} \quad \iint \frac{\partial^2 a_{01}(\zeta)}{\partial \zeta^2}\frac{d\zeta d\zeta}{B_0} = 0 \quad (6a)$$

can be calculated exactly without knowing[8,9,10,14] the dependence of a_{01} on ζ. In addition, there are the so–called fringing–field integrals of GIOS:

$$\int \frac{a_{01}^2(\zeta)d\zeta}{\chi B_0} \approx \Delta, \quad \text{and} \quad \iint \frac{a_{01}(\zeta)B_0 d\zeta d\zeta}{\chi^2} \approx \Delta^2. \quad (6b)$$

that depend on the detailed distribution of $a_{01}(\zeta) = B_\eta(0,0,\zeta)$. Note, however, that the magnitudes of these integrals are Δ^2 or Δ, respectively, with $\Delta \approx \overline{\Delta\zeta}B_0/\chi$ with $\overline{\Delta\zeta}$ being a measure for the extension of the fringing field. Here

$\chi = B_0\rho$ is the magnetic rigidity of the particles under investigation so that for all practical cases $\Delta \ll 1$. Thus[14], the integrals of Eqs. (6a) are of higher importance than the integrals of Eqs. (6b). However, note that the integrals of Eqs. (6a) rely on Maxwell's equations being valid throughout the fringing field.

2.2.3 THE RAY TRACING METHOD

For the fringing–field calculations in COSY specific fringing–field distributions must be assumed. The above defined $a_{i0}(\zeta)$ or $a_{i1}(\zeta)$ then are approximated as precisely as possible by modifying them until the measured or calculated fringing–field distributions were reproduced sufficiently accurate by $\vec{E} = -gradV_E$ or $\vec{B} = -gradV_B$. The thus obtained spatial fields $\vec{E}(\xi,\eta,\zeta)$ or $\vec{B}(\xi,\eta,\zeta)$ fulfill Maxwell's equations exactly for any ξ,η,ζ though the electrode or pole–face geometry that would produce these $\vec{E}(\xi,\eta,\zeta)$ or $\vec{B}(\xi,\eta,\zeta)$ is slightly different from the one that was used to measure or calculate $\vec{E}(\xi,\eta,\zeta)$ or $\vec{B}(\xi,\eta,\zeta)$ initially. The advantage of this relatively complex method of obtaining $\vec{E}(\xi,\eta,\zeta)$ or $\vec{B}(\xi,\eta,\zeta)$ is that the equivalents of the most important integrals of Eqs. (6a) are calculated exactly, i.e., to machine precision, while the equivalents of the smaller and thus not so important integrals of Eqs. (6b) are calculated with a precision that is equal or a little worse than the one existing in the original fringing–field distribution.

Employing the new differential algebra[11], then an "arbitrary particle" can be traced through this fringing–field distribution obtaining finally a transfer matrix that describes all fringing–field actions from ζ_a to ζ_b. Concatinating this transfer matrix with those of drift distances $\zeta_a - \zeta_{eff}$ and $\zeta_{eff} - \zeta_b$, i.e., drift distances of negative lengths, one obtains a fringing–field transfer–matrix that describes all the necessary bends and shifts an arbitrary particle should experience at the effective–field boundary[12].

3. SPACE–CHARGE FIELDS

For intense charged particle beams space–charge effects are not negligible and in some cases they dominate. In principle, a charged particle interacts with all its neighbours in the three-dimensional xyz–space. However, since the xy–cross-section of the beam varies only slowly with z, one can, as is done in GIOS, obtain good results from only two-dimensional space–charge calculations.

In detail, one must know the particle–density distribution at any z–value, determine from this distribution the internal space–charge field and add to it the external field before one solves the equations of motion. Since this particle-density distribution varies with z, the space–charge field varies with z also and must be determined in small steps in z–direction by some predictor–corrector procedure requiring knowledge of the external field and of the particle–density distribution at the preceding step. To obtain this distribution by a Monte–Carlo method requires that a large number of particle trajectories be traced, requiring intolerably long computation times[15]. For this reason GIOS uses two methods, both of which avoid the calculation of many individual particle trajectories:

1. In the first option, it is assumed that the current density is constant over any given xy–cross–section which corresponds to an even distribution of particles over the surface of the x, a, y, b–phase–space[16]. In this case the space–charge field increases linearly with the distance $r = \sqrt{x^2 + y^2}$ between an arbitrary point and the optic axis while its magnitude at the beam surface is inversely proportional to the beam envelope, the size of which can easily be determined[4].

2. In the second option[17], the particle beam density $\rho(x_n, y_n)$ at some position $x_n y_n$ is calculated, for any $z = z_n$, using the initial particle beam density $\sigma(x_0, a_0, y_0, b_0)$ in phase space at $z = z_0$. This option also allows the calculation of the effects of mirror charges in rectangularly shaped or cylindrical vacuum pipes.

Because of Liouville's theorem the local particle density σ is constant in phase space. Thus the final phase–space density σ_n is equal to the initial phase–space density σ_0

$$\sigma_n(x_n, a_n, y_n, b_n) = \sigma_0(x_0, a_0, y_0, b_0).$$

If "A" is a transfer matrix from z_0 to z_n, one finds $\sigma_n(x_n, a_n, y_n, b_n) = \sigma_0[A^{-1}(x_n, a_n, y_n, b_n)]$. The final particle–density $\rho(x_n, y_n)$ at the position x_n, a_n, y_n, b_n at $z = z_n$ thus is found[17] as

$$\rho(x_n, y_n) = \iint \sigma_n(x_n, a_n, y_n, b_n) da_n db_n = \iint \sigma_0[A^{-1}(x_n, a_n, y_n, b_n)] da_n db_n.$$

Though the second method is more accurate than the first one, it requires a much larger computational effort. Thus it is advisable for any GIOS calculation to use the first method to optimize an optical system until it seems to fulfill most requirements and then use the second option to optimize it further.

If the particle beam is bunched, the lengths of the particle packages usually shrink or grow in length monotonously. Thus the in GIOS assumed DC–current in principle varies. Knowing this variation GIOS can determine at least some of the in reality existing effects of longitudinally varying space–charge fields by accepting different currents for each calculated step.

Until now COSY has not been used routinely for space–charge calculations. However, the procedure is similar to the second method used in GIOS as described above.

4. GENERAL PROPERTIES OF GIOS AND COSY

Both GIOS and COSY determine optical properties of general particle–optical systems containing homogeneous and radially inhomogeneous magnetic and electrostatic sector fields which all have the same plane of deflection and possibly are assisted by magnetic or electrostatic quadrupoles, hexapoles or octopoles. GIOS computes all particle–optical effects up to third order while COSY can go to fifth order. Though in principle COSY is more precise and versatile GIOS is used more frequently since it contains more options. GIOS and COSY use an easy to remember mnemonic input form. As an example, the GIOS input is shown for a magnetic sector field analyzer preceded by a multipole doublet as shown in Figure 2., a graphical output of GIOS:

```
MAGNET SEPARATOR WITH PRECEDING ELECTROSTATIC MULTIPOLE DOUBLET
REFERENCE PARTICLE     0.06 100 1 ;
CALCULATION ORDER    2 2 ;
PARALLELOGRAM-LIKE X-DIRECTION    0.00001 0.037 ;
PARALLELOGRAM-LIKE Y-DIRECTION    0.003 0.01 ;
DEVIATION PARAMETER  0.02 3.5E-5 ;
FIT SIMPLEX ;
X = .01 ;
A = 20 ;
Y = 3 ;
B = 10 ;
DRIFT LENGTH    1.0 ;
ELECTROSTATIC MULTIPOLE    .3 0.81251 -0.01858V 0.0 0.05 ;
DRIFT LENGTH    .1 ;
ELECTROSTATIC MULTIPOLE    .3 -.73891 0.01168V -0.0 0.05 ;
DRIFT LENGTH    0.3 ;
FRINGING FIELD  1 0.0 ;
MAGNETIC SECTOR  1.4 90. 0.05 0.0 -0.2026 0.0 ;
FRINGING FIELD    1 30. ;
DRIFT LENGTH    2. ;
M = (((Y,Y)*Y)|2 + ((Y,B)*B)|2)|0.5 ;
FIT A M ;
END
```

Both GIOS and COSY allow the definition of algebraic quantities like $M = (((Y,Y)*Y)^2 + ((Y,B)*B)^2)^{0.5}$ where (Y,Y) and (Y,B) are first–order matrix elements. Those quantities may then be calculated and shown in the program output or they may be used in fitting routines. For the moment, there are two fitting routines installed in GIOS and four different ones in COSY. The fitting routines in GIOS are based on the method of conjugated gradients and on the simplex method. The first three fitting routines of COSY are also based on the method of conjugated gradients and the fourth one is a quadratic form algorithm.

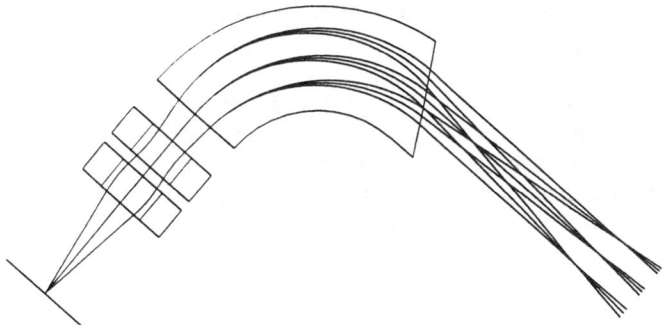

Figure 2. Plotter output of GIOS for particle trajectories as calculated for the listed example of a homogeneous magnet with a hexapole component $n_2 = -0.203$ preceded by an electrostatic quadrupole doublet. This doublet includes second order components as would be created by hexapoles of equal aperture radii and voltages of -0.0186 and +0.0117 kV Note that the image plane for particles of different masses is inclined with respect to the optic axis.

The standard output of GIOS and COSY is a listing of the optimized elements of the optical system under investigation as well as overall transfer matrices like the ones shown below:

(X,X)=-2.2224	(A,X)=-.50913	(T,X)=-.19155
(X,A)=-.23485E-04	(A,A)=-.44998	(T,A)=0.33028
(X,G)= 2.2751	(A,G)=0.78837	(T,G)=0.56467
(X,D)= 2.2751	(A,D)=0.78837	(T,D)=-.43533
(X,XX)=0.91217	(A,XX)=-.49158	(T,XX)=0.59467
(X,XA)= 5.4576	(A,XA)=0.88091	(T,XA)=0.87916
(X,XG)=-.88193E-01	(A,XG)=0.33295	(T,XG)=-.32062
(X,XD)= 2.1367	(A,XD)=0.34860	(T,XD)=0.42536
(X,AA)=-.36964E-02	(A,AA)=-.55342	(T,AA)=0.59482
(X,AG)= 2.1240	(A,AG)=0.50465	(T,AG)=0.12023
(X,AD)= 3.3859	(A,AD)=0.34327	(T,AD)=0.22920
(X,GG)=-1.8782	(A,GG)=-.79039	(T,GG)=0.31264E-02
(X,GD)=-1.4813	(A,GD)=-.79242	(T,GD)=0.63998E-02
(X,DD)=-1.8782	(A,DD)=-.79039	(T,DD)=0.39021
(X,YY)=-.94168	(A,YY)=-.36609	(T,YY)=0.16690
(X,YB)=-.16152	(A,YB)=-.59716	(T,YB)=0.75494
(X,BB)=0.10954	(A,BB)=-.15519	(T,BB)=0.82583
(Y,Y)=-1.0654	(B,Y)=-.64089	
(Y,B)=0.64973E-01	(B,B)=-.89952	

(Y,YX)= 1.3800   (B,YX)=0.19992
(Y,YA)=-.16182   (B,YA)=0.70246
(Y,YG)=0.23843   (B,YG)=0.23160
(Y,YD)= 3.4468   (B,YD)=0.48812
(Y,BX)=0.85938   (B,BX)=-.66037
(Y,BA)=-.32485E-01   (B,BA)=0.74247E-01
(Y,BG)= 3.5286   (B,BG)= 1.9072
(Y,BD)= 8.5191   (B,BD)= 2.1847

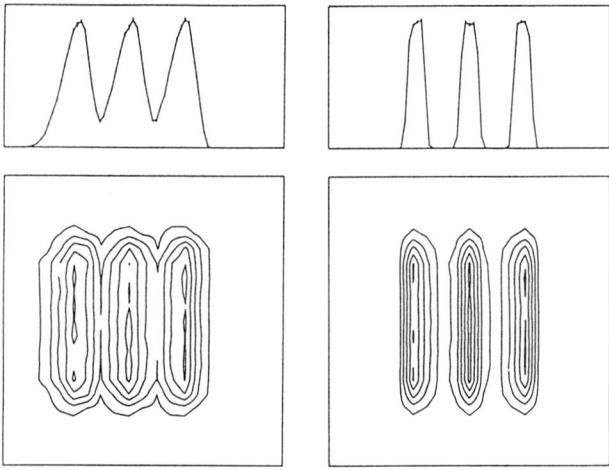

Figure 3. The intensity distribution of a particle beam containing particles of three different mass-to-charge ratios in the image plane of the mass analyzer shown in Figure 2. In Figure 3a the intensity distribution of the system is shown in which the hexapole components of the two quadrupoles are eliminated as is the hexapole component of the magnetic sector field. In Figure 3b the same intensity distribution is shown with all the hexapole components activated as indicated in the above GIOS input file.

In order to supply a quick visual survey, GIOS furnishes a printer plot that displays the envelopes of beams filling either elliptical or parallelogram-like phase-space areas. This printer plot also shows the cutoff effects by diaphragms that can be defined in GIOS. Parallel to this quick survey tool, GIOS can furnish graphical plots by use of the UGS-software package. Such graphical plots may display the geometry of the optical system under investigation as, for instance, that of in Figure 2, the one belonging to the above input example. However, also final intensity distributions can be supplied. As an example the intensity of three neighbouring mass-to-charge lines as obtained by the system of Figure 2 are shown in Figure 3.

Over the last ten years GIOS has proven to be a useful tool for designing almost any particle–optical system while over the last few years COSY has helped to answer specific questions mainly for the design of highly corrected optical devices.

5. REFERENCES

1. H. Wollnik, J. Brezina, M. Berz and W. Wendel;
   Proc. AMCO–7, GSI–Rep. THD–26 (1984) 679
2. H. Wollnik, J. Brezina and M. Berz; Nucl. Instrum. and Meth. **A258** (1987) 408
3. M. Berz; Nucl. Instrum. and Meth. **A258** (1987) 402
4. H. Wollnik; *"Optics of Charged Particles"* (1987), Acad. Press, Orlando
5. H. Wollnik and M. Berz; Nucl. Instrum. and Meth. **238** (1985) 127
6. H. Wollnik, T. Matsuo and H. Matsuda; Nucl. Instrum. and Meth. **102** (1972) 13
7. M. Berz and H. Wollnik; Nucl. Instrum. and Meth. **A258** (1987) 364
8. H. Wollnik and H. Ewald; Nucl. Instrum. and Meth. **36** (1965) 93
9. H. Matsuda and H. Wollnik; Nucl. Instrum. and Meth. **77** (1970) 40; **77** (1970) 283
10. H. Matsuda; Nucl. Instrum. and Meth. **91** (1971) 637
11. M. Berz, particle accelerator, in print
12. B. Hartmann; Thesis (1988) Universität Giessen, unpublished
13. A. H. Bucherer; Ann. Phys. (Leipzig) **28** (1909) 513
14. H. Wollnik; Nucl. Instrum. and Meth. **52** (1967) 250
15. M. Berz and H. Wollnik; Nucl. Instrum. and Meth. **A267** (1988) 25
16. I. M. Kapchinskij and V. V. Vladimirskij;
    Proc. Conf. High Energy Acc., CERN, Geneva, 274 (1959)
17. J. Brezina; Thesis (1986) Universität Giessen, unpublished
18. T. Mottershed; private communication (1988)

SOME PARTICLE BEAM COMPUTER PROGRAMS
ADAPTED FROM PLASMA PHYSICS RESEARCH

Brendan B. Godfrey
Mission Research Corporation, Albuquerque, NM   87106

ABSTRACT

Many of the computer programs developed for plasma physics and high current beam research may be applicable to high energy accelerators and their microwave power supplies. This paper describes briefly the codes used by Mission Research Corporation for addressing the physics of high power microwave and charged particle beam generation and propagation. The codes range from simple lumped-parameter models to detailed multidimensional PIC simulations. Both the capabilities and the underlying models are summarized.

1. INTRODUCTION

Over the past several years, Mission Research Corporation has developed two sets of general purpose computer programs for multi-dimensional, electromagnetic simulations of relativistic beams and plasmas. Applications have included magnetic insulation, electron diodes, modified betatrons, linear induction accelerators, collective ion acceleration, wakefield computations, electron-layer magnetic compression, ion-focused transport of electron beams, beam-plasma instabilities, electron beam transport in dense gases, and microwave generation. In addition, several simpler, faster running codes are available for investigating electron beam equilibrium and linear stability properties. The simulation codes, together with the more important equilibrium and stability codes, are listed in Table I.

This report summarizes the capabilities and describes the physical models of these codes, especially as they bear upon beam accelerator and microwave source investigations. Section 2 outlines the principal capabilities of each of the fourteen codes in Table I. Section 3 begins with an overview of particle-in-cell (PIC) simulation techniques and then enumerates the unique aspects of each of the general purpose PIC simulation codes. The models employed in the remaining computer programs are discussed in Sec. 4. Note that three programs footnoted in Table I are described in greater detail in other contributions to this Proceedings, as is the ARGUS simulation code by Science Applications International Corporation.

All programs listed in Table I except CPROP run on CRAY-1 computers under the CTSS operating system. As noted in Sec. 2, some run on other computers as well. Copies reside at the Los Alamos National Laboratory computer center and elsewhere. The

U. S. Government either owns or has a royalty-free license to use any of the programs.

Table I Major MRC computer programs for charged particle beam and microwave research

CODE	APPLICATIONS	REFERENCE
CCUBE*	2-D PIC Particle Beam and Microwave Simulations	1
IVORY†	3-D PIC Particle Beam and Microwave Simulations	2
ISIS*†	2-D PIC Particle Beam and Microwave Simulations	3
MAGIC	2-D PIC Particle Beam and Microwave Simulations	4
SOS†	3-D PIC Particle Beam and Microwave Simulations	5
CPROP	2-D PIC Beam Atmospheric Propagation	6
IPROP	3-D PIC Beam Atmospheric and IFR Propagation	7
SCRIBE	2-D Electron Diode Equilibria	8
ARCTIC	2-D Ion-Focused Transport Equilibria	9
BALTIC	Beam Transverse Stability in Accelerators	10
GRADR*	Laminar Beam Stability, Microwave Growth	11
ORBIT	1-D Beam Kinetic Equilibria	12
KMRAD	Kinetic Beam Stability, Microwave Growth	13
BTRSQ	Recirculating Accelerator Instabilities	14

*Developed by Los Alamos National Laboratory and used with its permission.
†Described in greater detail elsewhere in this Proceedings.

2. CODE CAPABILITIES

2.1 OVERVIEW

PIC codes treat the detailed electromagnetic interactions among one or more plasmas, charged particle beams, and metallic structures. They have been employed fruitfully in every branch of plasma physics from fusion energy to astrophysics; the linear and nonlinear evolution of instabilities is a common theme. In high-current electron accelerator research, PIC codes are used to treat pulsed-power flow in magnetically insulated transmission lines, electron beam generation in complex diodes, beam transport and inductive acceleration in vacuum drift tubes, and beam transport in low density ion channels. Beam propagation in partially ionized gases is a related application. High-power microwave sources are modeled effectively by PIC codes, which can follow simultaneously the electron beam nonlinear dynamics and the electromagnetic response of the microwave cavity. Likewise, particle beam interaction with cavity modes in RF accelerators can be treated in detail.

2.2 CCUBE

The PIC code CCUBE is used extensively to model microwave generation[15-18] and beam acceleration.[19-22] Although only two-dimensional, the code is very flexible in that it can run in any orthogonal coordinate system. Injection and extraction of particles and fields at boundaries allows it to treat complicated, open-ended devices. The code has extensive diagnostics, including particle and field plots, time histories of various quantities, and color movies. A graphics post-processor package, PEGASUS, is available.[23] CCUBE runs on VAX 780 and CONVEX computers, as well as on CRAY-1 computers.

2.3 IVORY

IVORY is a three-dimensional generalization of CCUBE. Field variation in the third dimension is treated by Fourier series of one to about six terms. As a consequence, the code is best suited for problems in which the particle dynamics are only weakly non-linear in the third dimension and boundary conditions are approximately two-dimensional. Examples include free electron lasers, many microwave sources, particle beam transport,[24-27] wakefield calculations, and plasmoid propagation across a magnetic field.[28] IVORY is very efficient for such problems, requiring computer resources only a few times greater than those needed for corresponding two-dimensional problems. The application of IVORY to transport of low energy negative ion beams in plasma columns is discussed elsewhere in this Proceedings.

2.4 ISIS

ISIS is a significantly enhanced, but still two-dimensional, version of CCUBE. Special features include Monte Carlo transport of high energy particles, volume creation of particles by high energy electrons and photons, and a hydrodynamic treatment of imploding resistive liners. The layout of complicated electrode geometries is largely automated, and a new algorithm for improved treatment of curved metallic surfaces has been developed. It and CCUBE are used at LANL in vircator,[29-30] electron diode, collective acceleration, ring compression, and beam-plasma interaction studies. The code runs on CRAY computers under the CTSS operating system. Another paper in the Proceedings describes ISIS in substantial detail.

2.5 MAGIC

MAGIC is a two-dimensional PIC code quite similar in capabilities to CCUBE but developed independently. It is employed at Sandia National Laboratories (SNL) and elsewhere for magnetic insulation, electron diode,[31] collective acceleration,[32] beam transport,[33] and microwave generation studies.[34]

2.6 SOS

SOS is a fully three-dimensional, general purpose simulation code. It accommodates essentially arbitrary boundary conditions in cartesian or cylindrical coordinates. Wave and particle emission and absorption, metallic structures, dielectrics, and air conductivity packages are included. Newly added features allow evaluation of wakefields in either the time or the frequency domain. Sophisticated data management techniques allow it to treat problems of immense size and complexity, limited only by the availability of computer time. A compact, "user friendly" data entry system simplifies problem specification. SOS runs on VAX and CONVEX computers, as well as on the CRAY-1 and CRAY-2 computers.

2.7 CPROP

CPROP is an enhanced and optimized version of CCUBE for axisymmetric investigations of electron beam propagation in partially ionized gas. Special features include a mesh which moves with the beam, a partially implicit field solver to permit large time steps, a multispecies air chemistry package, and a delta-ray creation and destruction procedure. CPROP is not limited by the paraxial particle and frozen field approximations used in many other propagation codes. The code has been employed in Nordsieck expansion,[35] beam front erosion, and hollowing instability studies supporting RADLAC,[36-37] IBEX,[38-40] PHERMEX,[41] and FX-100[42] experiments. CPROP runs on CDC-7600 computers under the LTSS operating system.

2.8 IPROP

IPROP is a three-dimensional electron beam propagation code obtained by adding the special features of CPROP to IVORY. Diamagnetic effects are included to treat annular beam rotation properly. The code can model all resistive instabilities—hollowing, hose, and filamentation—either separately or in combination.[36] Significant progress on electron beam steering and focusing in dense plasma columns has been achieved recently.[43,44] With the conductivity package omitted, IPROP also is used for research on electron beam transport in low-density ionized channels, the so-called ion-focused regime (IFR).[45] Research centers on beam front erosion, two-stream instabilities, and channel electron trapping.

2.9 SCRIBE

SCRIBE can be described as a two-dimensional, steady-state PIC code. It is based on Hermannsfeldt's Electron Trajectory Program.[8] It computes axisymmetric electron flow patterns in self-consistently determined static electric and magnetic fields. Being time-independent, the code is fast and relatively free of numerical noise. SCRIBE is ideal for studying electron equilibrium behavior in diodes,[46] accelerators and microwave tubes. A particularly noteworthy achievement with SCRIBE is the design of very low emittance electron diodes and transport systems for free electron laser studies at the Naval Research Laboratory.[47] SCRIBE runs on VAX 780 computers, as well as on CRAY-1 computers.

2.10 ARCTIC

ARCTIC is similar to SCRIBE but runs in the beam frame and solves the frozen field equations. It was written to determine electron and ion trajectories and associated steady-state fields for relativistic electron beam propagation in IFR channels and plasma wakefield devices. For such applications ARCTIC is far more efficient than alternative codes, especially when very long beam pulses and large drift tubes are involved.

2.11 BALTIC

BALTIC is a simple code used to follow the transverse oscillations of the centroid of a particle beam in an accelerator. Specifically, it computes the linear growth of the resistive wall,[48] image displacement,[24] and beam breakup instabilities[10,49] for long pulse beams in multigap induction accelerators. It accommodates wakefield data from BBU, IVORY, and related codes.

2.12 GRADR

GRADR is a general-purpose equilibrium and linear dispersion code for laminar, cylindrically symmetric, radially inhomogeneous particle beams. Conducting boundaries and resistive and dielectric mediums are allowed. Starting from user-supplied constraint equations, the code first computes the beam equilibrium current and field profiles. It then determines wave frequencies and growth rates for arbitrary, user-requested wave numbers. GRADR is interactive and exceedingly fast. Typically, the code is used in collective acceleration[19,50-52] and beam stability studies.[48,53-56] With modifications, it could be employed for microwave generation in slow wave structures.[57]

2.13 ORBIT

ORBIT and KMRAD are used in place of GRADR for nonlaminar beams, the more common situation. ORBIT determines the equilibrium radial profiles of the particle currents and fields, based on a user-provided subroutine specifying the particle distribution function in terms of constants of the motion. A family of distribution function subroutines for relativistic Bennett beams and for radial diodes is available.[58]

2.14 KMRAD

KMRAD is a three-dimensional linearized PIC code. A cylindrically symmetric, radially inhomogeneous particle distribution is specified by the user, possibly with the aid of ORBIT. (An interface between these codes is provided.) It then computes the temporal evolution of perturbed particle orbits and electromagnetic fields for arbitrary wave numbers chosen by the user. The code is interactive and moderately fast. KMRAD is used regularly for parametric studies of instability growth for propagating particle beams in air[36,59-61] and in IFR channels. Smooth-bore magnetron, Cherenkov masers, gyrotrons, and orbitrons also can be studied.

2.15 BTRSQ

BTRSQ is the most detailed of a family of dispersion-relation solvers for high-current betatrons.[25,62] The beam is treated in the rigid disk limit, while the fields are determined exactly for a cavity of rectangular minor cross section. The code is interactive and exceedingly fast. Many new properties of the negative mass instability in modified betatrons have been discovered recently with BTRSQ.[14,27,63]

Dispersion relation codes for microwave generation can be developed rapidly as the needs arise. A convenient shell, developed originally for GRADR, provides user interaction, error recovery, a Muller's method[64] complex root evaluater, and a graphics interface. The programmer merely writes a dispersion function evaluation routine for the new device. For instance, the recently written Pierce diode electromagnetic stability code PEM2D uses this shell.

3. PIC CODE ALGORITHMS

3.1 GENERAL FEATURES OF PIC SIMULATION CODES

PIC codes determine the time evolution of complex plasmas by computing the dynamics of many thousands of representative plasma particles (electrons and/or ions) moving in electromagnetic fields externally applied or produced by the plasma itself. Thus, PIC codes provide the most fundamental and detailed representation possible of plasma problems. In effect, they solve the Vlasov equation. Of course, this precision comes at the cost of substantial computer requirements, and for this reason PIC codes should be employed only when simpler numerical or analytical techniques are inadequate.

The electromagnetic fields are defined on a regular mesh in one, two, or three dimensions, depending on the symmetry of the problem to be solved. The mesh can be in rectangular, cylindrical, or other desired geometry. At each time step, new electric (E) and magnetic (B) fields are computed by advancing the finite difference approximations to Maxwell's equations,

$$\frac{\partial E}{\partial t} = \nabla \times B - J \qquad (1)$$

$$\frac{\partial B}{\partial t} = -\nabla \times E \qquad (2)$$

using currents (J) determined from the plasma particle motion on the previous time step. Alternatively, equations for the scalar and vector potentials can be solved. Boundary conditions are required to define spatial derivatives at the mesh edges. Wave reflecting (i.e., metallic) or periodic boundary conditions are common choices. More complicated boundaries allow electromagnetic waves to be launched into the computational region or to leave it. Metal structures at locations within the computational region are represented simply by setting electric field components to zero there. Representing dielectrics, resistors, and simple driven antennas is only slightly more difficult.

Particle momenta ($P = \gamma V$), positions (X), and energy (γ) are then advanced using the relativistic equations of motion with the newly computed fields.

$$\frac{dP}{dt} = E + V \times B \qquad (3)$$

$$\frac{dX}{dt} = V \qquad (4)$$

$$\gamma = \left(P^2 + 1\right)^{1/2} \qquad (5)$$

(Replace γ by 1 for nonrelativistic problems.) The fields appearing in Eq. 3 are those at the particle location, obtained from the fields at nearby mesh points by (typically, linear) interpolation. When a particle leaves the computational region, it is destroyed or it is returned to the mesh by some prescribed procedure (e.g., reflected). By the same token, particles can be injected from boundaries, a feature particularly useful in particle beam simulations. After a particle's new position and momenta have been determined, its contribution to the plasma currents is obtained by interpolating $V = P/\gamma$ to nearby mesh points.

This cycle of advancing fields based on particle currents and then advancing particles based on the new fields is repeated hundreds or thousands of times in a typical simulation. The time step is set by the smallest time scale in the calculation, which may be the plasma oscillation period, the electron cyclotron period, or the Courant time (of order the time for a light wave to cross a cell in the mesh). Progress has been made in the last few years at surmounting the Courant limit, which is numerical rather than physical in character.[65] Cell dimensions must be small compared to spatial scales of interest.

The PIC code running times and memory requirements are highly problem-dependent. The CPU times on a CRAY-1 computer typically range between fifteen minutes and four hours, although twenty hour runs are not unheard of. Corresponding central memory needs vary between 2×10^5 and 4×10^6 words. At least two fast, large capacity disks also are needed. Historically, the physics problems attempted with PIC codes have expanded to consume the maximum resources available in each generation of computers.

PIC codes usually have extensive graphics output capabilities and operating-system interfaces. Advances in PIC technique are described in several books.[66-67]

3.2 CCUBE

CCUBE[1] is a general-purpose two-dimensional PIC code, and as such operates along the lines described above. It does, however, have a number of special features which increase its flexibility. The Galerkin finite-element algorithm for advancing particle quantities[68] eliminates the numerical Cherenkov instability[69] and ameliorates some other numerical problems. The electromagnetic fields incorporate backward biasing to damp unwanted high-frequency field fluctuations and to further suppress possible numerical instabilities.[70] The MRC-developed time-biased field solver also can significantly relax the Courant time step constraint, reducing computing costs. Marder's procedure for reducing charge conservation errors is being added.[71] CCUBE is written for arbitrary orthogonal coordinates. The user can convert from, say, cylindrical to toroidal coordinates merely by revising routines which define metric elements. Nonuniform zoning of the spatial mesh is accomplished with equal ease.

A number of other features deserve mention. For compatibility with nonuniform zoning, variable particle weighting within particle type, with several particle species permitted, is implemented. The code supports periodic, wave transmitting, and inhomogeneous Dirichlet-Neumann field boundary conditions. Particles can be absorbed, reflected, or injected from surfaces not necessarily coinciding with edges of the mesh. The physics for wave launching and particle field-emission from surfaces, which is necessary to accurately model microwave and particle beam sources, is incorporated. Graphics output can be produced either directly from CCUBE or interactively from the PEGASUS[23] postprocessor. Options include microfiche plots of various slices through particle phase space, particle distribution functions, contour plots and one-dimensional cuts of fields and currents, and histories of particle and field energies and other selected quantities. Color movies of particle and field data can be generated concurrently. In addition to the more traditional simulation diagnostics just mentioned, CCUBE contains numerical equivalents of such experimental diagnostics as Faraday cups, calorimeters, compensated diamagnetic loops, Rogowski coils, and local probes for field and current measurements. Each is time dependent and can be Fourier analyzed.

3.3 IVORY

IVORY[2] is a three-dimensional generalization of CCUBE and shares with it most of the special features described above. For economy of operation and clarity of interpretation, variations of field quantities in the third (usually, azimuthal) coordinate are represented by spectral rather than finite difference methods; i.e., the fields are Fourier decomposed in that dimension. In this way the behavior of selected modes can be examined without wasting time and storage on other, uninteresting modes. It should be emphasized that the spectral method is not a linearization in either particle or field behavior. Of course, the greater the nonlinearity of a problem, the more azimuthal modes are needed to resolve it accurately.

Thus, IVORY is particularly suitable for configurations which depart only weakly from symmetry in one coordinate, such as a particle beam in an accelerator or microwave device. The code provides three-dimensional results at costs comparable to two-dimensional simulations in such cases.

3.4 ISIS

The two-dimensional ISIS code[3] also is a derivative of CCUBE, modified to allow more accurate treatment of particle beam interaction with materials. Electron- and photon-generated delta rays can be created, transported, and destroyed in the simulations. A Monte Carlo treatment of Moliere scattering is provided for particles passing through foils and other materials. Energy loss is modeled by Bethe's continuous-slowing-down approximation. A resistive magnetohydrodynamic representation of metallic shells is useful in magnetic compression studies. Bulk resistivity terms have been added to the field equations.

Several numerical enhancements are noteworthy. Bilinear interpolation of currents and fields is permitted in addition to the Galerkin scheme. A current lay-down procedure explicitly guaranteeing charge conservation is provided. The backward-biased field algorithm is inverted exactly rather than iteratively. The algorithm for space-charge-limited particle emission from electrodes is more accurate, and an improved treatment of curved electrodes is available. The procedure for specifying the locations of complex electrodes on the mesh is automated.

3.5 MAGIC

The two-dimensional code MAGIC[4] was developed almost entirely independently of CCUBE. Nonetheless, the two are very similar not only in their features but also in their internal algorithms.

Differences are noted here. Field interpolation in MAGIC is bilinear, and current interpolation is explicitly charge-conserving. Triangular zoning at conductor surfaces improves space charge emission accuracy there. Wave absorption at boundaries by graded electric and magnetic resistivity is accommodated. Input data processing is a bit more flexible. There is no post-processor.

3.6 SOS

SOS[5] is a general purpose three-dimensional PIC code. Unlike IVORY, it treats field variations in all three coordinates by finite differences. The code was developed originally for electromagnetic pulse computations but has been generalized to treat particle beam, microwave, and magnetic insulation problems.

A distinguishing feature of SOS is its data management procedure. At each time step, the code processes the three-dimensional mesh and the particles on it as a sequence of two-dimensional slices. Only a few contiguous slices are in computer memory at any instant, while the rest are stored on disk. (In most other PIC codes, fields and currents are retained entirely in computer memory, while particles are cycled between disk and memory.) In this way, extremely large computations can be performed, limited only by budget.

SOS has an X-ray-generated Compton electron creation routine. Conductivity generation by Compton electrons and other particles is determined using a three-species lumped air chemistry model with direct ionization, avalanche, and recombination. Bulk Compton currents and conductivity in dielectrics are treated by a special model. Exponential integrating factors in the field algorithm stably accommodate arbitrarily variable conductivity. A family of subgrid models is provided for structures too small to be represented accurately by standard finite differences. A cavity eigenmode capability was implemented recently. Other features of the codes are similar to those in MAGIC.

3.7 CPROP

CPROP[6] is a two-dimensional PIC code designed for investigating the axisymmetric dynamics of high-current relativistic electron beam propagation in air. It is based on CCUBE and thus possesses most of the diagnostic and other capabilities of that program. Three major enhancements were required to treat beam transport in air: an electromagnetic field solver accommodating arbitrary scalar conductivity and a moving coordinate mesh, a conductivity generation package, and a Moliere electron scattering routine. CPROP also contains a high energy delta-ray generation and transport package.

The CPROP field algorithm is unique in several respects.[40] Calculations are carried out in a frame moving axially at a user-specified, possibly time-varying velocity. The field evaluation is partially implicit, so that the Courant limit is set only by the axial cell size. The so-called frozen field approximation[43,72] to Maxwell's equations can be treated as a special case, if desired. Exponential integrating factors, similar to those in SOS, accommodate arbitrary scalar conductivities. Zoning is non-uniform both axially and radially.

Both the BMCOND[73] and the PHOENIX[74] air chemistry models are available. The former solves rate equations for gas electron density and temperature and for the densities of several ionic species. The latter is a lumped parameter model similar to that in SOS. Moliere scattering[35,75] is implemented in CPROP by applying a deflection to the beam electrons every few time steps. The deflection angle is chosen randomly from a set of previously computed small angles forming a truncated Moliere distribution or from an analytical expression for occasional large angles.

3.8 IPROP

IPROP[7] is a version of IVORY into which most of the special features of CPROP have been transplanted. A new multispecies conductivity model calibrated against the HICHEM chemistry code is used.[76] Straggling due to bremsstrahlung is incorporated, with correlation between bremsstrahlung emission and atomic scattering an option.

4. ALGORITHMS OF OTHER MRC CODES

The remaining computer programs are much faster than PIC codes. SCRIBE, ARCTIC, BALTIC, and KMRAD typically use no more than a few minutes of CRAY-1 computer time; GRADR, ORBIT, and BTRSQ only seconds. None of these codes needs more than 10^5 words of memory.

4.1 SCRIBE

When only a steady-state simulation of electron beam generation and transport is needed, great economy can be achieved with SCRIBE. It is a version of the Stanford Linear Accelerator (SLAC) code EGUN,[8] extensively modified by MRC for research on intense beam diodes. SCRIBE self-consistently solves two-dimensional problems (usually in (r,z) geometry) by an iterative method. First, a solution to Laplace's equation is obtained, then current emission is calculated, and the electron trajectories are integrated using the fields. The trajectories deposit space charge at the mesh points, and a solution to Poisson's equation is calculated. Current emission is then recalculated and the cycle repeated until the potentials and trajectories are unchanged from cycle to cycle. Improved algorithms assure rapid convergence. All relevant self and applied electric and magnetic fields (except the beam-generated axial magnetic field) are included in the calculations. Both Dirichlet and Neumann boundaries are available, and surfaces are not restricted to lie along the grids.

4.2 ARCTIC

ARCTIC is a newly developed two-dimensional code for determining the steady-state flow of channel electrons and ions for IFR beam transport problems. At present, the beam is treated as a rigid rod propagating at approximately the speed of light. Calculations are performed in the beam frame. Representative channel electron and ion trajectories are computed based on self-consistent fields, and the corresponding currents interpolated onto a spatial mesh. Because fields are evaluated in the frozen field limit,[43,72] no iterations are required to reach the solution. In many respects, ARCTIC resembles a one-dimensional PIC code in which the axial coordinate takes the place of time.

A three-dimensional version of ARCTIC is planned for the future. Representing the beam cross section by an envelope model also is attractive, although iterative solution of the resulting system of equations would then be necessary.

4.3 BALTIC

BALTIC[10] provides a less detailed but immensely faster treatment than IVORY of transverse beam instabilities in accelerators. Presently, it evaluates the simultaneous evolution of resistive wall,[48] image displacement,[24] and beam breakup[10,49] instabilities in multiple-gap linear induction and radio frequency accelerators. Arbitrary spatial variations in the beam energy and axial magnetic field strength are allowed. Provision exists for external forces. Adding centrifugal force and a vertical magnetic field for recirculating accelerator studies is straightforward.

BALTIC employs a number of simplifying assumptions for computational efficiency. The beam is represented as a chain of rigid disks propagating forward at a uniform speed. Electric and magnetic fields entering into the equations of transverse motion are computed in the long wavelength limit, and accelerating gaps are treated in the thin lens approximation. Coupling coefficients entering into various instabilities can be estimated analytically[77] or determined from experiments or more detailed numerical calculations. Frequency-transformed transverse wakefields can be convoluted with beam displacements by fast Fourier transverse techniques to treat very low Q interactions, where resonance models are inappropriate.[78,79]

E3WAVE, a computer code similar to BALTIC in many respects, has just been developed to investigate strong-focusing parametric instabilities in high current accelerators.[26,79]

4.4 GRADR

GRADR[11] determines the equilibrium profiles and linear eigenmodes of cylindrically symmetric, radially inhomogeneous, relativistic electron beams in the laminar flow approximation. The cylindrical equilibrium is evaluated from six nonlinear, coupled, algebraic and first-order differential equations. Four follow from Maxwell's equations plus the fluid equations, while two can be specified arbitrarily to select a desired equilibrium.[80] The two constraint equations should, of course, be physically realizable. One constraint almost always is appropriate: total particle energy, kinetic plus potential, must be constant across the beam and equal to the injection energy. As a second constraint, typically we let the current density profile of the beam be specified at injection. Conservation of canonical angular momentum then determines the current density profile within the accelerator drift tube. The resulting system of equations is solved iteratively.

Given an equilibrium determined in this or any other way, GRADR then solves the corresponding linearized (small amplitude wave) equations to obtain eigenmodes and eigenvalues. The linear equations form a fourth-order differential system in radius and are integrated by a Fehlberg fourth-fifth-order Runge-Kutta routine.[81] Muller's method is employed in finding wave frequencies and growth rates.[64] The options of computing wave energy and the adiabatic variation of wave amplitude with changing beam parameters are available.[51] Including background plasma conductivity in GRADR allows study of laminar beam resistive instabilities in air.[56] Modifying GRADR boundary conditions to accommodate slow-wave structures is straightforward[57] and would allow gain calculations for many microwave devices.

4.5 ORBIT

ORBIT[12] calculates one-dimensional, radially inhomogeneous, cylindrically symmetric beam and plasma equilibria. The code solves the relativistic Vlasov-Maxwell equations for user-supplied particle distribution functions. Distributions are specified in terms of particle constants of motion: canonical axial momentum, canonical angular momentum, and energy. The kinetic framework includes finite temperatures in a natural manner and allows investigation of phenomena associated with the detailed momentum-space distributions that cannot be treated using a macroscopic fluid description.

ORBIT has been used primarily for investigating equilibria in relativistic electron beams and magnetically insulated diodes. Relativistic electron beam equilibria with both sharp and smooth edges have been examined, and distribution functions for hollow beams identified. For magnetically insulated diodes, both axial and azimuthal magnetic fields have been examined. Finite temperature effects, gas prefill, and electron injection with nonzero energy have been investigated. The use of ORBIT has eliminated making many of the assumptions and approximations necessary in the analytic treatment of magnetic insulation. A catalog of distribution functions and their properties has been compiled for use with ORBIT.[58]

4.6 KMRAD

KMRAD[13] is a linearized but fully relativistic and electromagnetic linear stability code for arbitrary cylindrically symmetric plasma systems. It consists of two parts: a one-dimensional, radially resolved, nonlinear PIC code; and a three-dimensional, spectrally resolved, linear PIC code. The one-dimensional routines determine particle orbits and corresponding fields in a cylindrically symmetric, possibly slowly evolving, equilibrium. It can be initialized from ORBIT, and an interface exists between the codes.

The three-dimensional routines compute the temporal growth of small amplitude perturbations of the particle equilibrium orbits, and the fields arising from these perturbations. The three-dimensional fields are Fourier decomposed axially and azimuthally, and one mode at a time is analyzed. The instability growth rate of the fastest unstable wave of a selected mode-number pair (k_z,m) can be determined from the exponential growth of field energy in the simulation. User-specified radially inhomogeneous background conductivity and return current profiles allow KMRAD to treat resistive instabilities. The code is interactive and quite fast. In effect, it simulates three-dimensional linear dynamics but requires computer resources comparable to those of a one-dimensional code. KMRAD plus ORBIT constitutes a kinetic beam generalization of GRADR.

4.7 BTRSQ

BTRSQ[14] calculates the normal mode frequencies and growth rates, if any, of a particle beam in a modified betatron or similar device. It does so by numerically evaluating the roots of an analytical dispersion relation. The dispersion relation was derived by treating the beam as a string of rigid disks traveling around the toroidal acceleration cavity while executing small toroidal and poloidal oscillations. The electromagnetic fields which interact with these oscillations were evaluated exactly for a cavity of rectangular minor cross section by means of Green's functions.

Many features of BTRSQ are patterned after those in GRADR. It has interactive NAMELIST input, an optimized Muller's method root finder, error recovery and user-interrupt procedures, and graphical output. It is extremely fast and very easy to use. Other linear stability problems can be addressed in the same way simply by providing the appropriate dispersion function routines. This has been done in several instances.

ACKNOWLEDGMENTS

The computer programs CCUBE, ISIS, and GRADR were developed at Los Alamos National Laboratory (LANL). MAGIC, CPROP, IPROP, ORBIT, and KMRAD were written by MRC for Sandia National Laboratories (SNL), Albuquerque. SOS and BALTIC were written by MRC for the Air Force Weapons Laboratory (AFWL). IVORY was written by MRC for LANL. ARCTIC was written by MRC for the Strategic Defense Initiative Office (contract monitored by the Naval Surface Weapons Center, White Oak). BTRSQ was written by MRC for the Office of Naval Research (ONR). SCRIBE was adapted by MRC from the Stanford Linear Accelerator Center Beam Trajectory Program, EGUN.

The preparation of this report was supported in part by the Air Force Weapons Laboratory and in part by the Defense Advanced Research Projects Agency under a contract monitored by the Naval Surface Warfare Center.

REFERENCES

1. M. M. Campbell, D. J. Sullivan, and B. B. Godfrey, CCUBE User's Manual, AMRC-R-341, Mission Research Corporation, Albuquerque, NM, 1983.
2. M. M. Campbell and B. B. Godfrey, IVORY User's Manual, AMRC-R-454, Mission Research Corporation, Albuquerque, NM, 1983.
3. M. E. Jones, Los Alamos National Laboratory, Los Alamos, NM, unpublished.
4. B. Goplen, R. E. Clark, J. McDonald, and W. M. Bollen, User's Manual for MAGIC / Version - September 1983, MRC/WDC-R-068, Mission Research Corporation, Washington, D.C., 1983.

5. B. Goplen, R. J. Barker, R. E. Clark, and J. McDonald, User's Manual for SOS / Version - September 1983, MRC/WDC-R-065, Mission Research Corporation, Washington, D.C, 1983.
6. B. B. Godfrey, High Current Beam Propagation Study, AMRC-R-367, Mission Research Corporation, Albuquerque, NM, 1982.
7. B. B. Godfrey, The IPROP Three-Dimensional Beam Propagation Code, AMRC-R-690, Mission Research Corporation, Albuquerque, NM, 1985.
8. W. B. Hermannsfeldt, Electron Trajectory Program, SLAC-226, Stanford Linear Accelerator Center, Stanford, CA, 1979.
9. M. Mostrom, Mission Research Corporation, Albuquerque, NM, unpublished.
10. R. J. Adler and B. B. Godfrey, Radial Pulseline Electron Accelerator Study, AMRC-R-368, Mission Research Corporation, Albuquerque, NM, 1982.
11. D. J. Sullivan and B. B. Godfrey, GRADR: Interactive User's Manual, AMRC-R-309, Mission Research Corporation, Albuquerque, NM, 1981; B. B. Godfrey, GRADR: Batch User's Manual, AMRC-R-223, Mission Research Corporation, Albuquerque, NM, 1980.
12. L. A. Wright, ORBIT: Batch User's Manual, AMRC-R-249, Mission Research Corporation, Albuquerque, NM, 1980.
13. T. P. Hughes and B. B. Godfrey, KMRAD: A Linearized Particle Code for the Study of Resistive Instabilities, AMRC-R-364, Mission Research Corporation, Albuquerque, NM, 1982.
14. T. P. Hughes and B. B. Godfrey, Modified Betatron Accelerator Studies, Sec. 2, AMRC-R-655, Mission Research Corporation, Albuquerque, NM, 1984.
15. T. Kwan and B. Godfrey, IEEE Trans. Nuc. Sci. 26, 3933 (1979).
16. D. J. Sullivan, The VIRCATOR: A Tunable High-Power Microwave Generator, AMRC-N-166, Mission Research Corporation, Albuquerque, NM, 1981.
17. D. J. Sullivan, R. Adler, D. Voss, W. Bollen, R. Jackson, and E. Coutsias, Virtual Cathode Oscillator Study, AMRC-R-614, Mission Research Corporation, Albuquerque, NM, 1984.
18. D. Sullivan, G. Kiuttu, R. Adler, D. Voss, and J. Walsh, Resonant Vircator Study, AMRC-R-654, Mission Research Corporation, Albuquerque, NM, 1984.
19. R. J. Faehl, B. S. Newberger, and B. B. Godfrey, Phys. Fluids 23, 2440 (1980).
20. D. J. Sullivan and B. B. Godfrey, Laser Electron Acceleration, AMRC-R-281, Mission Research Corporation, Albuquerque, NM, 1981.
21. D. J. Sullivan and R. J. Faehl, Simulation of Collective Ion Acceleration in the Luce Diode, AMRC-R-353, Mission Research Corporation, Albuquerque, NM, 1982.
22. D. J. Sullivan and M. M. Campbell, Numerical Simulation of the Plasma Beatwave Accelerator Concept, AMRC-R-611, Mission Research Corporation, Albuquerque, NM, 1984.

23. C. Snell, Los Alamos National Laboratory, Los Alamos, NM, unpublished.
24. R. Adler, M. Campbell, B. Godfrey, D. Sullivan, and T. Genoni, Part. Accel. $\underline{13}$, 25 (1983).
25. B. B. Godfrey and T. P. Hughes, Phys. Fluids $\underline{28}$, 669 (1985).
26. T. P. Hughes and B. B. Godfrey, Phys. Fluids $\underline{29}$, 1698 (1986).
27. B. B. Godfrey and T. P. Hughes, High Current Electron Beam Transport in Recirculating Accelerators, AMRC-R-842, Mission Research Corporation, Albuquerque, NM, 1986.
28. W. R. Shanahan, Compilation of Interim Technical Research Memoranda, LA-10084, Vol. 1, Los Alamos National Laboratory, Los Alamos, NM, 1984.
29. T. J. Kwan, Phys. Rev. Lett. $\underline{57}$, 1985 (1986).
30. A. Kadish, R. J. Faehl, and C. N. Snell, Phys. Fluids $\underline{29}$, 4192 (1986).
31. J. W. Poukey, Electron-Beam Diode Simulations, 1982, SAND83-0173, Sandia National Laboratories, Albuquerque, NM, 1983.
32. J. W. Poukey, Electromagnetic Effects on the Formation of Potential Wells for Ion Acceleration, SAND81-1687, Sandia National Laboratories, Albuquerque, NM, 1981.
33. J. W. Poukey, Computer Studies of Oscillations in RADLAC, SAND82-0192, Sandia National Laboratories, Albuquerque, NM, 1982.
34. J. W. Poukey, Particle Simulations of BWO, SAND84-1242, Sandia National Laboratories, Albuquerque, NM, 1984.
35. T. P. Hughes and B. B. Godfrey, Phys. Fluids $\underline{27}$, 1531 (1984).
36. B. B. Godfrey, R. J. Adler, K. O. Busby, T. P. Hughes, G. Z. Hutcheson, G. F. Kiuttu, R. J. Richter-Sand, N. F. Roderick, and D. J. Sullivan, High Power Electron Beams and Microwaves, AMRC-R-864, Mission Research Corporation, Albuquerque, NM, 1986.
37. N. F. Roderick, "Hollowing Instability in Rotating Annular Beams," submitted to Phys. Fluids.
38. C. A. Ekdahl, J. R. Freeman, G. T. Leifeste, R. B. Miller, W. A. Stygar, and B. B. Godfrey, Phys. Rev. Lett. $\underline{55}$, 935 (1985).
39. B. B. Godfrey, Phys. Fluids $\underline{30}$, 570 (1987).
40. B. B. Godfrey, Phys. Fluids $\underline{30}$, 575 (1987).
41. L. A. Wright, B. B. Godfrey, and T. P. Hughes, PHERMEX Beam Propagation Studies, AMRC-R-445, Mission Research Corporation, Albuquerque, NM, 1983.
42. C. A. Ekdahl, L. A. Wright, T. P. Hughes, and B. B. Godfrey, FX-25 and FX-100 Propagation Experiments, AMRC-R-391, Mission Research Corporation, Albuquerque, NM, 1982.
43. D. R. Welch and B. B. Godfrey, Enhanced Channel Tracking Due to Beam Generated Magnetic Fields, AMRC-R-814, Mission Research Corporation, Albuquerque, NM, 1986.
44. B. B. Godfrey and D. R. Welch, Analytical Models of Conductivity-Channel Tracking, AMRC-R-899, Mission Research Corporation, Albuquerque, NM, 1987.

45. B. B. Godfrey, B. S. Newberger, L. A. Wright, and M. M. Campbell, IFR Transport in Recirculating Accelerators, AMRC-R-741, Mission Research Corporation, Albuquerque, NM, 1985.
46. D. E. Pershing, C. A. Sedlak, R. H. Jackson, M. M. Campbell, and B. B. Godfrey, Simulation of the IBEX Diode with Scribe, MRC/WDC-R-082, Mission Research Corporation, Washington, D.C., 1984.
47. R. K. Parker, R. H. Jackson, S. H. Gold, H. P. Freund, V. L. Granitstein, P. C. Efthimion, M. Herndon, and A. K. Kinkead, Phys. Rev. Lett. 48, 238 (1982).
48. B. B. Godfrey, Resistive Wall Instabilities in Radial Pulseline Accelerators, AMRC-R-345, Mission Research Corporation, Albuquerque, NM, 1981.
49. R. J. Adler and B. B. Godfrey, Radial Pulseline Electron Accelerator Study (Jan - May, 1981), AMRC-R-306, Mission Research Corporation, Albuquerque, NM, 1981.
50. R. J. Faehl, W. R. Shanahan, and B. B. Godfrey, "Nonlinear Characteristics of Cyclotron Waves in an ARA Configuration," in N. Rostoker and M. Reiser, Collective Methods of Acceleration (Harwood Academic, N. Y., 1979)
51. B. B. Godfrey and B. S. Newberger, J. Appl. Phys. 50, 2470 (1979).
52. D. J. Sullivan and B. B. Godfrey, Linear Theory and Adiabatic Wave Variation in the Converging Guide Accelerator, AMRC-R-353, Mission Research Corporation, Albuquerque, NM, 1982.
53. L. E. Thode, B. B. Godfrey, and W. R. Shanahan, Phys. Fluids 22, 747 (1979).
54. B. B. Godfrey, IEEE Trans. Plas. Sci. 7, 53 (1979).
55. B. B. Godfrey and B. G. Epstein, Diocotron and Resistive Wall Instabilities, AMRC-N-141, Mission Research Corporation, Albuquerque, NM, 1980.
56. T. P. Hughes and B. B. Godfrey, Cold-Fluid Model of Resistive Instabilities on Intense Relativistic Electron Beams, AMRC-R-285, Mission Research Corporation, Albuquerque, NM, 1981.
57. W. R. Shanahan, B. B. Godfrey, and R. J. Faehl, "Slow Cyclotron Wave Growth by Periodic Inductive Structures," in N. Rostoker and M. Reiser, Collective Methods of Acceleration (Harwood Academic, N. Y., 1979).
58. L. A. Wright and B. B. Godfrey, Distribution Function Directory for the Code ORBIT, AMRC-R-413, Mission Research Corporation, Albuquerque, NM, 1982.
59. T. P. Hughes and H. S. Uhm, Resistive Instabilities on a Thick, Hollow Electron Beam, AMRC-R-543, Mission Research Corporation, Albuquerque, NM, 1984.
60. T. P. Hughes and H. S. Uhm, J. Appl. Phys. 60, 577 (1986).
61. H. S. Uhm and T. P. Hughes, Phys. Fluids 29, 3074 (1986).
62. T. P. Hughes and B. B. Godfrey, Linear Stability of the Modified Betatron, AMRC-R-354, Mission Research Corporation, Albuquerque, NM, 1982.

63. B. B. Godfrey and T. P. Hughes, IEEE Trans. Nuc. Sci. 32, 2495 (1985).
64. J. F. Traub, Iterative Methods for the Solution of Equations (Prentice-Hall, N. Y., 1964) Ch. 10.
65. B. I. Cohen, ed., Multiple Time Scales, (Academic Press, Orlando, FL, 1985).
66. B. Adler, S. Fernbach, and M. Rotenberg, eds., Methods of Computational Physics, Vol. 9.
67. C. K. Birdsall and A. B. Langdon, Plasma Physics via Computer Simulation (McGraw-Hill, N. Y., 1985).
68. B. B. Godfrey, A Galerkin Algorithm for Multidimensional Plasma Simulation Codes, LA-7687-MS, Los Alamos National Laboratory, Los Alamos, NM, 1979.
69. B. B. Godfrey, J. Comp. Phys. 19, 58 (1975).
70. B. B. Godfrey, Time-Biased Field Solver for Electromagnetic PIC Codes, AMRC-N-138, Mission Research Corporation, Albuquerque, NM, 1980.
71. B. Marder, A Method for Incorporating Gauss' Law into Electromagnetic PIC Codes, SAND84-2051, Sandia National Laboratories, Albuquerque, NM, 1985.
72. E. P. Lee, The New Field Equations, UCID-17286, Lawrence Livermore National Laboratory, Livermore, CA, 1976.
73. R. L. Feinstein, BMCOND Model, SAI-U-080-8203, Science Applications International Corporation, Palo Alto, CA, 1982.
74. F. M. Chambers and D. M. Cox, Standard Test Case Runs for the Empulse Monopole Fieldsolver and Conductivity Generation Model, UCID-19213, Lawrence Livermore National Laboratory, Livermore, CA, 1971.
75. H. A. Bethe, Phys. Rev. 89, 1256 (1953).
76. R. L. Feinstein, Science Applications International Corporation, Los Altos, CA, private communication, 1985.
77. R. J. Briggs, D. L. Birx, G. J. Caporaso, V. K. Neil, and T. C. Genoni, Part. Accel. 18, 41 (1985).
78. B. B. Godfrey, Transverse Wake Potentials for Wide Radial Lines, MRC/ABQ-R-1046, Mission Research Corporation, Albuquerque, NM, 1988.
79. T. P. Hughes and B. B. Godfrey, Electron Beam Stability in Compact Recirculating Accelerators, MRC/ABQ-R-1040, Mission Research Corporation, Albuquerque, NM, 1988.
80. R. C. Davidson, Theory of Nonneutral Plasmas (Benjamin, N. Y., 1974) pp. 17-21.
81. L. F. Shampine, H. A. Watts, and S. M. Davenport, SIAM Rev. 18, 376 (1976).

RECENT APPLICATIONS OF SUPERFISH*

R.L. Gluckstern and F. Neri
Physics Department, University of Maryland, College Park, MD 20742

INTRODUCTION

The program, SUPERFISH,[1] obtains the frequencies and fields for azimuthally symmetric TM or TE modes in azimuthally symmetric cavities. The r-z plane is covered by a triangular mesh and the resulting difference equations for the field component H_ϕ (TM) or E_ϕ (TE) at the vertices of the mesh solved by direct matrix inversion.

In the present paper, we describe the following recent modifications of SUPERFISH:
1. A variational formulation for the frequency is used to greatly increase the accuracy.
2. The SUPERFISH calculation is modified to obtain the dispersion curve for a periodic structure by a mesh calculation in one cell.
3. The contribution of an azimuthally symmetric cavity or obstacle in a beam pipe to the longitudinal coupling impedance is obtained by a SUPERFISH-like mesh calculation in the region of the obstacle, with appropriate boundary conditions.

Each of these modifications is briefly described.

ACCURACY IMPROVEMENT

A variational form for the frequency ($\omega = kc$) of a cavity is

$$k^2 = \frac{\int dv \, (\nabla \times \vec{H})^2}{\int dv \, H^2} \qquad (1)$$

where the trial function $\vec{H}$ must be continuous within the cavity. The output of SUPERFISH for a TM mode is such a function, where H_ϕ is taken to be a linear function of r and z within each triangle of the mesh chosen to match the three values of H_ϕ at the vertices of the triangle.

It is clear the $\nabla \times \vec{H}$ is discontinuous across the triangle boundaries. From this fact, it can be shown that the accuracy of the SUPERFISH calculation is of order

$$\frac{\delta k}{k} \sim \mathcal{O}(h^2 k^2) \sim \mathcal{O}(1/N^2), \text{ linear } H_\phi \qquad (2)$$

where h is the mesh size and N^2 is the number of mesh points. Moreover, if H_ϕ had

---
*Work supported by DOE Contract DE-AS05080-ER-10666.

continuous derivatives, the accuracy would be

$$\frac{\delta k}{k} \sim \mathcal{O}(h^4 k^4) \sim \mathcal{O}(1/N^4), \text{ quadratic } H_\phi \ . \qquad (3)$$

Gluckstern, Ryne and Holsinger[2] therefore used the information from the output of SUPERFISH in nearby triangles to estimate the quadratic correction to H_ϕ in each triangle, which makes the derivatives of H_ϕ continuous across triangle boundaries. Contributions of this correction to the integrals in Eq. (1) were then calculated, and a modified frequency was obtained. In addition, a correction was applied to SUPERFISH'S polygon approximation to the curved boundary. The results were then tested on several TM and TE modes in cylinders and spheres, where the statements in Eqs. (2) and (3) were confirmed in detail. Improvements in frequency accuracy were typically a factor of about 20 to 50 using a 20×20 mesh.

This version of the post-processor supplement to SUPERFISH is available at LANL, Group AT-6.

PERIODIC STRUCTURE

The Floquet Theorem states that the fields in a structure with period L are related, for a traveling wave mode, to the fields in the previous period by

$$F(r, \theta, z + L) = e^{i\sigma L} F(r, \theta, z) \ . \qquad (4)$$

If one chooses a mesh which is exactly periodic, the difference equations at the boundary of a cell will relate to the field values at the mesh lines corresponding to z_{n-1}, z_n and z_{n+1} to one another, where n runs from 0 to $N-1$, with $z_o = 0$, $z_N = L$. The difference equations for $n = N-1$ and $n = 0$ can be written in terms of $F(z_n)$, $n = 0, 1, \ldots N-1$ by using Eq. (4) to express $F(z_{-1})$ in terms of $F(z_{N-1})$ and $F(z_N)$ in terms of $F(z_o)$. Thus the equivalent SUPERFISH problem is to determine the value of σ for a given ω which will make the determinant of the difference equations vanish. This is now a problem involving complex coefficients, and the result is the desired dispersion curve $\omega(\sigma)$ for either TM of TE traveling wave modes.

This program, designed and written by Gluckstern and Opp,[3] is available at LANL, Group AT-6.

LONGITUDINAL COUPLING IMPEDANCE

A current perturbation along the axis of a beam pipe will generate wake fields that are proportional to the driving perturbation, which can exchange longitudinal energy with the beam. The constant of proportionality between the resulting voltage gain in one revolution and the perturbing current is the longitudinal coupling impedance. Clearly the current limit in a ring will depend on the overall impedance.

Analytic results are available for the longitudinal coupling impedance for a beam pipe with conducting walls. The contribution of obstacles (pump ports, kicker magnets, bellows, beam position monitors, etc.) is more difficult to calculate. We have modified SUPERFISH[4] to yield the additional longitudinal coupling impedance due to an azimuthally symmetric obstacle. Specifically, we use Maxwell's equations for the field difference between the case of the beam pipe plus obstacle, and the beam pipe alone, requiring waves which are outgong from the obstacle.

The resulting impedance as a function of frequency has considerable structure in the vicinity of the cut-off frequency of the beam pipe, even for the simplest obstacle shape. The results of the program agree with the semi-analytic results of Henke[5] for a cylindrical resonator. The semi-analytic approach is then applied to a small cylindrical resonator in order to explain the structure of the coupling impedance in the vicinity of the cut-off of the first few allowed modes. This result is also used as a guideline for the fitting of the broad resonance near the cut-off for which numerical parameters are obtained.[6].

WORK IN PROGRESS

The program described in the previous section is now being used to examine:
1. the high frequency dependence of the longitudinal coupling impedance,
2. the interference between adjacent obstacles as a function of frequency and the separation of the obstacles, and
3. the dependence of the longitudinal coupling impedance on the shape of the obstacle for different frequency ranges.

Work is also being carried out[7] to use SUPERFISH to predict the parameters associated with the coupling between a cavity and a wave guide. Of particular interest is the external Q[8] for the coupling of the cavity modes into the wave guide.

We are also exploring the possibility of using either ULTRAFISH[9] or URMEL[10] to obtain the dispersion curves for transverse mode bands in a periodic structure, as well as to calculate the transverse coupling impedance, which is also important in setting current limits in proton storage rings. Some work along this line was reported earlier using URMEL to calculate the longitudinal coupling impedance.[11]

REFERENCES

1. K. Halbach and R.F. Holsinger, Particle Accelerators, **7**, 213 (1976).
2. R.L. Gluckstern, R.D. Ryne and R.F. Holsinger, Proceedings of the COMPUMAG Conference, Genoa, Italy, p. 141, May 1983.
3. R.L. Gluckstern and E.N. Opp, IEEE Transactions on Magnetics, Vol. **Mag-21**, No. 6, 2344 (1985).

4. R.L. Gluckstern and F. Neri, IEEE Transactions on Nuclear Science Vol. **NS-32**, No. 5, 2403 (1985); Proceedings of the 13th International Conference on High Energy Accelerators, Novosibirsk, USSR, August 1986.
5. H. Henke, Point charge passing a resonator with beam tubes, CERN LEP-RF/85/41.
6. R.L. Gluckstern and F. Neri, Proceedings of the Particle Accelerator Conference, Washington, DC, March 1987.
7. In collaboration with Rui Li.
8. See, for example, J.C. Slater, Microwave Electronics, Van Nostrand, 1950.
9. Gluckstern, Holsinger, Halbach and Minerbo, Proceedings of the 1981 Linac Conference, Santa Fe, NM, p. 102, October 1981.
10. T. Weiland, Nucl. Instr. and Meth. **216**, 329 (1983).
11. U. van Rienen and T. Weiland, Proceedings of the Linac Conference, SLAC, p. 298 (1986).

AN INTERACTIVE INTERFACE TO THE BEAM OPTICS CODE "TRANSPORT"

Richard R. Silbar
Theoretical Division, Los Alamos National Laboratory
Los Alamos, New Mexico 87545

ABSTRACT

The Fortran code, TRANSPORT, used for calculating beam optics properties, has been converted to run on a Symbolics Lisp Machine. Using the rich graphics and object-oriented programming environment provided by the KEE expert system shell, a beam designer can now use TRANSPORT in a manner similar to that of a user of a CAD/CAM tool or a spreadsheet. Beamline elements can be modified, added, or deleted at the user's screen, with the changes in the beam ellipses, envelopes, and transfer matrices being re-computed and re-displayed within a few seconds.

INTRODUCTION

In the ancient past, which for me was before there was a LAMPF accelerator, we used to design beamlines in a primitive way: input to the TRANSPORT code[1] was accomplished by means of punched card decks. With a typical turnaround-time for the mainframe batch job of half a day, a cut-and-try design process was a pretty tedious affair. With the advent of minicomputers, such as a VAX, the cards gave way to input and output files[2] and turnaround became much faster, of the order of minutes on a quiet day. More recently, there have been conversions of the TRANSPORT code to stand-alone micro-VAX computers,[3] which, in addition to still more rapid turnaround, allow for a more graphical display of results and for a more interactive environment in which the beamline designer can work.[4]

Today, I will tell you about an experiment in which we have embedded the TRANSPORT code in an expert system shell on a LISP machine. The purpose of this work is to see if we can provide an interactive environment for the beamline designer similar to that seen by the user of a spreadsheet or of a CAD/CAM system. The reason for doing this in a LISP environment is because, in the backs of our minds, there are some future applications using artificial intelligence that will build on the present exercise. These include an intelligent design advisor, a tutoring or training system, and an on-line controller for a beamline. However, I will not show you anything about those future applications today.

There are three major components in the prototype system I am about to show you:
1. The TRANSPORT code itself, which came to us as a 9000 line VAX-VMS Fortran source,
2. A Symbolics LISP machine, configured with a Fortran-77 compiler, and
3. The Knowledge Engineering Environment (KEE) expert system shell.[5]

© 1988 American Institute of Physics

The LISP machine provides a rich development environment for the LISP (and even the Fortran) programmer. The KEE shell provides a powerful set of graphics routines and images and, more importantly, an object-oriented programming system and active values. It is these latter two concepts that are the new ingredients in this prototype interface, and I will give below some examples to illustrate what these terms mean.

It is probably useful to say that a Symbolics workstation is not just an ordinary desktop computer; the present application needs a system with at least 4 megabytes (MB) of memory (12 MB gives much better performance) and considerable disk capacity (typically 2 x 190 MB). The KEE software itself is also not trivial; a first-copy license costs about $50,000. Nonetheless, it is reasonable to believe that in the future the front-end for TRANSPORT that we are developing can be ported to less expensive machines, such as Sun workstations or 80386 PC's, in run-time versions soon to be commercially available. That is, the present cost of the prototype I am discussing today will most likely not be a problem when this system becomes available as a working tool.

THE INTERFACE AND HOW IT "LOOKS AND FEELS"

The first task we faced was to convert the VAX-VMS Fortran code for TRANSPORT to compile with the Fortran-77 package provided by Symbolics Corporation. This was complicated by the use of machine-dependent routines ("VAXisms") beyond the scope of pure Fortran-77. The compiled TRANSPORT system then had to be integrated with the LISP-based KEE software. For those who are not computer scientists, let me just say that the number-crunching FORTRAN and the symbol-processing LISP languages differ considerably in the way they allocate computer memory, so the solution to this problem is not obvious. We have chosen, for now, to do the simplest thing; communicate between the two programming environments by writing and reading disk files. This does not give the fastest possible response -- the file manipulation probably takes as much time than as the actual calculation done by TRANSPORT -- but updates of the beam displays typically are made in a few seconds.

The major effort has been devoted to the construction of the interface with the beamline designer. There has been a conscious effort to keep it simple. The user selects program operations using a mouse, through menus and a standard input-output window. It is fairly easy for a person with some familiarity with a Sun work-station, for example, to sit down at the Symbolics console and begin constructing a new beam.

Figure 1 shows what the operator sees on the right half of the console screen.[6] The panel at the top displays a set of actuators or "digital knobs" for changing the parameters of certain selected beamline elements. The beamline under consideration consists of six quadrupoles and their associated drifts. The lower panels display the beam envelopes for that beamline in the horizontal (x-) and vertical (y-) planes, along with the positions of the quadrupoles

(dashed rectangles). For the case shown, the initial beam is
misaligned, in both planes, with respect to the principal axis in
both position and direction ("launch errors"). The apertures and
lengths of the quads are dynamic in the sense that the quadrupole
icons will change appropriately (and to scale) if these parameters
are changed.

The actuator boxes in the Beamline Elements Modifier Panel are
"mouseable." That is, one can move the mouse cursor into the box
labeled "Eye3's Field" and initiate a change of its value by
clicking the leftmost of the three mouse buttons. This brings up a
prompt in a standard input-output window (not shown in Fig. 1; it is
on the left side of the screen) asking for a new value of the field.
The user enters this from the keyboard, hits the carriage return,
and the number displayed in the "Eye3's Field" box is updated to the
new value. The program then automatically invokes an update of the
TRANSPORT input deck and a recalculation of the beamline. Upon completion of the TRANSPORT calculation, the beam envelope displays are
updated (with a new scaling, if necessary). The whole process takes
about ten seconds, depending on the complexity of the beamline.
This is not fast enough to allow these actuators to be truly
considered as "knobs," which one normally thinks of as responding

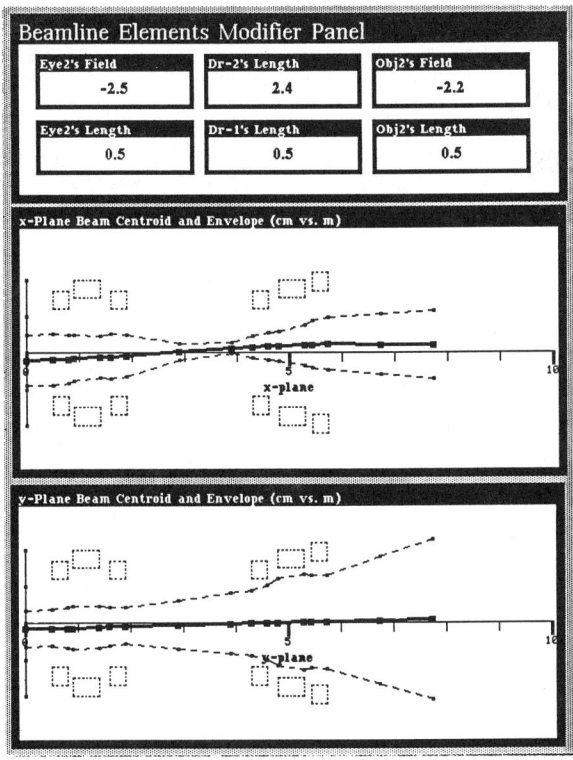

Fig. 1. Beamline elements modifier panel and beam envelopes.

to a tweak with an instantaneous response. It is fast enough, however, to give the operator a feeling, albeit less than kinesthetic, for how things change with respect to a given parameter.

Alternatively, instead of the Beamline Elements Modifier Panel, the operator can choose to display a set of actuators which display, and allow changes to be made in, the parameters describing the initial beam conditions. These are mouseable in exactly the same way as described above; again leading to a TRANSPORT calculation and an update of the various displays. In this regard, the operator can also choose to put other kinds of displays describing the beamline properties on the screen. Thus, instead of the beam envelopes, he or she can choose to look at phase-space ellipses or the transfer matrix.

At the top of the right-hand half of the screen (but not shown in Fig. 1) there are two "buttons" for bringing up two general menus, one for general utilities and the other for more general changes to the beamline. A left click of the mouse in the Modify-Beamline button-box brings up the cascading menu shown in Fig. 2. This menu shows exactly what elements are in the beamline and what kinds of operations can be done on them. Note the ability to display or change the parameters of a given element, to make or delete a "knob" in the Modifier Panel, and to rename, delete or add a new element to the beamline. (In the latter cases, the menu is recreated with the appropriate name changes.) This menu can be exited in one of two ways -- either with or without a TRANSPORT recalculation. (The latter would be desirable if only cosmetic changes were made.) When this pop-up menu is present on the screen, the actuator boxes in the "knob panels" are de-activated to avoid accidentally starting a TRANSPORT calculation.

For completeness, I should say that all areas of the display screen, including those in the lower half of Fig. 1, are mouseable. This includes "click-right" and "click-middle" mouse functions, such as for resizing, scrolling, and zooming of windows. In fact, the mouse is one of the chief ways the program developer communicates with KEE. In a run-time version of this tool, it would probably be useful to "hide" these functions from the naive user to keep him or her from getting into trouble by corrupting the knowledge base.

The pop-up cascading menu brought up by the Choose-Utility button looks similar to that shown in Fig. 2. The choices available to the user, at present, include:
1. Selection of which displays are to appear on the screen, such as beam envelopes vs. phase-space ellipses or beamline element knobs vs. intial beam knobs.
2. Display a (mouseable) "spreadsheet" of the beamline elements and their parameters.
3. Save the present beamline as a recallable TRANSPORT input deck.
4. Read one of the previously saved TRANSPORT decks and reconstruct the beamline from it.
5. Print the TRANSPORT output file.

6. Flush the beamline, i.e., remove all the elements except for one generic drift, so that one can begin anew.
7. Choose which of several windows useful to the developer to display, if any.
8. And, again, to exit, with or without a TRANSPORT calculation.

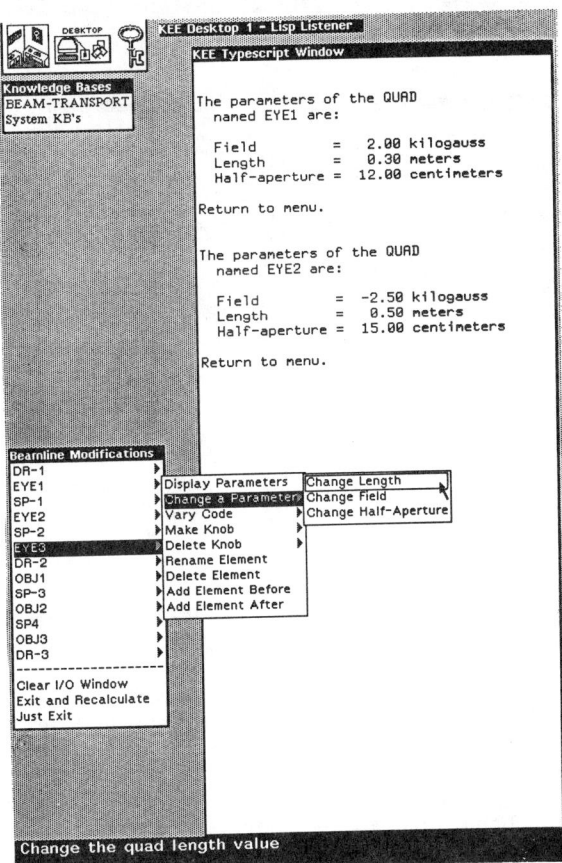

Fig. 2. Pop-up menu for modifying beamline elements.

OBJECT-ORIENTED PROGRAMMING AND ACTIVE VALUES

As stated earlier, the new ingredient in this interface is the use of object-oriented programming techniques (OOPS) and active values (AV's). There are the usual (valid) good-programming-practice reasons for using OOPS and AV's -- transparency, easier code maintenance and upgrades, etc. For our purposes, however, the main reason for using these techniques is that they provide the opportunity to expand the present software into new applications, such as on-line control.

In the OOPS approach taken here, every beamline element has its own "unit," or frame, containing some number of property slots, as shown in Fig. 3. The slots are used for storing parameter values or pieces of code, called "methods", which may be specialized to provide behavior appropriate for that particular kind of element. The values in the slots may be local or inherited from some parent-class containing this object.

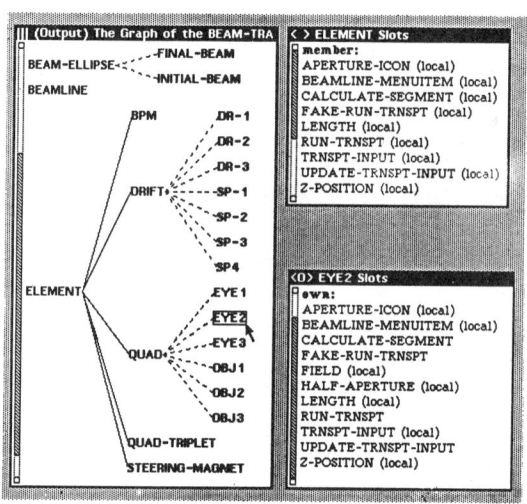

Fig. 3. Beamline element units in the knowledge base.

The class hierarchy for beamline elements is seen in Fig. 3 as the tree of devices descending from the general element called ELEMENT. As one goes from the left to the right down this tree, the units become more and more specialized, with the final leaves being the individual elements themselves (instances). The inheritance of properties and behavior can be seen here as well in the slot lists: drifts and quads each have a length slot, inherited from ELEMENT, but quads have in addition slots for field strength and half-aperture. The generic unit QUAD does not actually have any values in those slots, but an instance such as EYE3 does.

Methods can be invoked by "sending a message" to an object to carry out the action specified by the method. This can be different from one class to another. Thus, sending the message named UPDATE-TRNSPT-INPUT to a unit will give different results for quadrupoles and drifts, simply because the method named in that value-slot is different for the two types of object. The input deck "card' created by UPDATE-TRNSPT-INPUT will therefore have type-code 5 for the quad but 3 for the drift, as desired.

Active values refer to slots that, when their contents change or are accessed, invoke some sort of procedure. This style of data-driven program flow can be useful for many purposes, including the propagation of constraints through a given tentative design. (That has not yet been implemented in our prototype, however.) For example, the AVPUT active value which is invoked whenever any of the initial beam parameters is changed involves the following sequence of actions:

1. The initial beam ellipse is updated by sending the message CALCULATE-ELLIPSE to the unit called INITIAL-BEAM. If the ellipse is on-screen, that window is re-displayed.
2. The LISP routine FORM-TRNSPT-INPUT is called (which, in turn may send the UPDATE-TRNSPT-INPUT message discussed above to one or more units).
3. The message RUN-TRNSPT is sent, i.e., the TRANSPORT code is run with the new input deck.
4. The LISP routine READOUT-TRNSPT-RESULTS is called, which parses that output file, stores the results in appropriate value-slots, and generates updates of the displays of, say, the beam envelopes.
5. The final phase-space ellipse is updated (and redisplayed, if on-screen).

The change, as we have seen, may have occurred because the designer made it intentionally by mousing on the initial beam knob panel and entering a new value. Alternatively, in an on-line control situation in which this TRANSPORT model of the beamline simulates a real one, the change may occur because some device in the real beamline breaks.

FUTURE DIRECTIONS AND SUMMARY

As you can imagine from seeing the figures, this prototype interface between the beam optics code, TRANSPORT, and the beamline designer is still much in its infancy. It is changing rapidly as we experiment with different ways of presenting information (or extracting it) from the user. Some of the near-future enhancements will (or might) be:

- Ability to look at a small region of the beam-envelope plots in more detail (e.g., to "zoom" it).
- Filling in the knowledge base with other types of beamline elements, such as bends, sextupoles, etc., that TRANSPORT can also handle.
- Linking parameter values for similar devices and making sure the design satisfies layout constraints.
- Using the optimization facility that is already built into TRANSPORT.

Further in the future would be the extension of the software to support beam optics codes other than TRANSPORT, such as MARYLIE[7] or COSY.[8] At this point, one could imagine the system being transformed into an expert system for giving intelligent advice. For example, if the user declares "I want a final beam which is very nearly parallel", the expert system could look through its heuristic rules and reply with something like "To get a (nearly) parallel beam, put the eyepiece somewhere near the focal point of the objective lens. In the present case that would be near z=1.7, if 'objective' means the triplet near z=9."

To summarize, the work we have done so far in developing this prototype indicates that the interface between TRANSPORT and the beamline designer <u>can be much improved</u> by using a LISP workstation environment in conjunction with an expert system shell.

ACKNOWLEDGEMENTS

I thank Drs. E. A. Heighway, W. P. Lysenko, and C. T. Mottershead for their suggestions of features which would be useful to beam line designers during the design process. Also I thank Mary Fuka and Rozelle Wright for their encouragement. This work has been supported by the U.S. Department of Energy.

REFERENCES

1. D. Carey, this conference.
2. E.g., M. Hoehn, LAMPF version of FNAL TRANSPORT, 1981.
3. E.g., U. Rohrer, SIN version of TRANSPORT, 1982.
4. Similar developments have occurred for <u>other</u> beam optics codes. At this conference there are reports on TRACE-3D (K. Crandell) and LATTICE (J. Staples), which are also highly interactive and graphical. Also of note is the COMFORT code (M. Lee), which has been adapted to the Sun workstation and the SunTools package by V. Paxson.
5. A software product developed and sold by IntelliCorp, Mountain View, California.
6. The description of the interface given in this paper reflects changes made to the system since this talk was given.
7. A. Dragt, this conference.
8. H. Wollnick, this conference.

STATUS AND FUTURE OF THE 3D MAFIA GROUP OF CODES

F.Ebeling. R.Klatt. F.Krawzcyk. E.Lawinsky, T.Weiland, S.G.Wipf
Deutsches Elektronen-Synchrotron DESY, Notkestr.85, D-2000 Hamburg 52,
Germany

B.Steffen
Kernforschungsanlage Jülich KFA, D-5170 Jülich,Germany

T.Barts, J.Browman, R.K.Cooper, G.Rodenz
Los Alamos National Laboratory, Los Alamos, New Mexico 87544, U.S.A.

ABSTRACT

The group of fully three dimensional computer codes for solving Maxwell's equations for a wide range of applications, MAFIA, is already well established. Extensive comparisons with measurements have demonstrated the accuracy of the computations. A large number of components have been designed for accelerators, such as kicker magnets, non cylindrical cavities, ferrite loaded cavities, vacuum chambers with slots and transitions, etc. The latest additions to the system include a new static solver that can calculate 3D magneto- and electrostatic fields, and a self consistent version of the 2D-BCI that solves the field equations and the equations of motion in parallel. Work on new eddy current modules has started, which will allow treatment of laminated and/or solid iron cores excited by low frequency currents. Based on our experience with the present releases 1 and 2, we have started a complete revision of the whole user interface and data structure, which will make the codes even more user-friendly and flexible.

INTRODUCTION

The acronym MAFIA stands for the solution of **MA**xwell's equations by the **F**inite **I**ntegration **A**lgorithm. This is the name given to a set of fully three-dimensional codes used in the computer-aided design of accelerators. Rf structures, magnets and electrostatic devices as well as structures in which wake-field effects are of importance can all be calculated. The Finite Integration Technique (FIT) [1,2] produces a first order approximation of Maxwell's equations by replacing the line and surface integrals by mean field values multiplied by path lengths or areas. The algorithm produces a set of matrix equations which can subsequently be solved. The allocation of the field components to the rectangular grid is as shown in Figure 1, with the electric field components allocated at the mid-points of the sides of the rectangular cells and the magnetic field components at the centre of each face. This defines a dual grid where the origin of each dual cell lies at the centre of each original grid cell, and to which

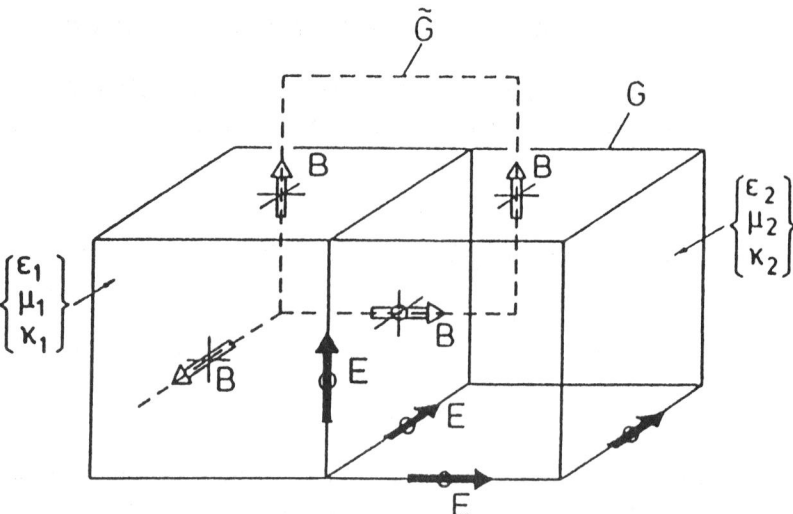

Figure 1: Geometry and Allocation of the Field Components for the FIT Method

the magnetic components are allocated in the same way as the electric components on the regular grid. Thus the transition from one cell to the next involves only continuous components, leading to the fact that Maxwell's equations are always satisfied at the points of allocation, even when different material fillings are involved. In addition, the analytical properties of the matrix operators are preserved on the discrete grid [2], in particular $divcurl \equiv 0$, $curlgrad \equiv 0$, which is not necessarily the case for all discretisation methods. Thus the FIT method has an advantage over other methods in that, after the eigenvalues of the problem are calculated, the numerical results can be tested for their physical correctness and any spurious solutions removed automatically.

DESCRIPTION OF THE MAFIA PROGRAMS

- M3 - is the mesh generator used by all programs, which translates the physical problem into mesh data and material distribution data.

  Figure 3 shows typical hidden line plots produced by M3.

- S3 - is the three-dimensional static solver for electro- and magneto-static problems.

  Figure 4 shows the structure and equipotential lines for the MAFIA emblem, the first and last letters are metal and electrostatically charged and the three central letters are dielectric.

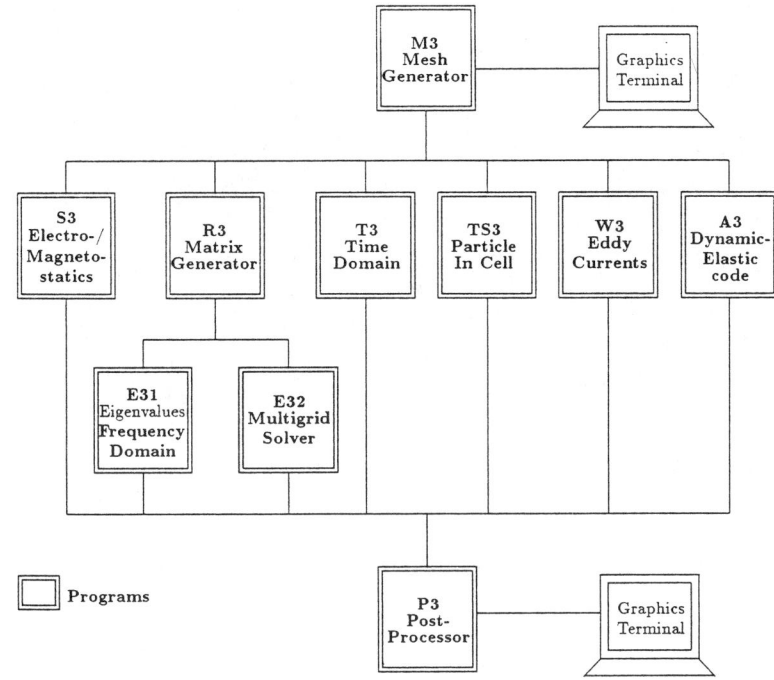

Figure 2: The MAFIA System with its Interrelationships

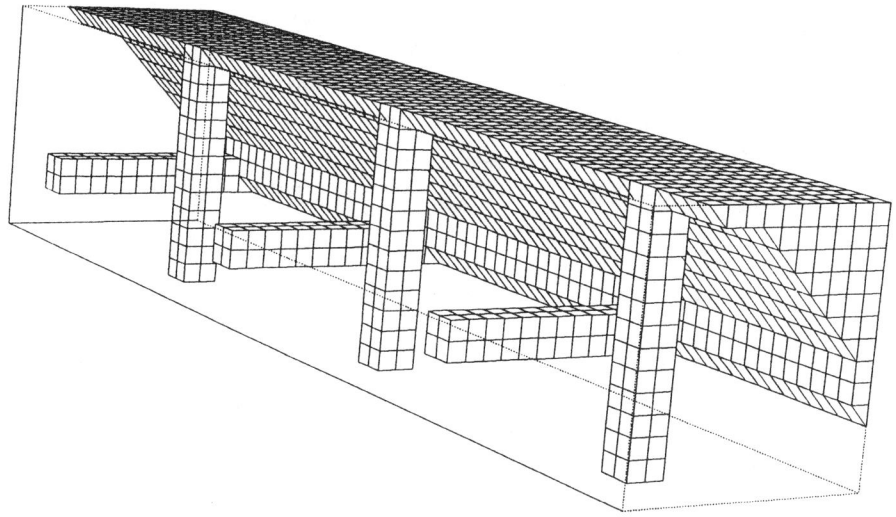

Figure 3: M3 Plots of Geometry and Mesh for the Jungle Gym Accelerating Structure

Figure 4: Equipotential plot from S3

- R3 and E31 (or E32) - these two codes, run in sequence, solve Maxwell's equations in the frequency domain. R3 sets up the matrices, adding the material properties and the boundary conditions to the mesh data from M3. E31 is the improved, more accurate version of the eigenvalue solver, which solves the matrix equations from R3 and writes the required eigenvectors onto the direct access file. E32 uses advanced multigrid methods and may be used instead of E31 for very large meshes.

  Figure 5 shows the structure of a proposed kicker magnet for injection of the proton beam from PETRA into the HERA ring. The magnet is carried on supports by which it can be rotated into the beam line. The computer simulations from the programs M3-R3-E3-P3 showed that the required fields were not economically obtainable by this method and the magnet was not built.

- T3 - solves Maxwell's equations in the time domain.

  Figure 6 shows a test example of a single bunch passing through a modified pill box cavity, the wakepotentials were calculated by T3 . The graph shows the three components of the wake potential and the bunch shape.

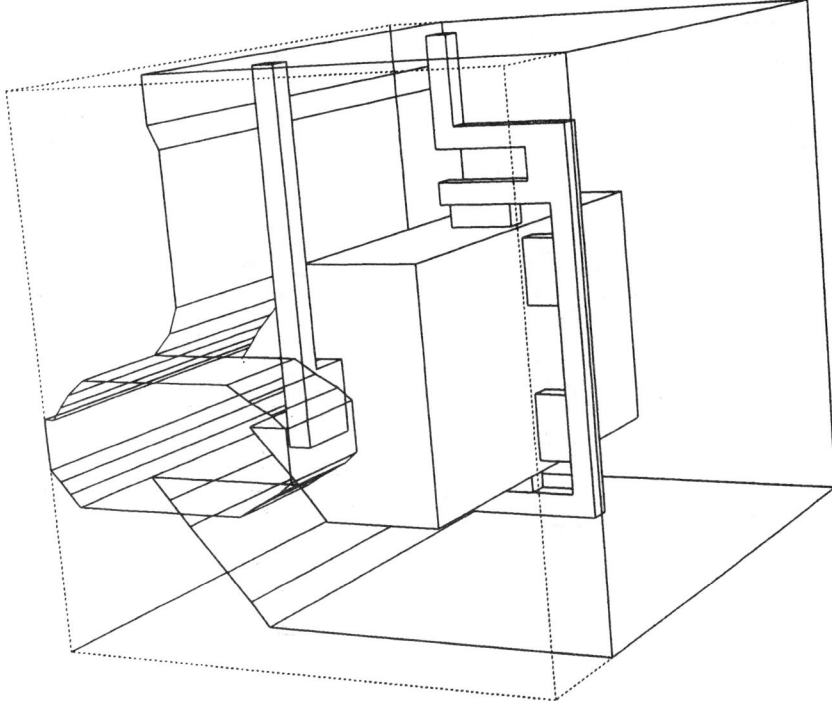

Figure 5: Structure of a Proposed Kicker Magnet for Injection into **HERA**

- TS3 - is a three-dimensional particle-in-cell code, not yet available for public release.

- W3 - will solve Maxwell's equations for eddy currents, not yet available for public release.

- A3 - will solve dynamic-elastic problems such as sound propagation, projected for future release.

- P3 - is the postprocessor for all the codes. Solutions can be displayed graphically, calculations carried out and file manipulations from within the program are possible. Figure 7 shows typical arrow plots of the electric and magnetic fields across the gap of the magnet from figure 5.

All the MAFIA codes for the third release can be used either interactively or in batch. Thus the former input files which were necessary for some of the codes will be replaced by a series of commands, either entered interactively, with a

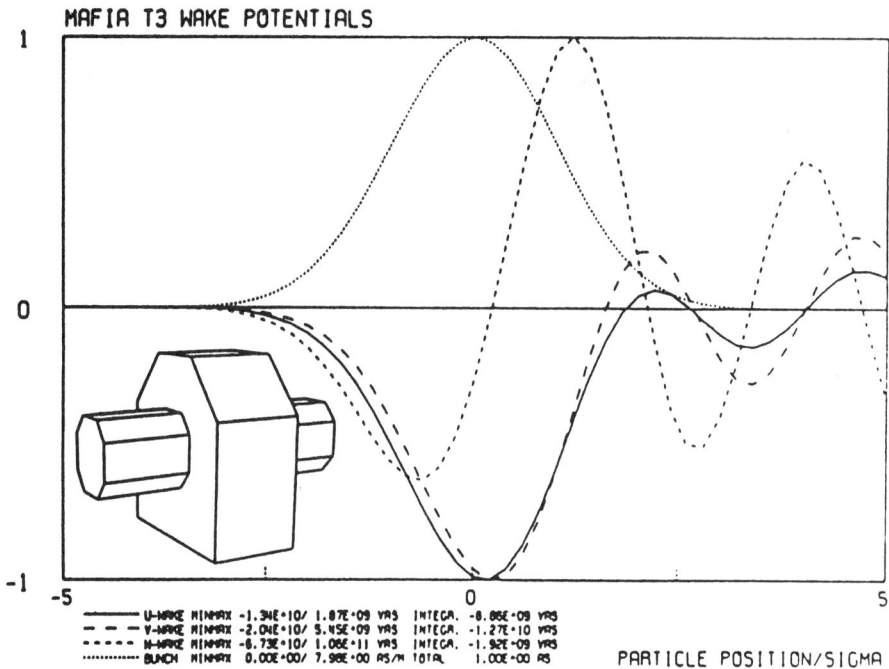

Figure 6: Wakefield Potentials versus bunch coordinates from T3, for the geometry shown in the insert

help facility to provide explanations where necessary, or read from a command file either interactively or in batch.

PRESENT STATUS

The second release of the MAFIA Programs, comprising M3, R3, E31, E32 and P3, has been distributed to over 75 installations, including MFE in the United States. The programs have already proved their worth through comparison with theoretical calculations and by successful design of accelerator components. For the third release, the existing MAFIA codes are being completely overhauled and new codes added.

NEW FEATURES OF THE PROGRAMS

Some of the new features are listed below:-

- A modular program structure has been adopted.
- A unified, menu-controlled, user interface has been created for all the codes.

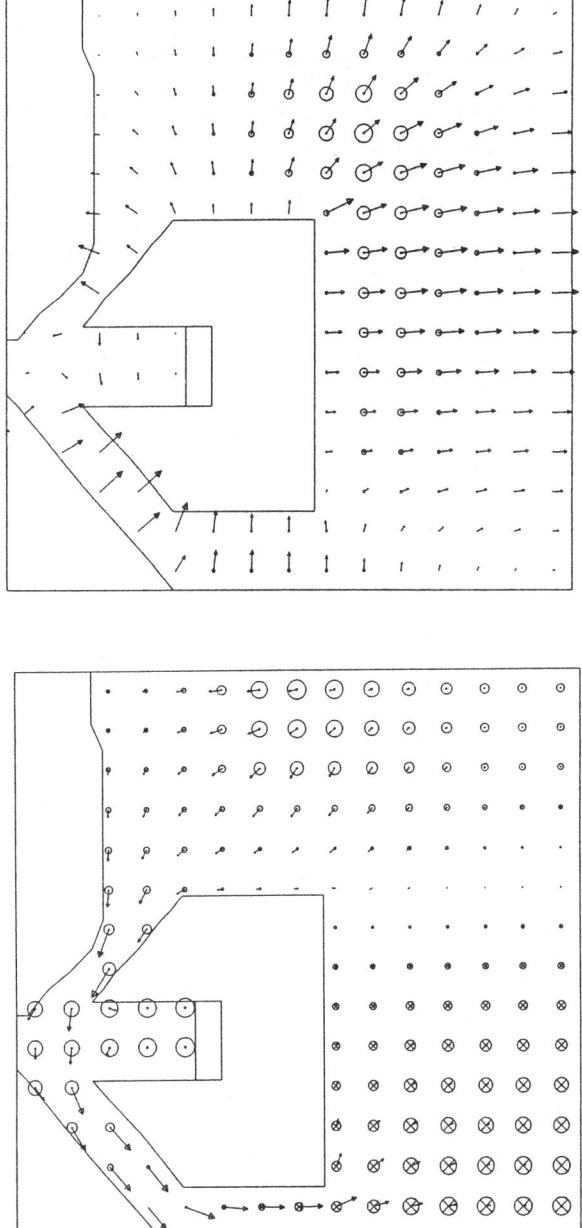

Figure 7: Arrow plots of electric(above) and magnetic(below) field, from E3 - P3

- A new command processor has been written.
- The data structure has been changed to accomodate all codes.
- A dynamic memory manager has been added to control both the file and the memory operations necessary for the new structure.
- Many new facilities have been included.

Program Structure

The programs have been restructured and divided into *sections*, each of which represents an independent function of the program, for example, the setting of the mesh or the drawing of an arrow plot.

Each program begins at an *Entry Level* where a menu is displayed that lists all the available commands. One can return to this *entry level* at any time and any *section* may be called from there .

User Interface

The new user interface is guided by menus. Each section prints a menu which lists all the commands available in that section, with their current settings or default values. This reduces the memorisation of command names necessary for running the programs, and displays the status information for the section. All sections have the same basic structure -

menu display
reading and processing of command lines
setting of appropriate variables
optional execution or exit
(execution may involve calls to additional subroutines)

Control remains with the current section until another section is invoked or the program is terminated.

It is also possible to run the programs without the menu display, which can be slow on some systems. In this mode, the information about the values set for certain commands can still be obtained on request.

Help system - The user interface includes a hierarchical help system to lead the beginning user through the functions of the program.

The help system is matched to the program structure. At the entry level the most basic help is available, addressing the questions -

What does this program do?
How do I use it?

What should I do next?
Where am I?
How do I get out of this?

These questions are answered by the command *help* Once this hurdle is passed, help is available for more specific areas, such as types of commands or any named section. A help menu is provided for each section which describes the function of the section, the meaning of the subcommands and gives any necessary instructions on its use. At any stage of the program the command *help* alone will give advice on how to proceed.

Command Processor

All commands are defined with unabbreviated names. This should relieve the user of the task of memorising specific command mnemonics. The command processor is capable of processing shortened versions of these commands; it checks for validity and ambiguity and will accept any unambiguous truncation.
The user has the freedom to enter and leave any section at will and, wherever possible, the order of operations carried out is not predetermined. However, certain sequences are not possible - a file must be opened before it can be read and must be read before anything may be calculated. In order to keep the programs as streamlined as possible, the idea of interactive prompting for missing information was rejected as impractical and time-consuming. Certain necessary checks will be made when the request to execute a section is received and warning messages issued.
A *macro* facility will be included and the command processor will be able to accept predefined macro commands, standing for a string of basic commands. This will enable experienced users to streamline terminal sessions and tailor the program to their particular needs.

Command Structure The commands themselves correspond closely to the structure of the program. For each section of the program there is a *branch* command by which that section is invoked. Each section has its own set of subcommands, its own menu listing these commands with their assigned values or defaults, and its own help menu describing the subcommands.
A certain subset of commands, which are necessary or desirable to have available at any time, have been selected and called *global* commands. Certain global settings are listed on every menu for convenience and a list of global commmands can be obtained at will.

Data structure of the direct access file

The structure of the direct access file has been made more flexible and can now be used by all programs. The new file has a transparent structure with a directory, organised like a database. The storage locations of named fields are recorded in the directory and the user has control of the reading and writing of quantities on the file. The directory is stored and updated in memory and written to the file when the program is terminated. Much general information stored on the file, such as the number of mesh points and the boundary conditions, are stored in readable form so that they can be listed on request from inside the program. It will be possible to allocate, open and close direct access files from within the MAFIA programs. The direct access file is designed to be used by all the MAFIA programs, see Figure 8. However, many systems do not support expandable direct access files. While it would be possible to work with one file which is large enough for all the programs, it is unwieldy to do so and often it is not obvious, when M3 is started, how large the final file should be. This is especially true for P3 where many new quantities can be calculated and stored. For this reason, the MAFIA programs are also designed to allocate their own files. Under this regime, see Figure 2, each program uses both an input and an output direct access file. The information stored by the previous program is read, the size of the output file calculated, a new file allocated and the information copied from the old to the new file.

Dynamic Memory Manager

One of the restrictions of three dimensional mesh codes in general is the number of mesh points which can be handled on any particular system. The more mesh points available, the greater the accuracy obtainable in the approximation of the structure. One of the aims of the new file structure and memory manager is to provide the maximum flexibility in the use of available memory. A combination of selective loading of quantities from the file and of selective deletion in memory of whatever is not needed for a particular operation, can make optimum use of the space available. Thus, in the post-processor, it will be possible to load a specific quantity from the file, or even to define a range and load only a part of that quantity if the field is very large. Then a calculation can be carried out and the results stored and lastly deleted from memory, freeing the space for further use. A load section is provided, where the functions of the memory manager can be invoked. Here one can load a quantity from one file, write something to another file, rename a quantity, or delete a field in memory or on a file; however caution is advised - any precious file should be copied before processing begins. If this has not been done the file can still be copied from within the program. Only the command 'delete all' will give you a warning message and a second chance.

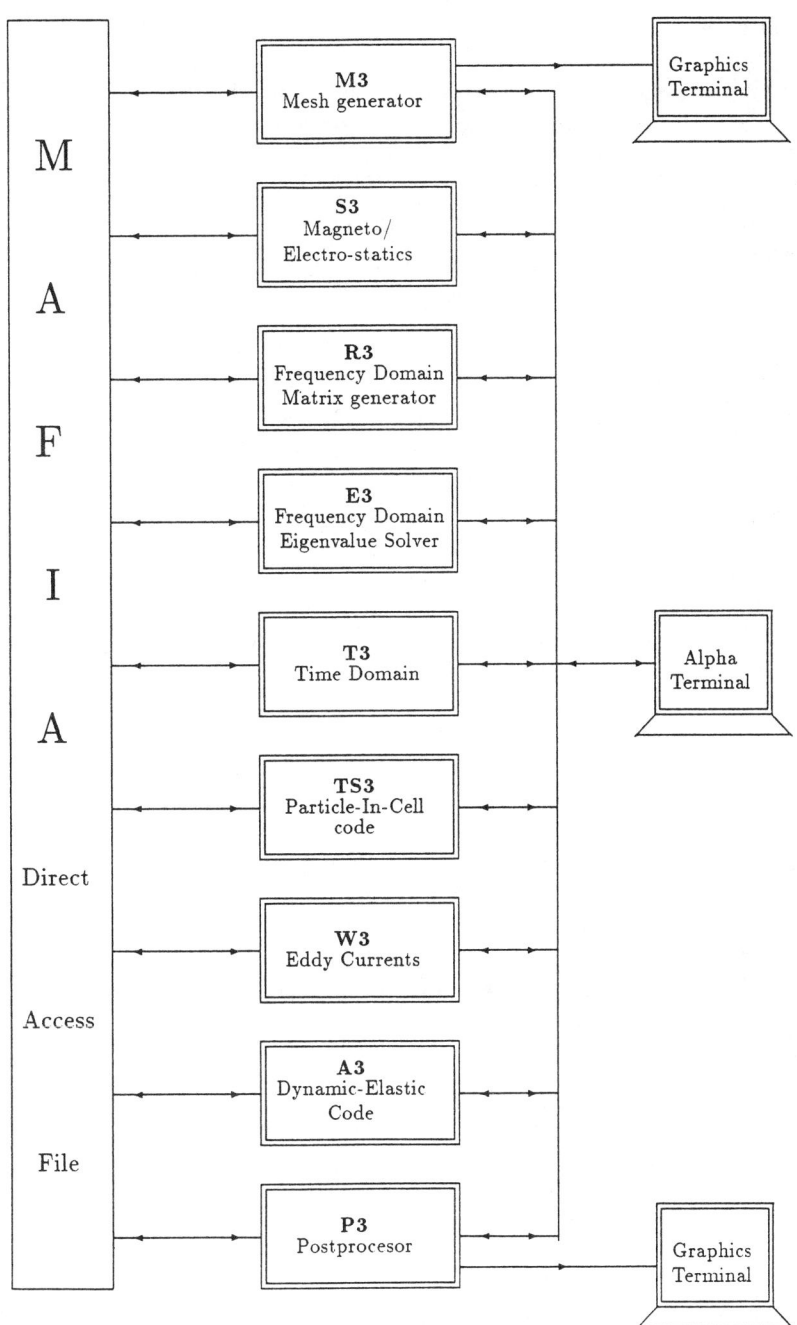

Figure 8: The Flow of Data between the the Programs and the Direct Access File

Sequential Files

The following sequential files are provided, all of which may be opened or closed by name from within the program and switched on or off at any time :

- Command file - Commands may be read from this file instead of from the terminal.

- Log file - When this file is open, all the commands entered will be written onto it. This is a useful method of creating a command file for future use as well as of recording the progress of a session. The commands used during a session are also automatically recorded on the direct access file when the program is terminated. This is to aid in the diagnostic work of following up on user-reported problems which are often not otherwise exactly reproducible.

- Print files - Two print files are provided, one of which is usually the interactive terminal. Output to the interactive terminal may also be turned off by a global command.

- Plot files - Similarly, there are two plot files. Graphic output may be directed to either, the graphic terminal or a metafile for later display. Global commands switch output on or off to these files.

There is also a section where the contents of any of these files can be listed. In this way a command file can be used several times to make a set of arrow and contour plots for the same geometrical values, or more than one command file may be used in a session to perform various tasks. The listing facility even enables an old command file to be opened to check its contents before using it. Also a logfile could be renamed as a command file, closed, and then reopened to repeat a complicated series of commands.

Other New Features of the Post-Processor, P3

Mathematical Operations - The original concept of the post processor was to provide the calculation of physical quantities of interest for accelerator design, each of which was separately programmed. This proved to be very inflexible and many users found that further calculations were necessary to adapt these quantities to their needs. Now these quantities will still be available but they will be programmed at the level of the individual mathematical operations and combined by means of internal macros. This facility will also be made available to the user. It will be possible, for example, to load a field from one direct access file, load the corresponding field from a file with the same mesh but slightly different geometry, difference the two and display the result graphically or perform further calculations as needed. This can be a very powerful tool in the design of accelerators and will enable the codes to be used more creatively. However this freedom also brings certain dangers and the full responsibility for the physical correctness and interpretation of the results lies with the user.

Windows - A zoom is available for all graphic output and calculations can be performed for a limited mesh window if desired. There is also a window section where the user has the possibility of presetting different windows for calculations or plots and for one, two or three dimensions. These windows are normally coupled so that changing one changes all the others but can be kept separate if required.

Range - A range is a limited set of values for which a quantity has been calculated or stored. The range section can set a reduced mesh or time range when a quantity is written to the file either in a previous program or in the post-processor. These ranges are recorded on the file and automatically checked whenever quantities are read.

Calculations - One, two and three dimensional integrals will be available, and calculations for impedance, charge and forces will be added.

AVAILABILITY

The MAFIA codes are written in Fortran 77 and currently run on IBM 3081, CRAY, VAX, Amdahl, Cyber and Apollo computers. The distribution centre for the codes is DESY, and the codes are available without charge to all non-profit organisations. For information contact T. Weiland at DESY.

CONCLUSION

Many new features will be added to the MAFIA group of codes. The additional programs will extend the scope of the codes while the new command processor, the conscious file manipulations and the additional calculations that will be possible, will be the features affecting users most directly. However the underlying restructuring of the programs themselves enables a much more efficient management of the support of the codes and is flexible enough to accomodate virtually any future extensions. The newest release of the MAFIA codes will be more user friendly than before, while the new flexibility will allow users to adapt it to their own particular needs.

References

[1] T.Weiland, Part.Acc. 15 (1984), pp 245-292 and references therein.

[2] T.Weiland, Part.Acc. 17 (1985), pp 227-242.

[3] K.S.Yee, IEEE, AP-14, 1966, pp 302-307.

[4] MAFIA User Guide, The Mafia Collaboration, DESY, Los Alamos National Laboratory, KFA-Jülich February 23,1987.

MAGNUS-3D: ACCELERATOR MAGNET CALCULATIONS IN 3-DIMENSIONS

S. Pissanetzky

Texas Accelerator Center, 2319 Timberlock, The Woodlands, Texas 77380. U.S.A. Phone (713)363-7925.

ABSTRACT

MAGNUS-3D is a professional finite element code for nonlinear magnetic engineering. MAGNUS-3D can solve numerically any general problem of linear or nonlinear magnetostatics in three dimensions. The problem is formulated in a domain with Dirichlet, Neumann or periodic boundary conditions, that can contain any combination of conductors of any shape in space, nonlinear magnetic materials with magnetic properties specified by magnetization tables, and nonlinear permanent magnets with any given demagnetization curve. MAGNUS-3D uses the two-scalar-potentials formulation of Magnetostatics and the finite element method, has an automatic 3D mesh generator, and advanced post-processing features that include graphics on a variety of supported devices, tabulation, and calculation of design quantities required in Magnetic Engineering. MAGNUS-3D is a general purpose 3D code, but it has been extensively used for accelerator work and many special features required for accelerator engineering have been incorporated into the code. One of such features is the calculation of field harmonic coefficients averaged in the direction of the beam, so important for the design of magnet ends. Another feature is its ability to calculate line integrals of any field component along the direction of the beam, or plot the field as a function of the z coordinate. MAGNUS-3D has found applications to the design of accelerator magnets and spectrometers, steering magnets, wigglers and undulators for free electron lasers, microtrons and magnets for synchrotron light sources, as well as magnets for NMR and medical applications, recording heads and various magnetic devices.

There are three more programs closely associated with MAGNUS-3D. MAGNUS-GKS is the graphical postprocessor for the package; it supports a number of output devices, including color vector or bit map devices. WIRE is an independent program that can calculate the field produced by any configuration of electric conductors in space, at any point in space, when iron is not present. HARMON performs the spherical harmonic analysis of magnetic fields in 3D, and is used for the design of passive and active shims and correction coils for high-precision magnets.

This paper deals with the latest extensions of MAGNUS-3D. Descriptions of the features and internal libraries of MAGNUS-3D are included. Emphasis is placed on the new features recently developed, which will become available to users in the next update.

INTRODUCTION

MAGNUS-3D solves Maxwell's equations $\nabla \times \mathbf{H} = \mathbf{j}$, $\nabla \cdot \mathbf{B} = 0$, with the constitutive equation $\mathbf{B} = \mu \mathbf{H}$ for soft ferromagnetic materials or $\mathbf{B} = \mu \mathbf{H} + \mu_0 \mathbf{M}_0$ for permanent magnets, where $\mu = \mu[H(x, y, z)]$ is a function of H

at each point. Thus, nonlinearity is fully accounted for.

The partial-scalar-potential formulation of Magnetostatics has been recognized as inaccurate in magnetic materials [1]. If $\mathbf{H} = \mathbf{H}_M + \mathbf{H}_S$, such that $\nabla \times \mathbf{H}_S = \mathbf{j}$ and $\nabla \times \mathbf{H}_M = 0$, then $\mathbf{H}_S$ is due to the electric currents alone and $\mathbf{H}_M$ is due to the iron. $\mathbf{H}_S$ can be calculated by integration over the conductors, and $\mathbf{H}_M$ can be written as $\mathbf{H}_M = -\nabla \Phi$, where Φ is the partial scalar potential. In permeable iron, $\mathbf{H}_M$ and $\mathbf{H}_S$ are of comparable magnitudes and opposite directions. $\mathbf{H}$, the small difference between two large numbers, is affected by a large round-off error, leading to inaccuracy and even loss of significance. For this reason, MAGNUS-3D uses the two-scalar-potentials formulation [2], or TSP formulation, where Φ is used only in regions with electric current (Φ-regions) where a total scalar potential cannot be defined. In the region without current, $\nabla \times \mathbf{H} = 0$, and $\mathbf{H} = -\nabla \Psi$, where Ψ is the total magnetic scalar potential. Thus the total $\mathbf{H}$ in the iron is directly calculated from Ψ with full accuracy, affected only by the normal discretization errors. No large errors appear inside the iron, and $\mu = \mu(H)$ or $\mathbf{M} = (\mu - 1)\mathbf{H}$ can be accurately obtained at each iteration. This is necessary for the solution to be accurate, because $\mathbf{M}$ is one of the two sources of magnetic field. The theory and other details of the implementation of the TSP method in MAGNUS-3D have been reported elsewhere [3]. The TSP method is considered to be state-of-the-art for both computational accuracy and efficiency in 3D Magnetostatics.

AUTOMATION

Automation, accuracy and efficiency are the three most important considerations in the design of professional scientific software. A very important step towards automation has recently been achieved and is reported in this Section. This step completely eliminates any reference to the two-scalar-potentials theory from the input files. There are two extremely important advantages: first, the program operator no longer needs to be knowledgeable in the intricacies of the theory of the method, and second, the simplifications in the input file make the program much easier and agreeable to use and reduce the risk of human errors. The theory of the TSP method requires that the domain of solution be divided into one Ψ-region that is singly-connected, and one or more Φ-regions enclosing the conductors. It further requires that the quantities

$$(\Phi - \Psi)_Q = \int_P^Q \mathbf{H}_S \cdot d\ell$$

be calculated for every Q, where P is some starting point and Q is any mesh point on the boundaries between the Ψ-region and the Φ-regions. In previous versions of MAGNUS-3D, only one Φ-region was allowed, and the user was required to specify this Φ-region, its boundary S, the location of point P, the definition of the Dirichlet boundaries, and their relation to the boundary S. Such a detailed specification was found to be inconvenient because it required an advanced knowledge of the TSP method, not always available to program operators.

These restrictions have been completely eliminated. Several Φ-regions are now allowed, making it possible to solve complicated problems with multiple electric circuits, even if the circuits are not all clustered together. The user no longer has to specify the boundaries of the Φ-regions, or the locations of point P, or the relations between S and the Dirichlet boundaries. The specification of the Φ-regions is conceptually simple: they are regions that enclose the conductors.

The definition of a Dirichlet boundary is also conceptually simple: a symmetry plane such that **B** and **H** are normal to it. Of course Dirichlet boundaries must be defined by the user in order for the problem to be mathematically complete.

It would not be appropriate here to enter in a detailed explanation of the computational algorithms that achieve these results. It will suffice to say that the program uses topology, graph theory and advanced set manipulation techniques [4] to identify the different boundaries and their topological relations, and the trajectories along which the line integrals have to be calculated. All this is done automatically and the correct results are obtained in all possible cases because only theoretically proved, completely general properties are used by the algorithms.

PROGRAM DESCRIPTION

MAGNUS-3D consists of the programs KUBIK, MAGNUS and EPILOG. Programs MAGNUS-GKS, WIRE and HARMON are also closely associated with the package.

The codes KUBIK, MAGNUS, EPILOG and MAGNUS-GKS are closely associated, and as a group perform the single task of solving problems of nonlinear magnetostatics. KUBIK is the pre-processor, used to describe the geometry and generate the finite element mesh. KUBIK is a semi-interactive, command operated, easy to use program that employs the modular concept. Simple modules of mesh are defined and then assembled together into a three dimensional structure of any degree of complexity. Plots of the structure seen from any point of view or zoomed into can be obtained and displayed on any of the devices supported by MAGNUS-GKS. KUBIK was described elsewhere [5].

MAGNUS is the solver. MAGNUS accepts the output from KUBIK containing the description of the mesh, and a user's file with the definition of the conductors, boundary conditions and magnetic properties of the materials and permanent magnets, and uses this information to assemble the nonlinear finite element equations and to solve them iteratively. MAGNUS uses the preconditioned conjugate gradient method and sparse matrix technology [4] to solve the system of linear equations in each iteration. This guarantees state-of-the-art accuracy, efficiency and storage economy. The input to the solver is now divided into four command modes. The commands in each mode are interchangeable, and many are optional and have default values. The general command mode contains commands to specify some general parameters for the run, like the maximum number of iterations, the desired accuracy, the relaxation coefficient and whether this is a new run, a restart or just a test run. Some limited output options are provided, as well as a feature to save intermediate results during the run.

In the boundary command mode, the user specifies the Dirichlet boundaries and the periodicity conditions of the domain, if any. Any unspecified boundary is treated as a Neumann boundary. There is no longer need to specify P, S, or the relations between S and the Dirichlet boundaries. Some limited output options are again provided.

The conductor command mode is used to define the electric conductors in the problem. MAGNUS has a library of 3D conductor elements, that includes infinitely thin conductors, solid conductors and strips of a variety of shapes, that are completely parameterized and independent of the mesh and of each other, and can be placed anywhere in space. The conductors existing in the problem are approximated by elements taken from the library. The operations of reflection, displacement and rotation are available to generate more conductors when symmetries exist. There is a command to specify the Φ-regions, and a command to generate a plot of the conductors that is stored into a device independent metafile.

Finally, in the magnetization command mode, the user can specify the magnetic properties of materials. There is a library of magnetization tables for the most common magnet steels, including some US and some japanese steels, and ideal materials such as pure iron or pure nickel. The user can enter his own table. Permanent magnets can be specified in a number of ways, including a constant $\mathbf{M}_0$ of a given magnitude and direction, a demagnetization curve determined by the coercive force H_c and the residual induction B_r, or a detailed non-linear demagnetization curve as provided by the manufacturer. There are commands to print the tables either as they are stored or as they are interpolated, and to specify the user's choice of interpolation algorithm.

THE POST-PROCESSOR EPILOG

The most important and useful feature of this program package is the post-processor EPILOG. After iterations have been completed and convergence has been achieved by MAGNUS, a complete solution to the problem is available. The solution consists of the two potentials, Φ and Ψ, which are continuous functions defined at every point of the respective domain by a finite element approximation. The solution can then be used by EPILOG for a variety of purposes, like print tables, obtain plots or calculate quantities that depend on the solution and are useful to the magnet designer or the Magnetic Engineer. EPILOG is command operated. The commands are interchangeable and can be executed repeatedly if so desired, because the solution obtained by MAGNUS is not affected and remains in storage. The commands currently available are briefly described.

Command ENERGY will calculate the total magnetic energy stored in the magnetic materials and the non-magnetic part of the solution domain. The energy is important for the design of power supplies, and also for quench protection systems of superconducting magnets.

Commands FORCE and TORQUE will calculate the total magnetic force or torque acting on any given region of space. The user specifies the region by defining a surface that encloses the region. The force or torque is calculated by integration of Maxwell's electromagnetic stress tensor over the given surface, and is actually acting on the conductors or magnetic materials contained inside the region. Force and torque are important for mechanical design.

Command FLUX can calculate the magnetic flux through a given surface in space. The surface is specified by the user, and there is a small assortment of surface types that can be used. The surface can be such that its perimeter coincides with an electric circuit, in which case the flux concatenated by the circuit is obtained and can be used to calculate the self inductance or mutual inductance of the circuit. Note, however, that the inductance of a nonlinear system is not a constant, and that it can also be defined in terms of the magnetic energy. The two definitions have different meanings, and their properties must be understood by the user.

Command LINE INTEGRAL will calculate any one of a variety of line integrals of potential or field components. The line is any line in space specified by the user. A line integral of the type $\int \mathbf{H} \cdot d\boldsymbol{\ell}$ is important because it has a physical meaning. Integrals like $\int B_y dz$ along the beam line are frequently needed in accelerator physics.

Command MULTIPOLES is a rather unique feature of EPILOG. It can calculate the harmonic coefficients of the field of an accelerator bending or steering magnet averaged over an interval along the beam direction. In this way, end effects, effective length and field errors in 3D geometries are all taken into account.

Command PRINT has many forms that meet practically every need of the Magnetic Engineer or the Accelerator Physicist. For example, it is possible to print a table of field values at points on a line or arc in space specified by the user. If the line is the beam, the table gives directly the values of the field acting on the beam. Such a table could be used as input for a particle tracking program. If the line coincides with a conductor, the values in the table can be used to determine the maximum field at the conductor, important for estimating the critical current in the case of a superconductor. PRINT can print a table of values of μ at points in the iron or in a selected region inside the iron; such a table indicates the effectiveness of shields and yokes, and the linear or nonlinear behavior of pole pieces and other components. PRINT can also be used to obtain a verbose description of all the electric conductors existing in the problem.

Command PLOT is very powerful. It allows the user to obtain plots or views of the mesh, the magnetic bodies, or the conductors, seen from any point of view in space. It is possible to zoom into particular details, and to combine or overlay different views. All plots go into device independent metafiles, which can then be displayed by running MAGNUS-GKS, in black and white or in color on any of the supported graphical devices. There is partial hidden line elimination, and an algorithm for full hidden surface removal is in the works and will be made available to users soon.

A further unique feature is the ability to plot curves representing a function, using command PLOT FUNCTION. The function can be a component of the field or of the magnetic induction, or the magnetic scalar potential or the total field, plotted as a function of a line coordinate along any line in space. The plots are appropriately labelled, and can be displayed on any graphical device. Plots of the transverse field along the beam, or of the total field along a conductor, can be obtained in this way.

Another unique feature of EPILOG is command FIELD LINES. This command obtains sets of field lines in 3D and generates a plot, where the field lines can be seen from any point of view. Field lines do not generally stay on a plane and must be seen and understood in space. Because of the complexity of structures involving field lines, it is recommended that they be displayed in color. Field lines, by simple examination, provide a degree of insight into the behavior of the field that is very difficult to achieve by other means, and we have not heard of any other program that has the capability of plotting them in three dimensions.

The output from EPILOG can easily be interfaced to programs that calculate particle trajectories or perform other tasks of Magnetic Engineering. EPILOG is an essential tool for the magnetic designer and the accelerator physicist. Further information on the use and capabilities of EPILOG will be published[6]

MAGNUS-GKS, WIRE AND HARMON

MAGNUS-GKS is the graphical post-processor for the package. It interfaces with all the programs, and is used to display a graphical metafile generated by one of the programs on one of the supported devices. The devices currently supported by MAGNUS-GKS are: VT-240 color terminal, LA-100 dot matrix printer, LN-03 laser printer, HP-7580 color plotter, TEK-4014 terminal and TEK-4125 color terminal. Different plots can be overlaid on the same view, appropriately scaled, and lines can be shown in different patterns or colors where available, or eliminated from the view. These devices include both bit mapped devices and vector devices. The list of supported devices grows continuously in response to requirements from the users.

WIRE is an independent program that can calculate the field of any configuration of electric conductors in three dimensions, at any given set of points in space, in cases where no magnetic materials are present. WIRE has a library of conductor elements, that includes a fully parameterized set of filaments, strips, and solid bars, solenoids and arcs. This library is similar to the one used by MAGNUS-3D. Conductors are approximated with elements of the library, and additional conductors can be defined by the symmetry operations of rotation, reflection or displacement. Plots and views of the conductors in space can be obtained and displayed on any of the devices supported by MAGNUS-GKS. WIRE is not a finite element program and does not require a finite element mesh. WIRE calculates the field of electric currents by direct integration of Biot and Savart law.

HARMON is a program for the spherical harmonic analysis of magnetic fields in three dimensions. The main application of HARMON is the design of passive and active shims and correction coils for high precision magnets, like NMR and MRI magnets for medical uses. It uses as input field values either calculated by MAGNUS or obtained from measurement.

CONCLUSIONS

We have presented an overview of the program package MAGNUS-3D for 3D nonlinear magnetostatics. We have emphasized the recently developed improvements and extensions, that include input-simplifying automation, better graphics and new post-processing features. We have also described the associated programs WIRE and HARMON. We conclude that MAGNUS-3D is a very useful design tool for the Magnetic Engineer and the Accelerator Physicist.

REFERENCES

1. J. Simkin and C. W. Trowbridge, Int. J. Num. Meth. Engng. 14, 423 (1975).
2. J. Simkin and C. W. Trowbridge, Proc. IEEE 127, 368 (1980).
3. S. Pissanetzky, Comp. Electromagnetics, Z. J. Cendes (editor), Elsevier Science Publishers B. V. (North-Holland), IMACS, 121 (1986).
4. S. Pissanetzky, Sparse Matrix Technology (Academic Press, London, 1984).
5. S. Pissanetzky, Int. J. Num. meth. Engng. 17, 255 (1981).
6. M. Fan and S. Pissanetzky, Nucl. Instr. Meth. in Phys. Research, to appear (1988).

THREE-DIMENSIONAL
ELECTROMAGNETIC PARTICLE CODES
AND APPLICATIONS TO ACCELERATORS

Alan Mankofsky
Science Applications International Corporation
McLean, VA 22102

ABSTRACT

We discuss the principles of three-dimensional electromagnetic particle-in-cell plasma simulation codes. Applications of these codes to present-day accelerator problems involving high intensity particle beams and cold testing of complex RF structures are presented.

INTRODUCTION

The numerical simulation techniques developed by the plasma physics community over the past 25 years are now finding broad applicability in many areas of accelerator design. A wide class of problems can be studied with a high degree of confidence using well-tested, robust, user-friendly codes. For example, fully three-dimensional cold testing calculations for cavities and structures can be performed in either the time domain or the frequency domain. Iterative equilibrium calculations can be applied to gun physics problems such as collector design and steady-state beam transport; secondary particle trajectories may be included in a self-consistent fashion. Full transient simulations can be used to study self-consistent beam dynamics including space charge, revealing the temporal and spatial evolution of unstable modes.

Numerical simulation codes are powerful tools both for the study of complex physical phenomena and for practical device design. Simulations expand the scope of both theoretical and experimental understanding of physical systems in a number of ways. For example, they allow the theorist to study systems with nonlinearities or those with irregular boundaries, configurations that cannot normally be treated analytically. The experimentalist can use a simulation to provide a window into a configuration for which accurate measurements are difficult to perform, or for which the process of making a measurement generates a perturbation that actually renders the measurement invalid. Furthermore, these tools have become more readily accessible to the researcher over the past decade, as the availability of supercomputers has grown. The traditional categorization of physics research as either theoretical or experimental has therefore been broadened to include a computational branch as well.

One- and two-dimensional codes have been successfully used in this way for a number of years. As supercomputer power has increased over the past five years or so, three-dimensional codes have now come into their own as essential research tools. While much of the impetus behind these developments has come from the traditional class of plasma physics problems, these techniques are also directly applicable to the study of accelerators and charged particle beams.

This paper is intended to serve several functions. First, we will discuss the basic ingredients required for building a simulation code capable of treating a variety of beam and field configurations. We will also describe techniques for combining these ingredients into a working code. We will then present a number of examples to illustrate present-day capabilities in simulation of accelerator physics problems. Finally, we will conclude with a brief summary.

PHYSICS MODELS AND CODE ORGANIZATION

For any simulation code to be useful, it must include some subset of the basic physics models contained in a list such as the following:

- electromagnetic fields
- particle flows
- material properties
- surface physics
- chemistry
- radiation
- complicated boundaries and geometries
- disparate time scales
- high resolution

The first two categories, electromagnetic fields and particle flows, are the fundamental building blocks for any simulation code. As we will see, however, it is not always necessary to solve Maxwell's equations in the full electromagnetic limit. There are many problems for which electrostatic, magnetostatic, or magnetoinductive field solutions are appropriate.

The inclusion of some combination of the remaining categories is highly problem-dependent. It is certainly possible to perform useful simulations with only fields and particle flows; however, the more accurately the true physics of a problem is represented in a simulation, the more realistic the simulation will be. For example, in performing a simulation of a gun with an irregularly shaped emitting surface, the ability to compute electric fields in the vicinity of a complicated boundary is crucial. The surface physics and material properties of the cathode are important if realistic electron emission is to be done. If the formation of a cathode plasma is to be considered, the code must also be able to treat the widely disparate time scales caused by the simultaneous presence of ions and electrons. One should therefore consider the above list to be a menu, from which the physics appropriate to each specific problem may be drawn and included in the simulation model.

There are a number of well-known methods for the discretization and numerical solution of the equations arising from the physics models described above. In this paper, we will be concerned only with the particle-in-cell, or PIC, technique.[1] This is a statistical technique wherein electrons and ions are represented by an ensemble of test particles. Each particle carries physical attributes such as mass, charge, position, and momentum. The ensemble evolves under the influence of both external and self-consistent electromagnetic fields, in the following fashion.

All of the vector and scalar fields involved in the calculation are organized on a lattice known as a <u>finite-difference mesh</u>. A fundamental unit, or cell, of a three-dimensional mesh might appear as in Figure 1. The lattice would be created by simply replicating this structure in all three directions. The staggered mesh positions of the various electromagnetic field components, chosen in this way to guarantee an accurate numerical solution when standard second-order finite difference operators are used, are also shown in Figure 1.

These quantities are updated by a <u>field solver</u>, which is a collection of routines that advance Maxwell's equations in time. As discussed above, it is often sufficient to solve only for the electrostatic potential via a generalized Poisson equation,

$$\nabla \cdot \varepsilon \cdot \nabla \phi = -\rho . \tag{1}$$

The presence of a dielectric tensor in Eq. (1) allows great generality in the type of problem being treated. Alternately, the Maxwell curl equations,

$$\nabla \times \boldsymbol{E} = -\frac{\partial \boldsymbol{B}}{\partial t} , \tag{2a}$$

$$\nabla \times \boldsymbol{H} = \frac{\partial \boldsymbol{D}}{\partial t} + \boldsymbol{J} , \tag{2b}$$

may be advanced in time for a fully electromagnetic solution, in which the divergence equations,

$$\nabla \cdot \boldsymbol{D} = \rho , \tag{3a}$$

$$\nabla \cdot \boldsymbol{B} = 0 , \tag{3b}$$

are taken as initial conditions and/or constraints. Notice that the

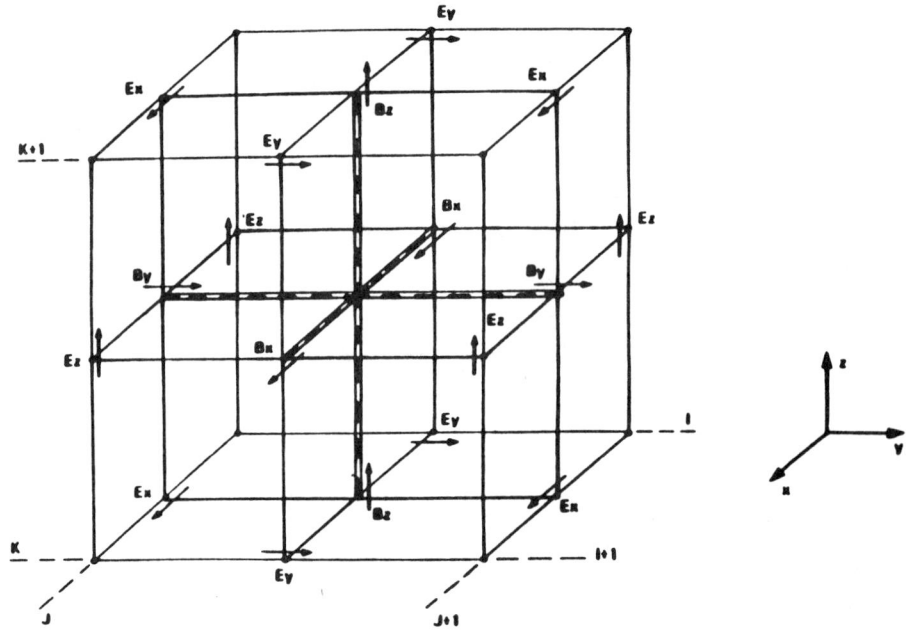

Fig. 1. A fundamental mesh cell in three dimensions, showing centering of the field quantities.

use of the ***D*** and ***H*** vectors allows us to model material properties: dielectrics, magnetic materials, and so on can be treated with the appropriate use of ε and μ (which may be time-, frequency-, and/or field-dependent), and the constitutive relations,

$$\boldsymbol{D} = \varepsilon \cdot \boldsymbol{E}, \quad (4a)$$

$$\boldsymbol{B} = \mu \cdot \boldsymbol{H}. \quad (4b)$$

The numerical solution of field equations such as (1)-(4) is a subject within itself, and a variety of techniques exist. The choice of the proper technique may even be problem-dependent, based in part on the physics that must be resolved. Detailed presentations of many of these techniques may be found in the references.[2,3]

As an example, two approaches exist to the cold testing of a device to determine its resonant frequencies and mode structures in the absence of particles or beams. In a time-domain calculation, one introduces an impulsive noise signal into the simulation. Since this drive signal acts as a delta function in both space and time, it excites all the modes that make up the signal, limited only by the grid resolution. Field time histories are collected at a number of sample points and are Fourier decomposed at the end of the run, giving the spectrum of modes present. Once the mode frequencies are known, another set of simulations can be performed in which the system is driven at one particular frequency, thereby providing a detailed picture of the spatial structure of the mode. Alternately, in a frequency-domain calculation,[4] the Maxwell curl equations are Fourier transformed in time, yielding an operator equation. The spectrum of this operator gives the resonant frequencies of the system, while the mode structure is given by the eigenvectors. The accuracy of this procedure can be increased as desired simply by using more basis modes in the calculation. In either case, one can also use these methods to determine integrated field quantities such as cavity Q and stored energy.

Particle interactions occur through the intermediary of the field quantities on the mesh; this is far cheaper than computing the $N(N-1)/2$ direct interactions that would be required by the force equation for N particles, since a typical simulation might involve several million particles. (Compare the approximate floating point operation count for a direct calculation, $10N^2 - N$, with that for a PIC calculation on an $M \times M \times M$ mesh, $20N + 5M^3 \log_2 M^3$. The scaling of the PIC calculation with the number of particles is clearly far more favorable.) The motion of the particles on the mesh is determined by a set of <u>PIC routines</u>, which performs three major functions. First, the fields are interpolated from the mesh to the particle positions. Using these interpolated fields, the particle quantities are then advanced in time using a set of equations of the form

$$\frac{d\boldsymbol{x}_k}{dt} = \boldsymbol{v}_k , \qquad (5a)$$

$$\frac{d\boldsymbol{u}_k}{dt} = \frac{Z_k e}{m_k}(\boldsymbol{E} + \boldsymbol{v}_k \times \boldsymbol{B}) - \sum_\alpha [v_s^{k\alpha}(\boldsymbol{v}_k - \boldsymbol{v}_\alpha) + \boldsymbol{S}_{k\alpha}] , \qquad (5b)$$

$$\boldsymbol{u}_k = \gamma_k \boldsymbol{v}_k , \qquad (5c)$$

where k is a particle index and α represents other species in the system. In Eq. (5b), the three terms on the right-hand-side represent the Lorentz force, slowing-down collisions, and pitch-angle scattering collisions, respectively.[5] The inclusion of these terms, along with the use of the relativistic gamma in Eq. (5c), means that much complex physics can be included in a PIC representation. As an example, with a proper model for the collision frequency in Eq. (5b), one could easily treat the slowing-down of a relativistic electron beam as it passes through matter. After Eqs. (5) are used to update all the particle velocities and positions, the PIC routines calculate moments of the particle distribution for use as source terms by the field solver.

Figures 2 and 3 show two possible ways to organize the above operations. The classical time-stepping simulation is diagrammed in Figure 2, where one complete circuit around the loop represents one time step. Given the source terms $\boldsymbol{J}$ and ρ, the field solver updates the values of $\boldsymbol{E}$ and $\boldsymbol{B}$ on the mesh. Using these new field values, the PIC routines perform the particle advance, and, if a fluid specie is present, the variables associated with this specie can be advanced as well. Updated source terms are then computed, the independent time variable is advanced from t to $t+\Delta t$, and the process begins again.

The above scheme should be contrasted to the equilibrium approach presented in Figure 3. A field solver updates the values for the electromagnetic fields, as before. However, one now integrates the particle orbits completely through the system using these fixed fields, accumulating charge and current densities along the way. Once the integration is complete (in other words, after all the particles have left the system), the fields are updated, the particles are reinitialized, and the cycle is then repeated. This type of iteration will converge (to a state where another orbit integration and source attribution will result in no further change in the fields) as an equilibrium is approached, and is therefore suitable in situations where such an equilibrium exists and is accessible given the chosen initial conditions.[6]

A comment about code architecture is appropriate at this point. Much of the basic code physics as described above is relatively straightforward; however, the huge amounts of data

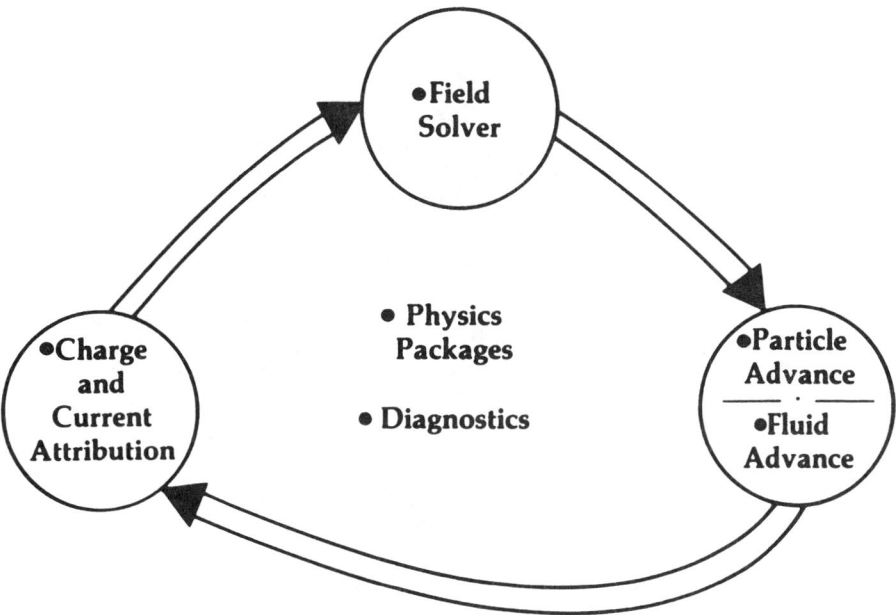

Fig. 2. Flow of a classical time-stepping simulation.

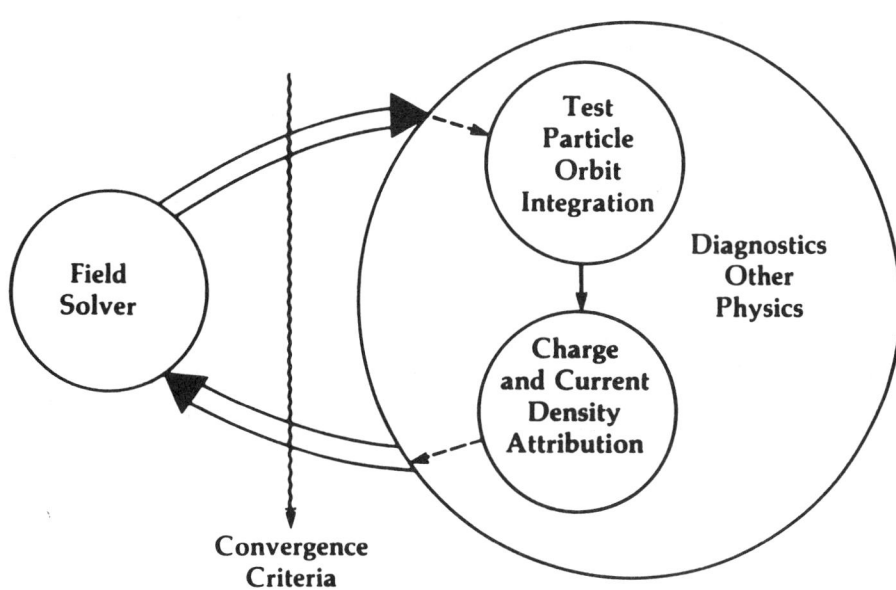

Fig. 3. Flow of an iterative equilibrium simulation.

involved in a three-dimensional simulation place a new set of requirements on the code. Even a marginally-resolved simulation may involve 10^6 cells (with 10 data/cell) and 5×10^6 particles (with 10 data/particle), numbers which strain the largest of today's supercomputers. As a consequence, a great deal of the effort that goes into three-dimensional code design centers on efficient memory and data management. Domain decomposition techniques, wherein the physical region being simulated is broken up into smaller subregions, each of which is processed separately, are among the methods that have been successfully applied to such problems.[7] Furthermore, effective visualization of this massive quantity of three-dimensional data is crucial if the code results are to be understood by the physicist. One therefore finds that the great majority of the code in such a simulation is devoted to data management and diagnostics, with only the remainder being devoted to the actual physics.

We conclude this section with a brief mention of the available literature on PIC techniques. Particle simulation is now a mature, well-established field, one in which the numerical properties of the algorithms have been analyzed and understood. There is a great deal of confidence in what may be called this first generation of simulation codes. Several excellent books and journal articles may be consulted on topics ranging from elementary presentations of standard methods through code architecture for multiprocessing vector supercomputers.[2,3,8,9,10,11,12] Furthermore, over the past five years, there has been interest in second generation techniques, which are advanced methods for doing an entirely new class of problems. One of the most promising of these is a group of multiple time scale methods for solving problems in which a number of physical processes occur, each with its own particular scale.[13] Results from multiple time scale codes are just now becoming available.

ACCELERATOR ISSUES AND EXAMPLES

In this section we go through a number of problems of current interest to the accelerator community and show by example how each can be approached with three-dimensional numerical simulation. It is our intent to give the reader a flavor for what is now the state-of-the-art in three-dimensional modeling. There are quite a few codes presently in active use in this area; we list the names and developers of several of the major ones:

- ARGUS (Science Applications International Corporation)
- IVORY (Mission Research Corporation)
- MAFIA [fields only] (DESY)
- QUICKSILVER (Sandia National Laboratories)
- SOS (Mission Research Corporation)

Since we are most familiar with ARGUS, the examples we present below are taken from work done with this code. However, we emphasize that similar simulations have been performed with most of the other codes listed above.

Cavity and structure design

One of the most successful applications of three-dimensional simulation codes to accelerator design has been the detailed study of the eigenfrequencies and related properties of cavities and gaps. For many years, engineers and physicists had to rely on intuition and tedious trial-and-error experimentation for developing new geometries. The process becomes far simpler when numerical modeling is employed.

As described above, these calculations can be performed in either the time domain or the frequency domain. In both cases, the simulations yield a wealth of accurate information about the device under study: in addition to the resonant frequencies and mode patterns, there is sufficient data to determine cavity Q, coupling impedances, wake fields, and other integrated field quantities.

We present two sample ARGUS cold testing simulations showing the complexity of the configurations that can be treated numerically. The first is taken from a study of 1 GHz RF separators used by CEBAF for beam splitting. Biperiodic disk loaded cavity structures are used for this purpose. The coupling cavity is circular; since the cylindrical symmetry results in a two-fold degeneracy in the deflecting mode, the designer must find some way of introducing an asymmetry into the separator. One solution would be to add bars or holes to the structure. As an initial measure, however, the deflecting cavities are simply stretched into an elliptical shape. The object is then to determine the minimal eccentricity for achieving the desired frequency splitting of somewhere between 20 and 40 MHz. The configuration used for this calculation is shown in Figure 4a. Each ARGUS run was performed on a half-grid of 123,255 mesh points, and required about 11 minutes of Cray X-MP CPU time to compute ten modes. Typical spatial structure of the lowest mode (661.753 MHz) is shown in the side view of Figure 4b and the end view of Figure 4c. For purposes of comparison, the results of a parallel series of calculations performed with MAFIA and ARGUS are presented in Table I. The two codes are seen to agree quite well.

As a second example, an ARGUS simulation of a double cavity output gap for a modified 5045 S-band klystron used by SLAC is shown in Figure 5. The double cavity structure, designed to facilitate high power operation of the device, is clearly shown in Figure 5a. In this set of calculations, 89,999 mesh points represented half of the structure. Approximately 6.5 minutes of

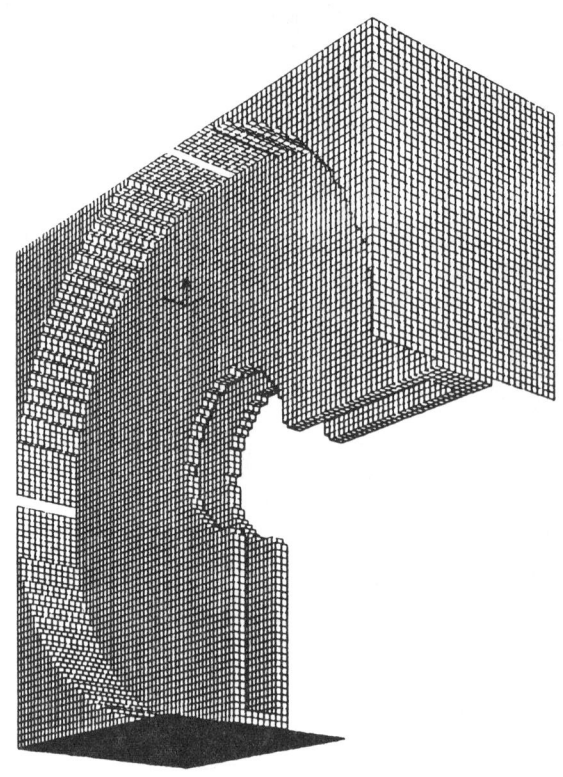

Fig. 4. ARGUS cold testing calculation for the CEBAF coupling cavity.
(a) Device configuration.

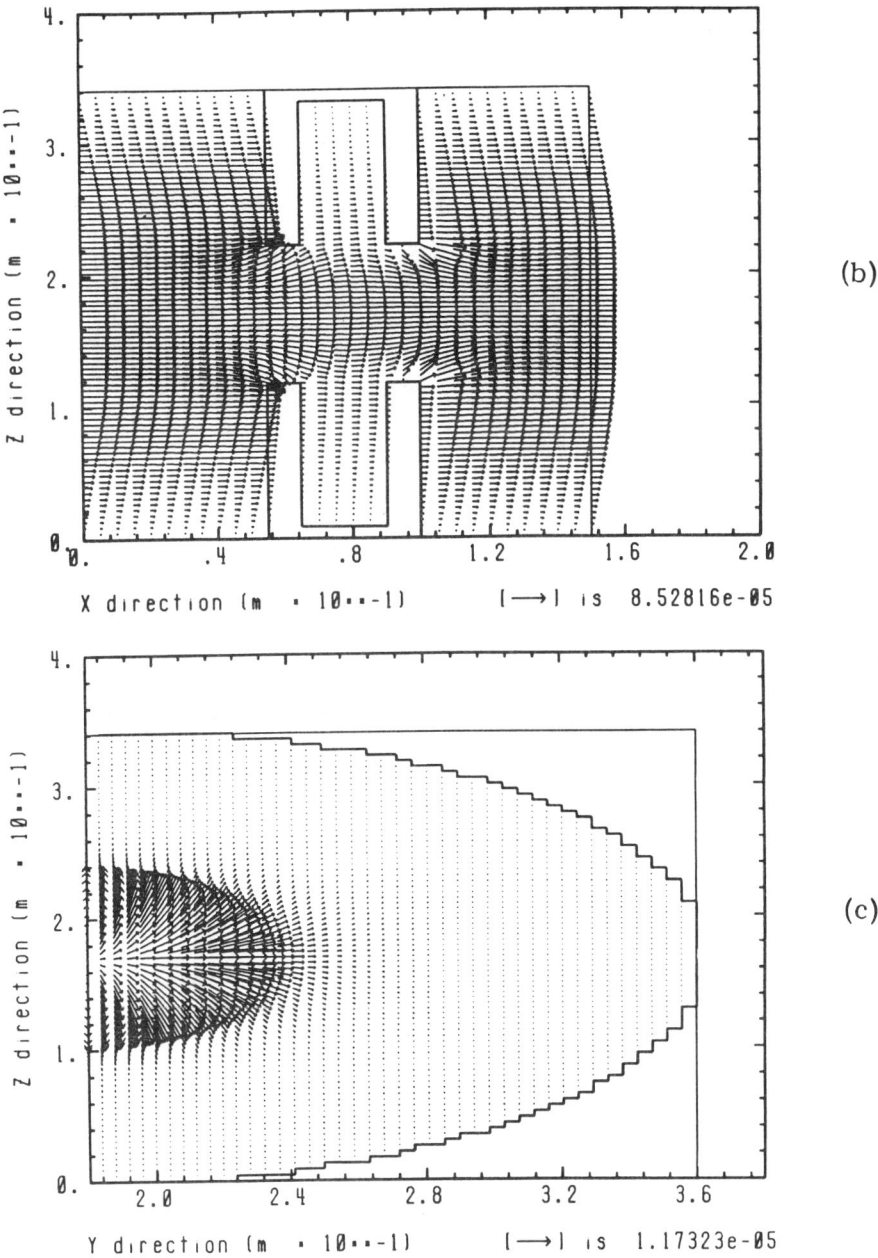

Fig. 4. ARGUS cold testing calculation for the CEBAF coupling cavity. (b) Normal mode pattern at 661.753 MHz (side view).
(c) Normal mode pattern at 661.753 MHz (end view).

Table I. MAFIA and ARGUS results for the CEBAF coupling cavity (eigenfrequencies are in MHz).

MODE	MAFIA	ARGUS	%DIFF.
1	662.413	661.753	.0996
2	667.105	667.824	.1077
3	766.228	757.714	1.111
4	992.353	983.726	.8693
5	995.116	986.267	.8892
6	1011.27	1011.27	.0000

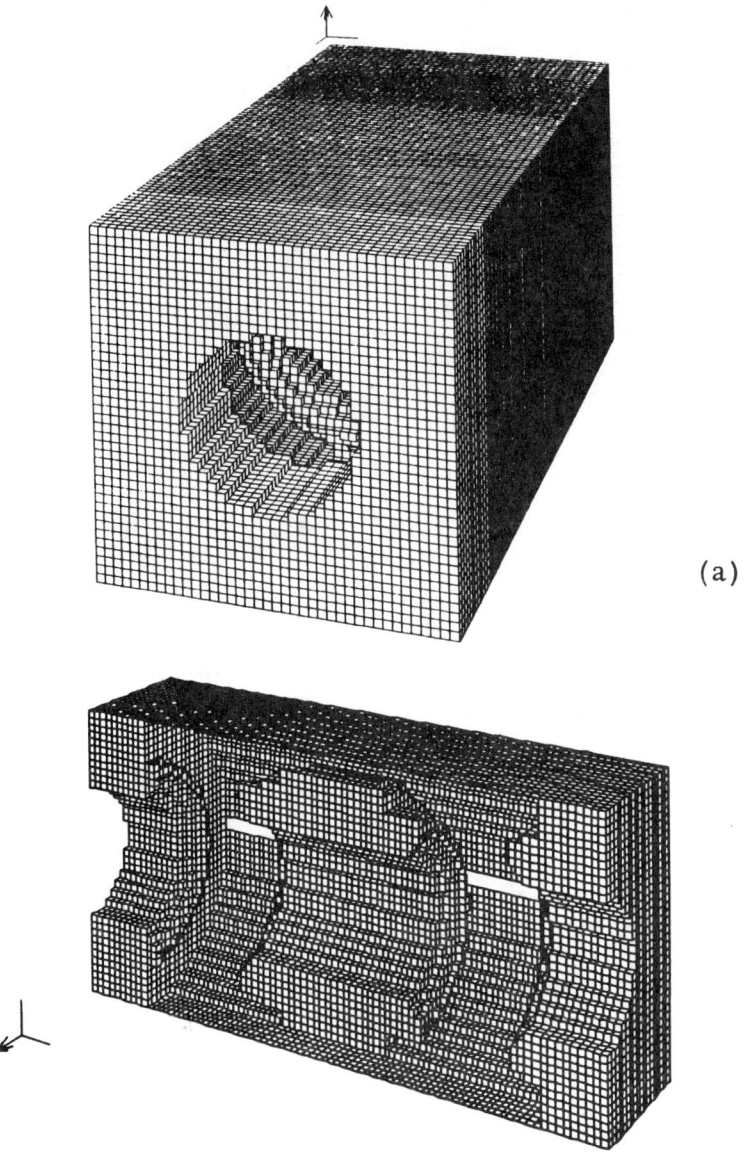

Fig. 5. ARGUS cold testing calculation for the SLAC double cavity output gap.
    (a) Device configuration.

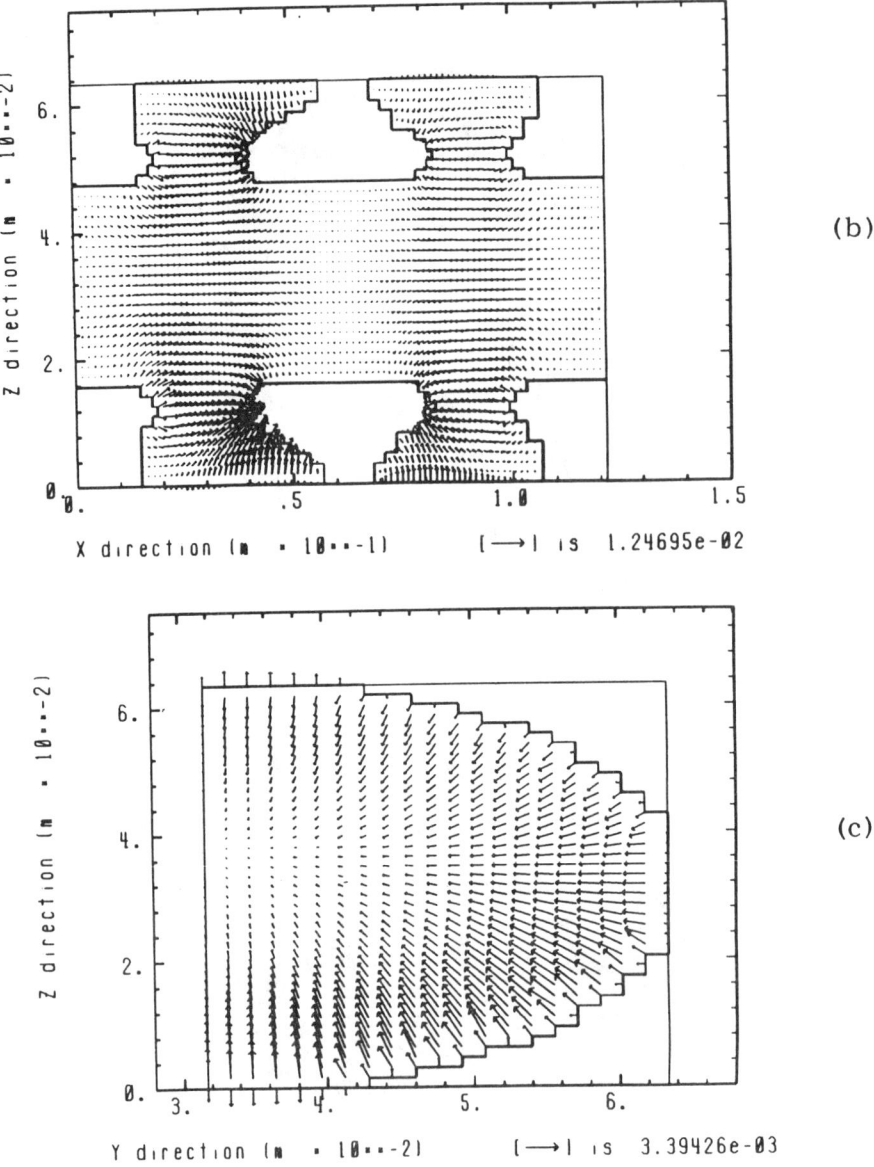

Fig. 5. ARGUS cold testing calculation for the SLAC double cavity output gap.
(b) Normal mode pattern at 2.536 GHz (side view).
(c) Normal mode pattern at 2.536 GHz (end view).

Cray X-MP CPU time were required to compute the first ten modes. In Table II we present a comparison between the eigenfrequencies of the three lowest modes as computed by ARGUS and the actual experimental measurements. The accuracy of the numerical results is seen to be excellent. Side and end views of the mode pattern for the lowest mode (2.536 GHz) are shown in Figures 5b and 5c, respectively.

Design of guns and collectors

Three-dimensional electromagnetic PIC simulations are extremely well-suited to the study of guns and collectors. Time- and/or frequency-domain cold-testing calculations can be used to examine the response of the gun region. Static calculations are useful for determining field patterns and equilibrium particle orbits. Finally, dynamic calculations can be used to study power flow, particle dynamics, and transients and fluctuations in the system.

As an example, we describe a series of ARGUS simulations of a depressed electron collector. This is a device that essentially acts as a beam dump. Such a configuration is shown in Figure 6. It is composed of four cylindrically symmetric collector plates of decreasing potential along the beam line. The symmetry is broken by the presence of vanes on the plates, as mentioned below. The incoming beam energy for this particular design is about 11 keV, and so the voltages on the plates start at slightly over 11 kV, decreasing to about 1 kV on the last plate.

As electrons enter the device at the high-voltage end, they are decelerated as their kinetic energy is transferred to the fields, until they ultimately strike the plates. The initial position and velocity of a given electron determines which plate it impacts: the lower energy particles will impact on the higher voltage (earlier) plates. Thus, the impact velocity of an electron with the plate is low, thereby minimizing effects such as noise due to radiation effects and secondary electron emission from the surface. In addition, energy can be retrieved from the device by collecting the current from the charge deposition on the plates.

As part of an ongoing study, electrostatic ARGUS simulations of this device are being performed. The design goal here is to determine the optimal placement of the vanes for improved performance of the device. Large space charge densities accumulate in regions where the electron velocity is nearly zero; new electrons entering these regions may be completely reflected by this space charge, and may actually exit the collector and reenter the accelerator. The vanes are placed so as to shape the field where space charge accumulation is highest, thereby causing the particle trajectories to become more like those of single particles and

Table II. Experimental and ARGUS results for the SLAC double cavity output gap (eigenfrequencies are in GHz).

MODE	EXPERIMENT	ARGUS	% ERROR
1	2.587	2.536	1.9
2	2.856	2.876	0.7
3	3.455	3.478	0.6

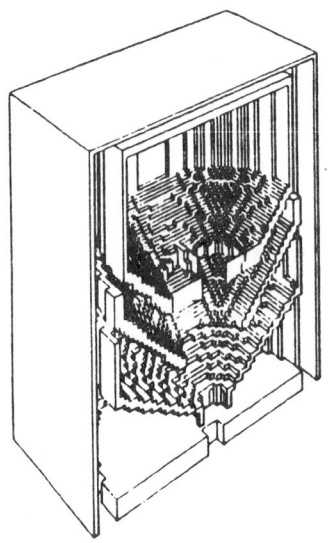 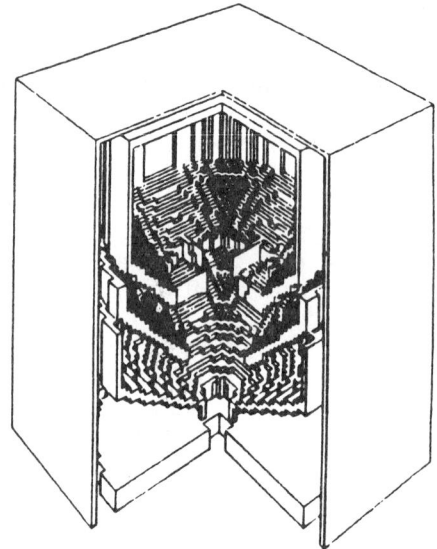

Fig. 6. Depressed electron collector with vanes.

reducing the probability of electron reflection. This increases the efficiency of the collector.

Initial results from the ARGUS simulations, done in the iterative equilibrium mode of operation, are presented in Figures 7 through 9. Figure 7 shows computed equipotentials for collector designs with and without vanes. Corresponding orbits of test particles after two iterations are shown in Figures 8 and 9. While these results are preliminary, the essential features of the device are verified. The lower energy electrons are indeed absorbed on the earlier plates, while the most energetic electrons reach the uppermost plate. Most importantly, the vanes are seen to greatly reduce the number of reflected particles. The designer is now in a position to optimize the performance of the device by adjusting the plate voltages and vane locations for a given initial distribution of electrons in the beam.

Beam Transport and Transient Effects

Three-dimensional simulation of charged particle beam transport, equilibrium, and stability is perhaps the most "traditional" use of PIC codes. The applications to accelerator problems are many. The coupling of a beam to a particular accelerating device can be studied in great detail, and the beam and accelerator parameters can be easily varied to optimize the design. Assuming the beam has already entered the accelerator, one can then study the interaction between the beam and the accelerating cavities or gaps. For example, an off-center beam can be injected to simulate shock excitation of cavities. Output from the simulation can then be used as input for existing models of the beam breakup instability.

As an illustration, we describe a series of preliminary ARGUS simulations of transport, compression, and quadrupole focusing of medium-to-heavy ion beams. The usual approach to such a problem is to assume that the beam can be described by a Kapchinskij-Vladimirskij (K-V) distribution[14] so that a set of so-called envelope equations can be derived. These equations are then propagated forward in time to determine the beam's behavior as it is transported through various field topologies. However, this approach breaks down for a non-ideal (i.e., non-K-V) beam or when the space charge distribution is nonuniform, since the envelope equations no longer apply. Thus, one must use a three-dimensional PIC code to study such effects as phase mixing of envelope oscillations or axial bunching and debunching of the beam.

The selection of a particular quadrupole focusing arrangement is driven by the application under study. For a beam focusing system, where transport and acceleration are the key issues, a series of electrostatic quadrupoles may be used. However, for inertial confinement fusion applications a system of magnetic

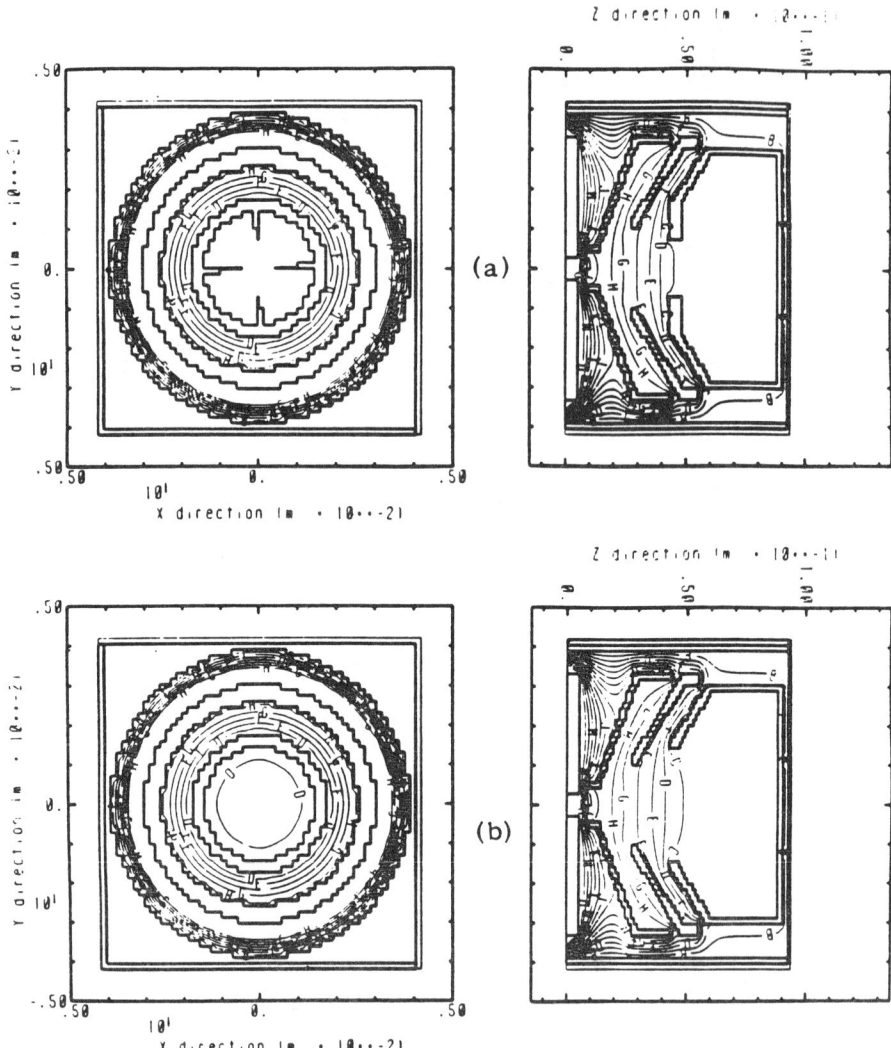

Fig. 7. ARGUS field solutions for the depressed electron collector.
(a) With vanes.
(b) Without vanes.

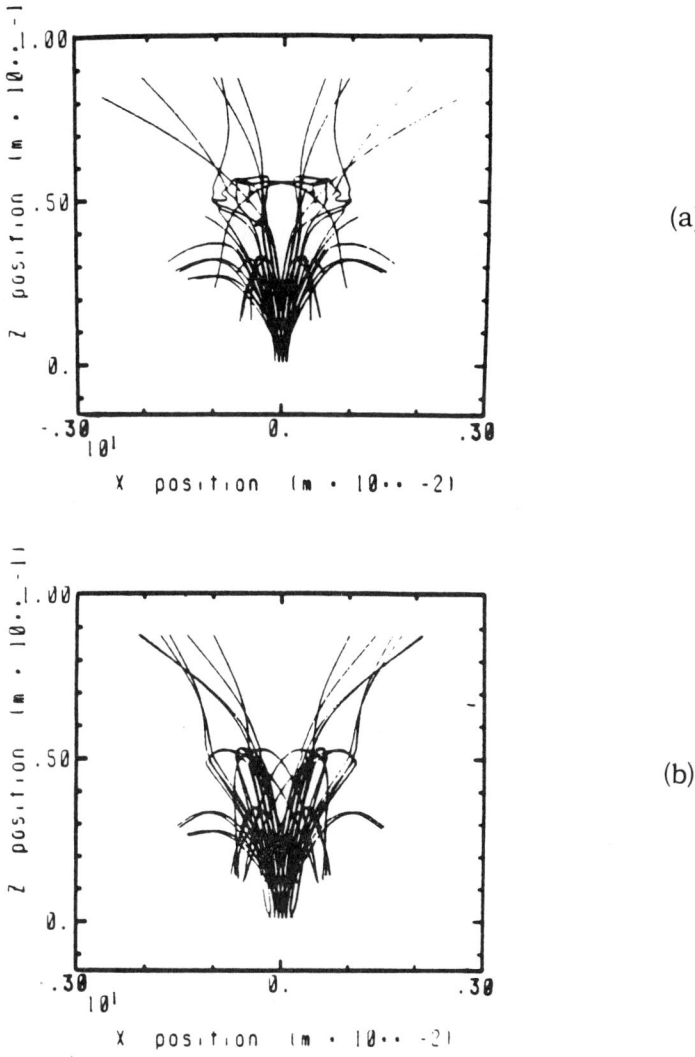

Fig. 8. ARGUS test particle trajectories (x-z) for the depressed electron collector after two iterations.
    (a) With vanes.
    (b) Without vanes.

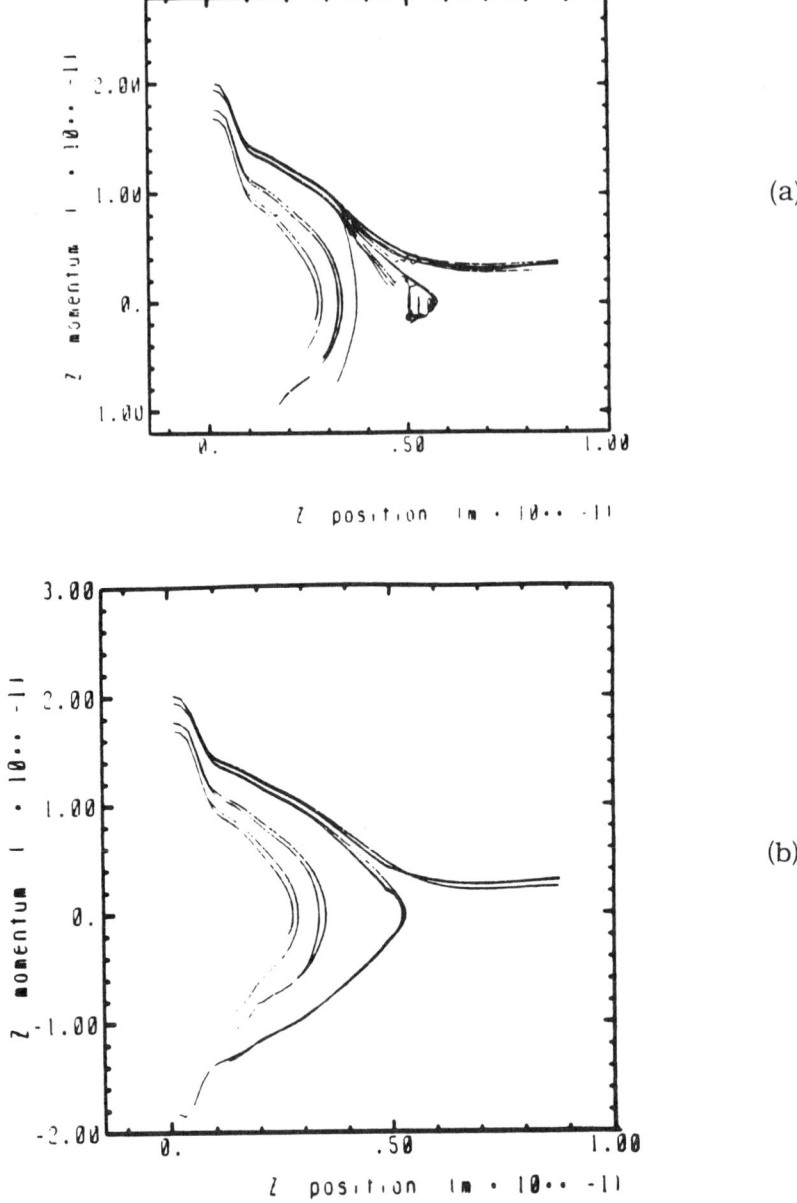

Fig. 9. ARGUS test particle trajectories (z-$\dot{z}$) for the depressed electron collector after two iterations.
    (a) With vanes.
    (b) Without vanes.

quadrupole windings is often the first choice. The ARGUS simulations we describe here utilize the latter scheme. In either case, the quadrupoles are placed so as to provide a FODO focusing array.

In ARGUS, each magnetic quadrupole doublet can be independently specified. At this time, a periodic set is chosen for convenience. Thus the axial field structure consists of a vacuum region of length 0.3927 m, followed by a field region of length 0.19635 m. The quadrupole magnetic field strength is 21.9 T/m. A cold 10 A beam, composed of singly charged 10 MeV C^{12} ions, is injected into the system; the individual ions are given initial velocity vectors such that the beam is convergent. The simulations are performed in a moving window in the lab frame. This window follows the longitudinal motion of the beam centroid.

Figure 10 shows a three-dimensional perspective plot of the beam ion positions early in the simulation. The direction of propagation is left-to-right. The alternate focusing and defocusing of the quadrupole windings in the two transverse directions is clearly visible. In Figure 11 we show the transverse cross-section of a well-matched beam later in the simulation, after transport over a distance of 3.2 m. The beam is seen to have retained its sharp boundary quite well. Compare this plot to the one shown in Figure 12, which is for an unmatched beam. The ragged edges in the latter case are caused by beam betatron oscillations due to mismatch.

These simulations are part of an ongoing effort to use numerical models for matching a given beam to a given transport configuration. In the near future, effects such as off-axis displacements, finite temperature, beam bending through curved sections, and fringing fields will be incorporated, yielding a more complete understanding of the underlying beam dynamics.

CONCLUSIONS

Numerical simulation is a mature, established technique for studying a wide variety of physical systems and devices, particularly those of interest to the accelerator community. A number of well-tested simulation codes are currently in use at national laboratories, private companies, and universities. These codes are being applied to many diverse problems, including the design, interpretation and understanding of experiments and the expansion of theoretical pictures.

The field continues to move forward. Over the next few years, as new techniques mature and computers become even more powerful, we can expect to see dramatic progress in numerical modeling capability. The obvious improvements will come in the areas of speed and resolution, but we will also see significant advances in treating complicated boundaries and geometries and in problems involving disparate time scales. Finally, new personal

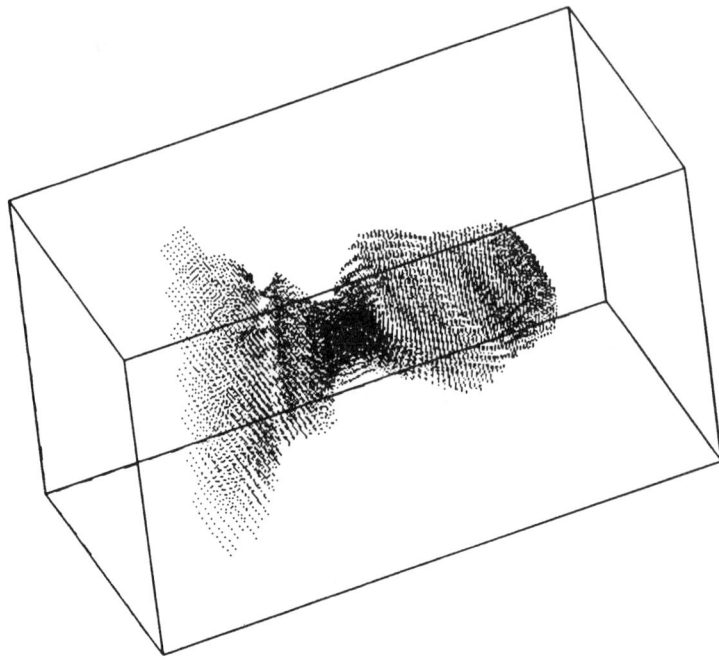

Fig. 10. Three-dimensional perspective plot of an ARGUS beam transport simulation, showing the effect of quadrupole focusing magnets.

computer and workstation technologies will greatly enhance the user's ability to set up and diagnose large, complex problems.

ACKNOWLEDGEMENTS

We wish to thank Chia-Lie Chang, Kwok Ko, John Petillo, Al Mondelli, Adam Drobot, Chris Kostas, Larry Seftor, and Dave Chernin of SAIC, Jan Moura, Scott Brandon, Bill Aimonetti, Dale Nielsen Jr., Alex Friedman, and Jim Mark of LLNL, and Bill Herrmannsfeldt, Harold Hanerfeld, and Ken Eppley of SLAC for their many contributions to this work.

Various aspects of this work were supported by the SAIC Independent R&D Program, Lawrence Livermore National Laboratory, Los Alamos National Laboratory, Sandia National Laboratories, the Stanford Linear Accelerator Center, the Naval Research Laboratory, Hughes Aircraft Company, the Tri-Services/NASA Program on Advanced Computational Modeling, and the U.S. Department of Energy.

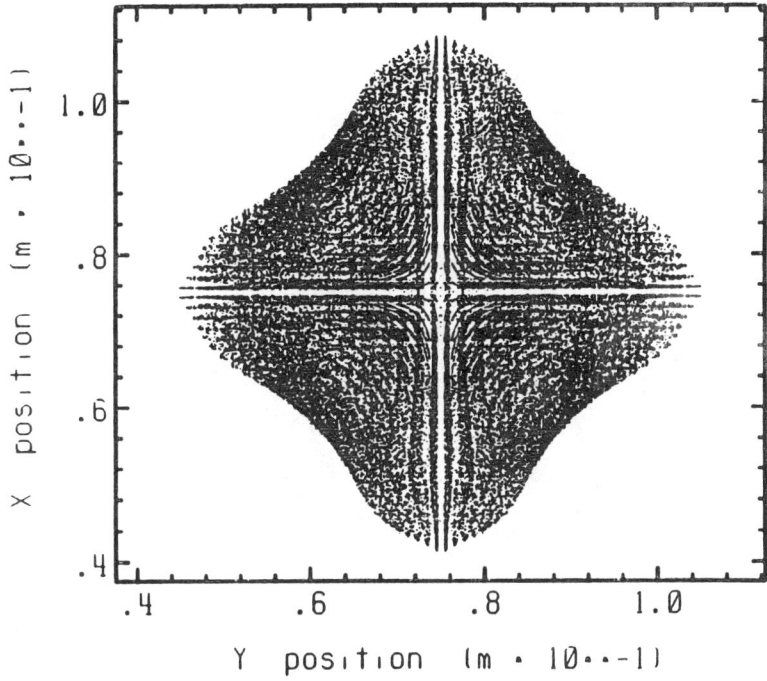

Fig. 11. Transverse cross-section of a well-matched beam.

REFERENCES

[1] See, for example, O. Buneman, *Phys. Rev.* **115**, 503 (1959), and J.M. Dawson, *Phys. Fluids* **5**, 445 (1962).
[2] R.W. Hockney and J.W. Eastwood, *Computer Simulation Using Particles* (McGraw-Hill, New York, 1981).
[3] C.K. Birdsall and A.B. Langdon, *Plasma Physics via Computer Simulation* (McGraw-Hill, New York, 1985).
[4] T. Weiland, *Part. Accel.* **15**, 245 (1984).
[5] B.A. Trubnikov, in *Reviews of Plasma Physics*, edited by M.A. Leontovich (Consultants Bureau, New York, 1965), Vol. 1, p. 105.
[6] W.B. Herrmannsfeldt, Stanford Linear Accelerator Center Report No. SLAC-226 (1979).
[7] See, for example, A. Mankofsky, J.L. Seftor, C.L. Chang, K. Ko. A.A. Mondelli, A.T. Drobot, J. Moura, W. Aimonetti, S.T. Brandon, D.E. Nielsen Jr., and K.M. Dyer, *Comp. Phys. Commun.* **48**, 155 (1988).
[8] *Proceedings of the Fourth Conference on Numerical Simulation of Plasmas* (U.S. Naval Research Laboratory, Washington, DC, 1970).
[9] J.M. Dawson, *Rev. Mod. Phys.* **55**, 403 (1983).

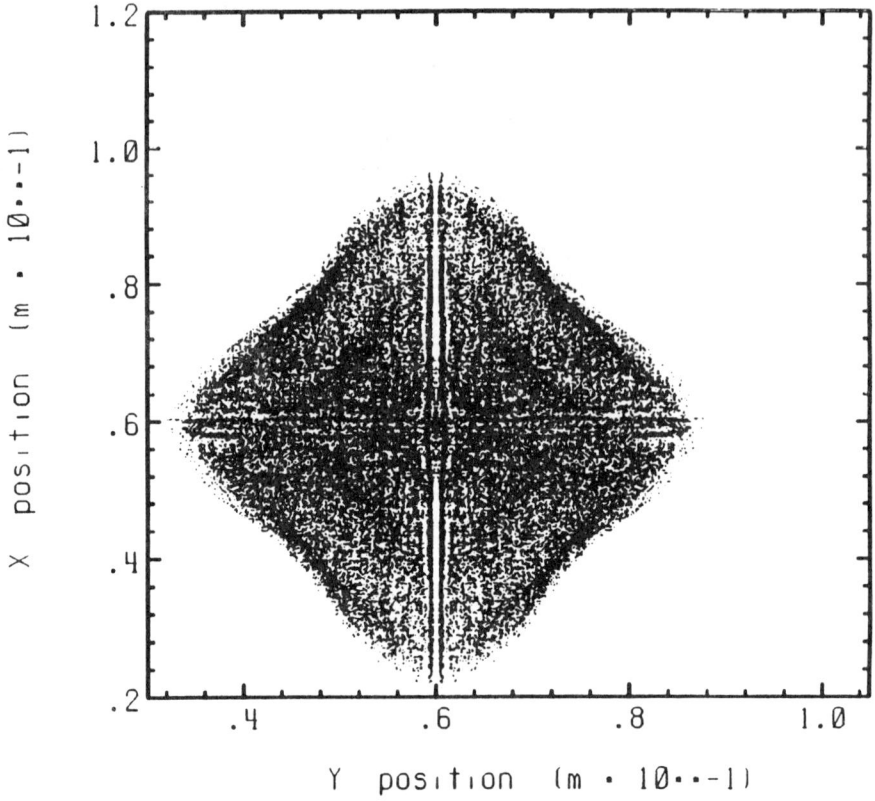

Fig. 12. Transverse cross-section of a poorly matched beam.

[10]J.M. Dawson and A.T. Lin, in *Handbook of Plasma Physics*, edited by M.N. Rosenbluth, R.Z. Sagdeev, A.A. Galeev, and R.N. Sudan (North-Holland, Amsterdam, 1984), Vol. 2, p. 555.
[11]*Particle Methods in Fluid Dynamics and Plasma Physics*, edited by J.U. Brackbill and J.J. Monaghan (North-Holland, Amsterdam, 1988).
[12]*Methods in Computational Physics*, edited by B. Alder, S. Fernbach, M. Rotenberg, and J. Killeen (Academic, New York, 1970 and 1976), Vols. 9 and 16.
[13]*Multiple Time Scales*, edited by J.U. Brackbill and B.I. Cohen (Academic, New York, 1985).
[14]See, for example, J.D. Lawson, *The Physics of Charged-Particle Beams* (Clarendon, Oxford, 1978), Chap. 4.

NUMERICAL LIMITS ON P.I.C. SIMULATION OF LOW EMITTANCE TRANSPORT

I. Haber
Naval Research Laboratory, Washington, DC 20375

H. Rudd
Berkeley Research Associates, Springfield, VA 22150

ABSTRACT

Use of P.I.C. codes to simulate low emittance, high current transport can be limited by the collisional emittance growth which is a consequence of the relatively small number of particles in a simulation, in contrast to the much larger number in an actual charged particle beam. Simulations are presented which show examples of the scaling of this collisional emittance growth. A knowledge of this scaling can be useful in discriminating between the collective physics which is under investigation and effects which are numerical in origin, without doing extensive numerical tests for each new parameter regime.

INTRODUCTION

Applications such as heavy ion fusion, free electron lasers and particle beam weapons systems have generated a requirement for beams with both high current and low emittance. This requirement, as well as recent experimental and theoretical progress in generating and transporting such beams, has substantially increased interest in understanding the detailed behavior of the accelerators needed to produce these very high beam intensities. At these intensities, space charge forces become increasingly dominant, so that the beam evolution can depend not only on averaged quantities such as beam current and emittance, but can also be strongly influenced by details of the six-dimensional distribution function.

Measurement of detailed information about the distribution function of a low emittance beam is usually very difficult. In addition, the complexity of individual particle orbits, which are influenced by the complicated set of focusing elements in a transport system, as well as by strong space charge forces, can render a detailed theoretical treatment exceedingly difficult. Numerical simulations have therefore emerged as an important tool in understanding the dynamics of high intensity particle beams, as well as the interpretation of experimental results, and as an aid in the design of intense beam accelerators. At the same time, the simulation of beams at such high intensities can also require substantial care in order to avoid problems which may be introduced by limits in the numerical methods.

A numerical limit, which can become particularly important as the beam intensity in a simulation is increased, is caused by the relatively small number of particles employed in particle simulations of the beam dynamics. Several examples are presented here to illustrate the quantitative consequences, primarily to the emittance evolution. By examining the scaling of the emittance growth as a function of the number of particles in the simulations, some guidelines are established for differentiating between those causes of emittance growth which occur as a consequence of collective space charge nonlinearities in the beam dynamics, and those caused by the choice of numerical parameters in the simulation.

© 1988 American Institute of Physics

BACKGROUND

A primary technique for the simulation of intense charged particle beams, where the space charge forces can not be represented adequately as a small perturbation to the single particle dynamics, is to numerically follow the orbits of a large number of simulation particles in their self-consistent electromagnetic fields. Particle in cell (P.I.C.) techniques have been extensively employed for this purpose by plasma physicists[1] because they use an amount of computer time which is proportional to the number of particles in the simulation, rather than the square of that number, as is the case when calculating the individual interactions between each of the particles. It is still not possible, however, using current computer technology, to calculate the orbits of a number of particles that approaches the number contained in even a tenuous beam. This smaller-than-physical number of particles results in a level of fluctuations, or a granularity in the distribution function, which can affect the behavior of the simulations in a way which will not occur for a physical beam with a much larger number of particles.

The collisionless Vlasov equation which is appropriately used to describe the dynamics of charged particle beams, assumes that the forces acting on each particle are dominated by the electromagnetic fields which are the collective result of the ensemble of particles in some vicinity of that particle rather than the effects of its nearest neighbors. The consequences of the small scale granularity in the distribution function, which would be caused by a small number of particles, is usually estimated by the addition of a collision term to the Vlasov equation. The collision frequency, measured in terms of the plasma frequency, v_{col}/ω_p, is inversely proportional to the number of particles in a Debye sphere, or a Debye cylinder of unit length in the two-dimensional cases to be discussed here.

P.I.C. simulations of a charged particle beam, as a result of the limited number of particles that they employ, generally have a collision frequency which is much higher than the beam being modeled. Because of the nature of the numerical algorithm, where the particle density is accumulated on a grid, the fields are found numerically on that grid, and the fields are then interpolated back to the particle positions, field gradients on a subgrid scale are generally not resolved. This effective suppression of subgrid scale fluctuations provides a lower limit to the fluctuation wavelength in the simulation. The particle collision frequency is therefore determined by the number of particles either in a Debye cylinder, or grid area, whichever is greater. In the particular instance of a finite beam of particles in force balance with external forces, as will be discussed here, the particle velocities go through a minimum as they are turned around near the beam edge. There is, therefore, usually a region near the beam edge where finite grid size smoothing, rather than the Debye limit, affects collisional behavior. The SHIFT-XY code[2], employed for the numerical simulations discussed here, imposes additional smoothing by using simulation macroparticles which are Gaussian in shape. That is, each simulation particle can be thought of as being a clump of actual point particles with a total charge density which is Gaussian in shape. This shape is generally chosen so that the charge density of a macroparticle centered in a cell would be down by 1/e at the cell edges.

From a knowledge of the collision frequency, however, it is still not a simple matter to predict the consequences of particle collisions on the macroscopic evolution of the beam. In order to aid in the differentiation between any collective physics which can be important to the beam evolution in an accelerator system, and numerical effects caused by particle collisions, a series of simulation experiments was run. These runs examine the numerical scaling of the collisional emittance growth by varying only those numerical parameters which should ideally have no effect on beam evolution. The data which have been obtained should be useful in choosing a numerical parameter regime which avoids unexpected collisional behavior.

Two series of simulations are presented here. They have been chosen to illustrate the consequences of numerical collisions in examining the thermalization of mismatch oscillations, as well as in the determination of the marginal stability region for space-charge-driven instabilities in alternating gradient transport systems.

COLLISIONAL EMITTANCE GROWTH ENHANCEMENT BY BEAM MISMATCH OSCILLATIONS

It has been observed both experimentally, on the Single Beam Transport Experiment at LBL[3], and in simulations[4], that the conversion of mismatch energy to emittance behaves differently depending on the symmetry of the mismatch. An asymmetric mismatch appears to couple the energy of the envelope oscillation into emittance much more quickly than a symmetric mismatch of the same magnitude. This behavior is observed in simulations of uniformly focused solenoidal systems, as well as in simulations of interrupted solenoid and alternating gradient transport systems.

Figure 1 is a plot of emittance evolution versus propagation distance down a uniform solenoidal focusing system, measured in magnet betatron periods, for an asymmetrically mismatched beam. The initial beam distribution is a semi-Gaussian, or thermal, distribution, which is uniform in configuration space and Gaussian in velocity distribution with a uniform temperature. The beam space charge is sufficient to depress the phase advance by a factor of six. Initial beam shape is an ellipse formed by increasing the x-dimension of the beam by 30 percent, and decreasing the y-dimension by 30 percent, compared with the matched radius. Examination of the behavior of the rms envelope oscillations shows that the growth of

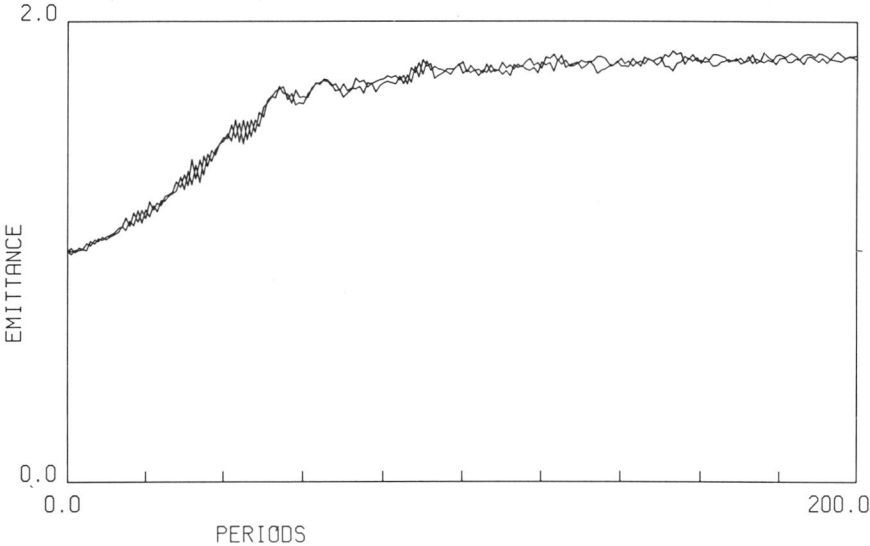

Fig. 1. Evolution of the x and y rms emittances, plotted vs. propagation distance in betatron periods, of an asymmetrically mismatched beam ($x_0 = 1.3\, x_{match}$, $y_0 = y_{match} / 1.3$), propagating down a uniform solenoidal focusing channel. The initial distribution is semi-Gaussian with a factor of six tune depression.

emittance is accompanied by a damping of the envelope oscillation on the same time scale and provides strong evidence that the emittance growth is driven by a thermalization of the mismatch energy. The simulation shown was run with 32K particles on a 256 by 256 grid. Periodic boundary conditions were used and the initial major radius of the beam, when matched, was approximately 50 cells, chosen to be small enough so that any nonlinear forces, which might result from a beam edge near the boundary, would not substantially affect beam evolution.

Figure 2 is a plot of the evolution of the rms emittance of a beam with the same initial parameters but which is mismatched symmetrically by increasing both the x- and y-dimensions the same 30 percent. Emittance growth is on a slower scale, but as in the asymmetric case, the growth appears to be driven by the thermalization of the mismatch energy. A significant difference in behavior between the symmetric and asymmetric mismatch behavior is observed, however, when the number of particles in the simulation is varied. Whereas the behavior in the asymmetric case is virtually unchanged when the number of particles is increased, the emittance growth in the symmetric case decreases as the approximate inverse square root of the number of particles.

Figure 3 shows the fractional growth in rms emittance, plotted against the number of simulation particles, on a log-log scale. Data for grids of 128x128 and 256x256 are plotted, as are parallel lines with a slope of $n^{-1/2}$. The lack of any apparent saturation in the $n^{-1/2}$ decrease in emittance growth with increase in number of particles, appears to indicate that the thermalization of mismatch energy in the symmetric case is, for the parameter regime studied, entirely numerical.

Also plotted in Figure 3 is data showing the much lower emittance growth observed in simulations of a matched solenoidal transport system with similar numerical parameters. A characteristic of the initial semi-Gaussian distribution[5] is that the rms emittance will initially

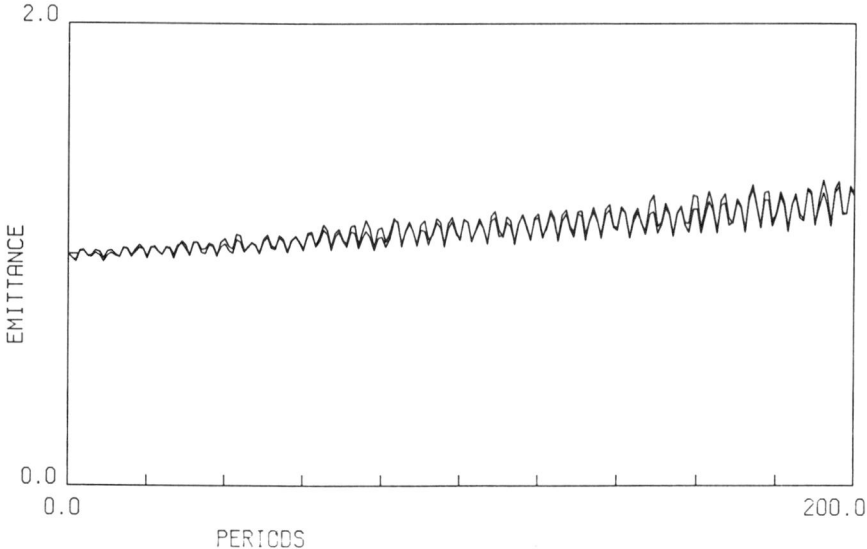

Fig. 2. Emittance versus propagation distance in betatron periods for parameters similar to those in Fig. 1, but with a symmetric match, $x_0 = 1.3\ x_{match}$, $y_0 = 1.3\ y_{match}$.

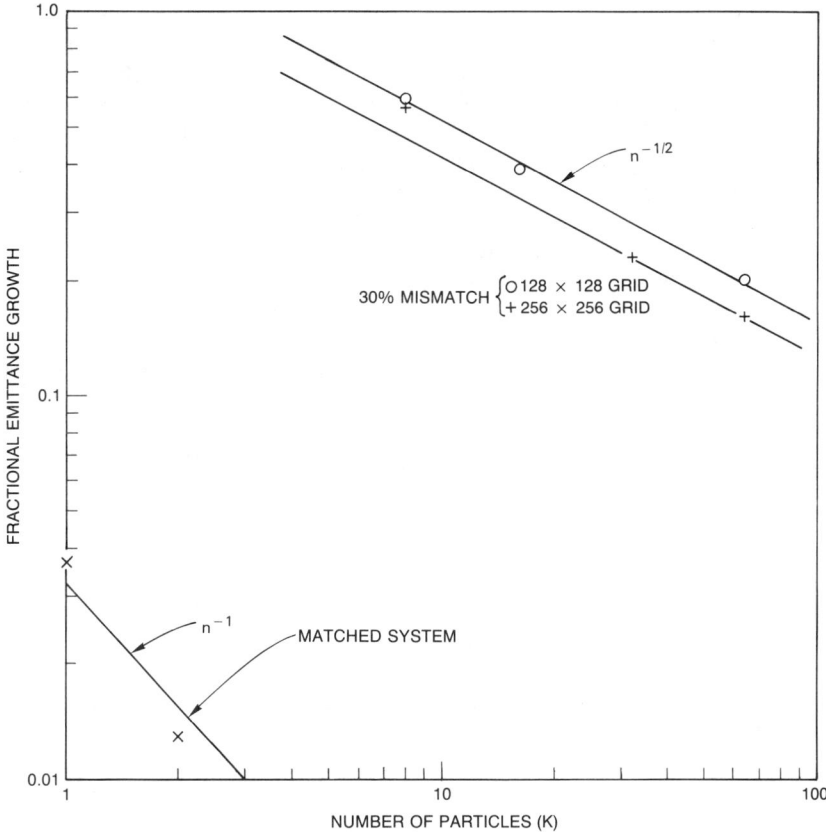

Fig. 3. Fractional growth in emittance versus number of macroparticles in the simulation for the symmetrically mismatched beam after propagating 100 betatron periods. Plot is on a log-log scale to show the $n^{-1/2}$ decrease in emittance growth. Also plotted, at lower left, is a matched beam to show the much lower level of collisional emittance growth and the more rapid n^{-1} decrease.

drop, in the first fraction of a plasma period, as the beam relaxes to its minimum energy state. Because of the small emittance growth in these runs, this initial drop becomes significant by comparison, so that approximately 2 percent was added to the emittance plotted to compensate for this initial emittance decrease.

An additional source of emittance growth which depends on the number of particles can become significant when a small number of particles is used in a simulation. Random density nonuniformities from the loading of the initial distribution cause an initial emittance growth as the field energy in these density variations is converted into thermal energy, and the beam relaxes toward a more uniform distribution. This behavior, in the same way as the redistribution inherent in a semi-Gaussian distribution, is easily separated from the much slower collisional behavior shown in Figure 2. However, in combination with the large fluctuations in rms emittance that characterize the evolution of a beam system with a small number of particles and with large tune depression, it becomes difficult to measure the slower scale growth. Both the initial increase and the continuing fluctuations are often larger than the small, slower time scale, emittance growth after a hundred periods of propagation. Therefore, no attempt will be made here to proceed further, because collisions do not appear to be a problem when modeling beam systems in this parameter range with an acceptable number of particles. Beyond a certain point, reducing the number of particles does not substantially reduce running time, since the computational effort becomes dominated by the field solver.

Nevertheless, if the matched solenoid curve is accepted merely as an upper limit on collisional growth, rather than as a quantitative estimate of collisional behavior, the much lower collisional emittance growth rates in the matched case, as well as the n^{-1} scaling, compared with the $n^{-1/2}$ observed when a strong mismatch is present, suggest a strong enhancement of the collisional emittance growth caused by the envelope motion during mismatch oscillations.

COLLISIONAL EMITTANCE GROWTH IN PERIODIC FOCUSING SYSTEMS

When a beam propagates down a periodically focused channel, the beam envelope is modulated at the magnet period, even when it is matched. Since this modulation increases as the phase advance per focusing period (σ_0) increases, it might be expected that any enhancement in collisional emittance growth by this modulation would also increase as σ_0 gets larger. It has also been observed in a large number of simulations,[6] as well as experimentally[7], that there are collective space charge instabilities which cause emittance growth as the phase advance approaches 90 degrees. Furthermore, details of the emittance growth dependence on phase advance can be important to the design of periodic quadrupole systems for use as heavy ion fusion drivers, because there is an economic advantage to operating as close to 90 degrees as possible, yet it is necessary at the same time to avoid any of the instability-caused emittance growth which can be encountered near 90 degrees.

Figure 4 shows the emittance growth, after the beam has propagated 100 periods, as a function of phase advance per focusing period, σ_0. The two curves shown correspond to runs with 16K and 32K macroparticles. The beam used in these runs has an initial semi-Gaussian distribution, about a 50 cell major radius, and the phase advance (σ) is depressed a factor of ten by space charge. The factor of ten tune depression was chosen to test the importance of collisions because, from the scaling of the matched equilibria, the beam area measured in units of Debye length goes as the tune depression ratio (σ^0/σ) squared. The number of particles in a Debye cylinder therefore decreases rapidly, and collisional emittance growth is most likely to become a serious problem, at substantially depressed phase advances. Since the increase in emittance growth as the phase advance increases can be attributed either to collective effects or numerical collisions, further investigation of the numerical behavior of these simulations is warranted, especially in view of the difference which can be seen between the two curves run with different numbers of particles.

The lowest value of the phase advance, σ_0, shown in Figure 4 is 60 degrees. This value was chosen because, from a variety of considerations (primarily the extrapolated behavior of systems with less depressed phase advance), no substantial collective-space-charge-instability caused emittance growth is expected at 60 degrees, at least for transport systems which are

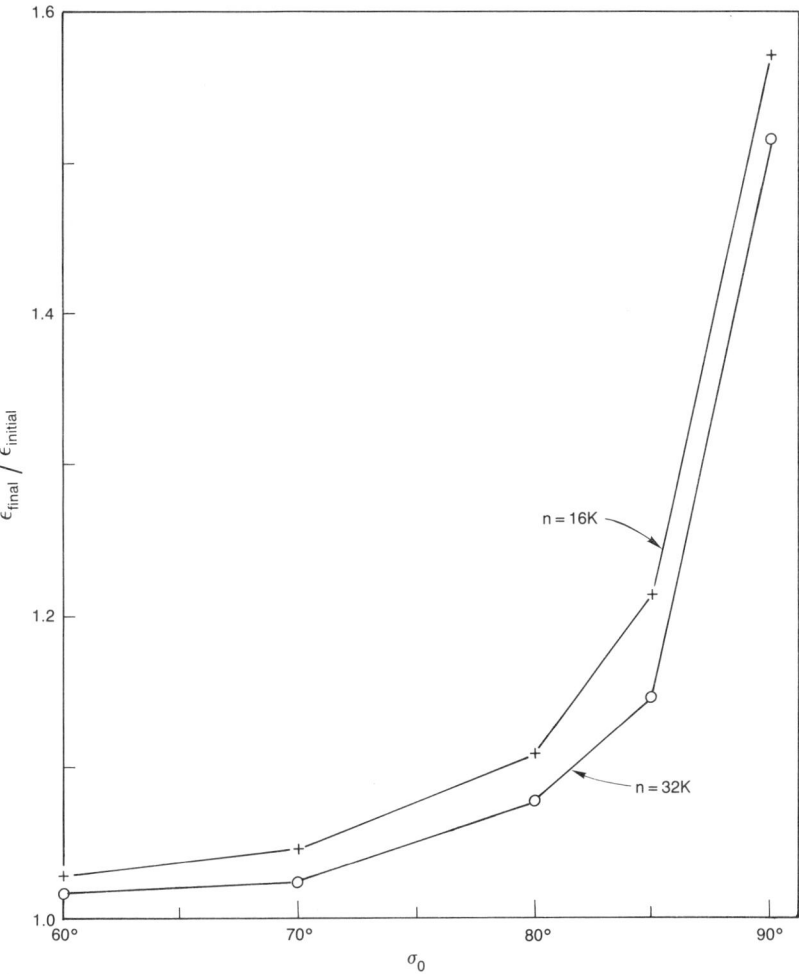

Fig. 4. Plot of the ratio of the emittance after 100 periods to the initial emittance, as a function of the phase advance per cell, for an initially semi-Gaussian distribution with the phase advance depressed by a factor of ten. Differences in the emittance growth between the simulations performed with 16K and 32K particles, especially at the lower phase advances, suggest that at least some of the emittance growth is a consequence of numerical collisions.

only a few hundred periods long. Figure 5 is a plot of the emittance growth after 100 magnet periods, versus number of simulation particles, of a 60 degree magnet system depressed by space charge to 6 degrees ($60°/6°$). Both alternating gradient and interrupted solenoidal transport systems are plotted. Both of these systems show the n^{-1} scaling of emittance growth with number of particles seen in the uniform solenoidal system, although the absolute growth is somewhat higher.

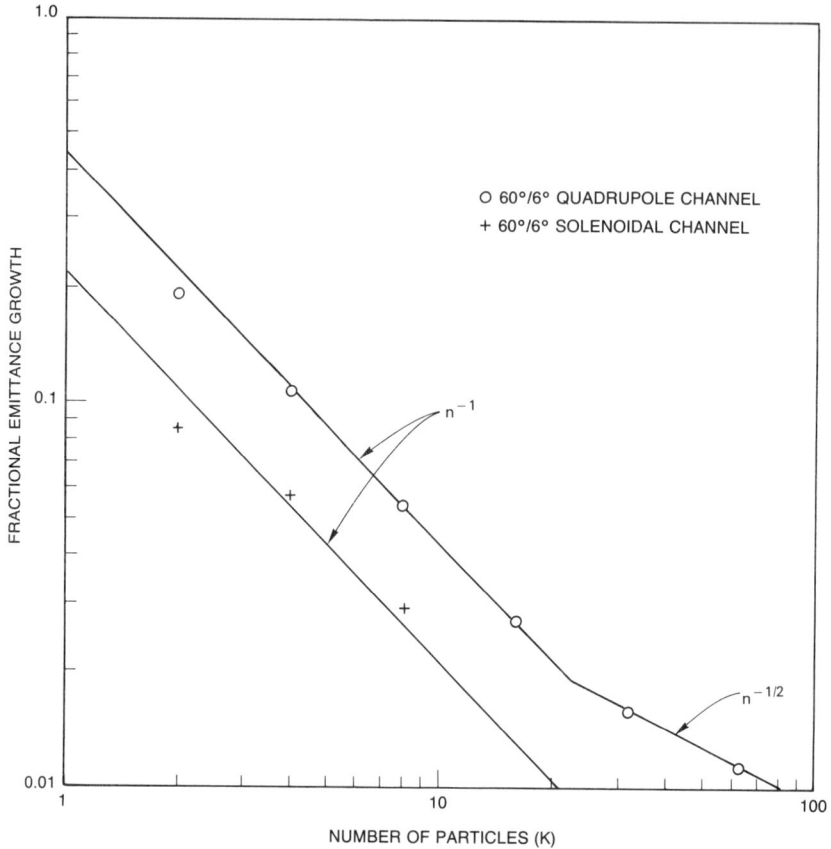

Fig. 5. Fractional emittance growth after 100 magnet periods versus number of simulation particles, for 60° interrupted solenoid and alternating gradient transport systems. The initial distribution in both cases is semi-Gaussian, with a factor of ten tune depression. Data for the quadrupole channel show the transition to an $n^{-1/2}$ dependence of collisional emittance growth on number of particles, as n becomes large.

The curve for the quadrupole magnet transport system, which extends to runs that employ a greater number of particles, also exhibits n^{-1} scaling, but makes a transition to the $n^{-1/2}$ behavior seen previously, in the case of the mismatched beam, to be characteristic of the movement of the beam edge. This is what would be expected from the sum of two effects scaling as n^{-1} and $n^{-1/2}$ respectively, where the inverse square root behavior will eventually dominate as n increases. Also noteworthy in the data is that, if we assume that the $n^{-1/2}$ behavior dominates at large n, the curve will extrapolate to zero as the number of particles increases.

The curves corresponding to both the symmetrical mismatch and matched continuous solenoid, as well as to the interrupted solenoid, and quadrupole systems with 60 degree phase advance per cell, are seen to extrapolate to zero in this way as the number of particles increase. In an attempt to determine whether this behavior "turns over" as we get closer to 90 degrees, so that there is a component to emittance growth which does not extrapolate to zero, but remains relatively unchanged as the number of particles increases, a large number of numerical tests were also conducted on systems with phase advances of 80 degrees and 85 degrees.

Figure 6 is a plot of the emittance growth after 100 periods, as a function of number of particles, for an $80°/8°$ semi-Gaussian distribution. Curves are plotted for a 128x128 numerical system with an initial beam major radius of approximately 50 cells, as well as for systems which are 256x256 and 512x512, with the beam major radius measured in cells doubled accordingly. For each resolution, the n^{-1} behavior goes into $n^{-1/2}$ for large n, and the

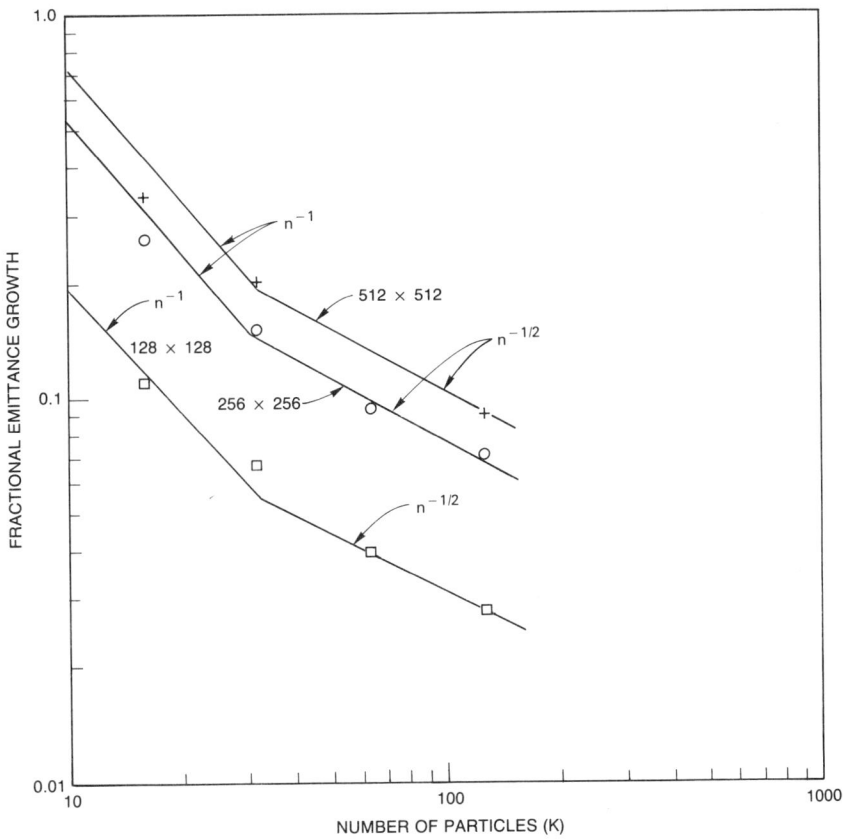

Fig. 6. Fractional emittance growth, after 100 magnet periods, for an $80°/8°$ semi-Gaussian beam. Curves of 128x128, 256x256 and 512x512 systems show the increase in collisional growth rates as the numerical resolution is made finer.

curves appear to go to zero as $n^{-1/2}$ in the same manner as the previous curves. An increase in collisional emittance growth with numerical resolution is expected because, as the number of cells increases, the beam radius at which the effective Debye distance is determined by the cell size rather than by beam thermal energy, moves out in the beam. Fewer particles in the beam then experience grid size smoothing. In effect, the numerical system is able to resolve finer scale fluctuations, and these are significant near the beam edge.

Figure 7 is a plot of the emittance growth as a function of the number of cells along a side of the simulation grid. As the number of grid cells increases, a larger portion of the collisional emittance growth should be limited by the thermal velocity, so that a saturation of the increase in collisional growth is expected.

Figure 8 is a plot of the dependence of collisional emittance growth on number of particles for an $85°/8.5°$ semi-Gaussian system. The major differences between this system and the one in Figure 6 are a higher level of emittance growth, and the leveling off of the

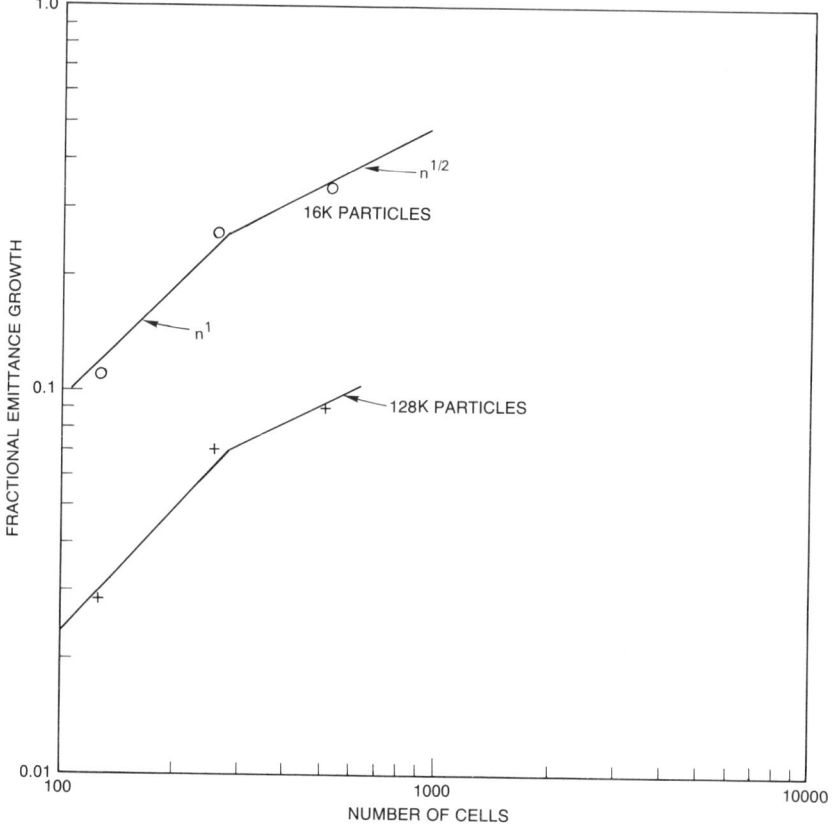

Fig. 7. Fractional emittance growth after 100 periods of the $80°/8°$ system versus the number of cells in a side of the numerical system. The rate of increase of the collisional emittance growth saturates as the numbers of cells increases and an increasingly large part of the beam can be numerically resolved on a Debye length scale and is therefore not affected by grid scale smoothing of the fine scale fluctuations.

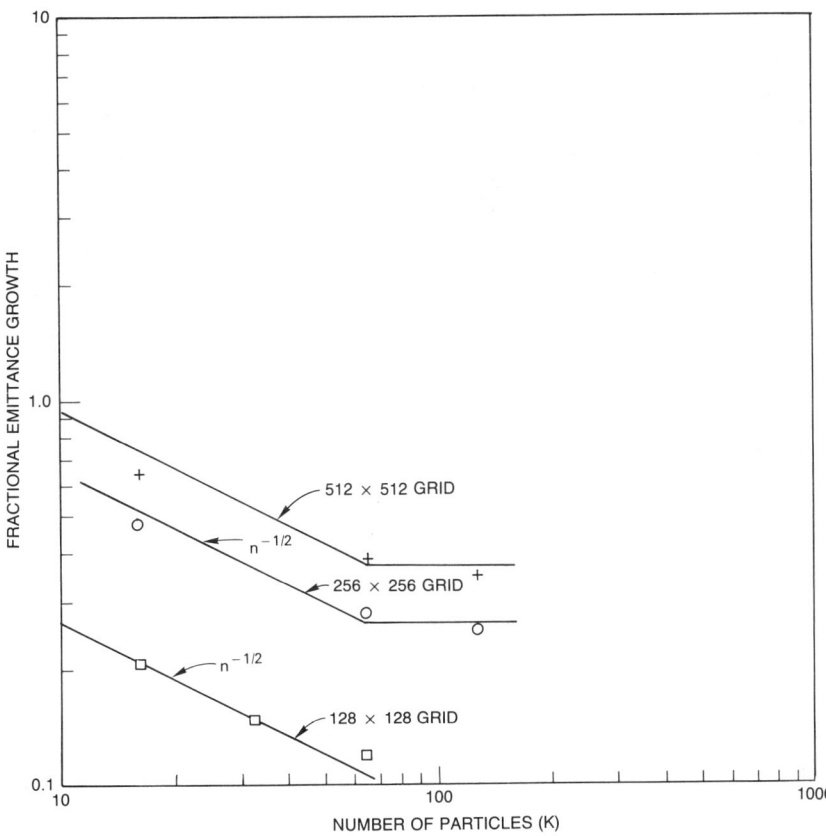

Fig. 8. Fractional emittance growth after 100 periods versus number of simulation particles for an 85°/8.5° semi-Gaussian beam. Unlike the previous examples, the emittance growth does not continue to decrease as the number of particles is increased.

decrease in collisional emittance growth as the number of simulation particles is increased. This leveling off occurs at all resolutions shown, but at a different level. It would be much more comforting if the actual level of emittance growth were independent of resolution. It is, however, not unreasonable to postulate, as a resolution-dependent cause of the emittance increase, some short wavelength phenomenon which depends on the details of the sharp gradients near the beam edge, or whose inherent wavelength is very short. It does appear, however, that at 85 degrees some new phenomenon is being observed which, for a sufficient number of particles, does not depend on the number of particles, so that the emittance growth is not primarily collisional. It should also be mentioned that halving the time step of the 256x256 run with 128K particles had virtually no effect on the amount of emittance growth.

CONCLUSIONS

While it would be nice to feel secure that the emittance growth in a simulation is independent of the details of the numerical model, and there indeed is a parameter range for which this appears to be true, there is also a parameter range where care in choosing the numerical parameters is important. Furthermore, the parameter range where simulation results strongly depend on the numerical parameters can be one of practical interest. We have attempted here to present some data on the scaling of collisional emittance growth to aid in avoiding the attribution of simulation results to an erroneous cause. For example, it is easily possible in a P.I.C. simulation to be in a parameter range where increasing both the number of particles and the numerical resolution in the same ratio would not result in any change in collisional emittance growth. Even though the simulation behavior was in fact dominated by collisions, it would then be tempting to conclude that the lack of change in emittance growth, after changing the numerical parameters precluded numerics as the cause of this growth. As a general conclusion, it is probably worthwhile to do a relatively small number of numerical experiments, when examining low emittance beam systems, to help reduce the possibility of a largely numerical cause for any behavior observed.

ACKNOWLEDGMENT

This work was supported by the United States Department of Energy under contracts DE-AI05-83ER40112 and DE-AI05-84ER40156.

REFERENCES

1. Charles K. Birdsall and A. Bruce Langdon, <u>Plasma Physics via Computer Simulation</u>, (McGraw Hill, New York, 1985).
2. I. Haber, "High Current Simulation Codes," <u>High Current, High Brightness, and High Duty Factor Ion Injectors</u>, AIP Conf. Proc. 139, Ed. by George H. Gillespie, Yu-Yun Kuo, Denis Keefe & Thomas P. Wangler, 107 (AIP, New York, 1986).
3. M. G. Tiefenback, "Experimental Measurement of Emittance Growth in Mismatched Space-Charge Dominated Beams," 1987 IEEE Particle Accel. Conf., IEEE Catalog No. 87CH2387, 1046 (1987).
4. R. A. Jameson, private communication.
5. T. P. Wangler, K. R. Crandall, R. S. Mills, and M. Reiser, "Relation between Field Energy and RMS Emittance in Intense Particle Beams," IEEE Trans. Nucl. Sci. <u>NS-32</u>, 2196 (Oct. 1985).
6. I. Hofmann, L. J. Laslett, L. Smith, I. Haber, "Stability of the Kapchinskij-Vladimirskij (K-V) Distribution in Long Periodic Transport Systems," Particle Accelerators, <u>13</u>, 145 (1983).
7. M. G. Tiefenback and D. Keefe, "Measurements of Stability Limits for a Space-Charge-Dominated beam in a Long A.G. Transport Channel," IEEE Trans. Nucl. Sci. <u>NS-32</u>, 2483 (Oct. 1985).

HOLLOW BEAM DYNAMICS IN THE DESY WAKE FIELD ACCELERATOR

P. Schütt
University of Maryland, College Park, Maryland 20742

W. Bialowons, F.-J. Decker, F. Ebeling, R. Wanzenberg,
T. Weiland and Xiao Chengde[1]
Deutsches Elektronen-Synchrotron DESY,
Notkestr. 85, 2000 Hamburg 52, West Germany

ABSTRACT

The characteristic requirements for studies of an intense hollow electron beam do not permit the use of simulation codes usually used for linac or storage ring design. At DESY, two special codes have been developed for this purpose: WAK-TRACK, a tracking code which includes arbitrarily shaped external fields and collective effects and TBCI-SF, a 2-1/2 dimensional, fully relativistic particle-in-cell code used for studies of the ring shaping and bunching process. TBCI-SF also includes arbitrarily shaped static or dynamic fields. Both computer codes will be discussed and applications for the Wake Field Transformation Linac will be indicated.

INTRODUCTION

The principle of the Wake Field acceleration mechanism[1,2] and the Wake Field Transformer experiment at DESY[3,4,5] have been described in detail in other papers. Here we will focus on the computational needs and methods which are used to study the dynamics of the hollow electron beam driving the Wake Field Transformer. Two simulation programs will be discussed and some example output will be presented. More detailed descriptions of the programs and more results will be published elsewhere.[6]

 - The hollow electron beam must have certain characteristic qualities in order to drive the transformer efficiently. Figure 1 shows a schematic view of the transformer geometry. In the left part, the electric field created by the hollow driving beam is shown. After the hollow beam has passed the structure, the field is compressed as it travels towards the center and bunch II on axis is accelerated by a higher field.

The accelerating gradient on the axis of the transformer is proportional to the charge in the driving beam. Therefore, it is desirable to use charges as high as possible. The design charge in the DESY experiment is 1 μC per hollow beam bunch. Even stronger is the influence of the bunch length and the wall thickness of the hollow beam. Preferably the charge should be concentrated within a torus and the minor radius of this torus should be smaller than 1 mm.

---
[1]On leave from Tsinghua University, Beijing, People's Republic of China.

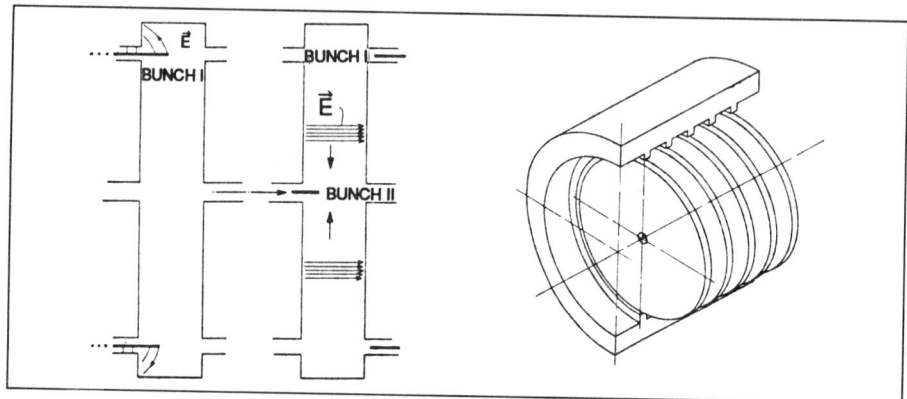

Figure 1. The Wake Field Transformer.

This high charge density causes strong space charge forces, especially in the low energy region, where the torus is formed. It also causes wake fields outside the Wake Field Transformer which influence the dynamics of the hollow beam. In order to shield partly the longitudinal space charge forces by image charges in the beam pipe, the radius of the hollow beam is designed to be only slightly smaller than the beam pipe radius.

A disadvantage of this large radius is the fact that higher multipole moments of the external fields are much stronger here than near the axis. Therefore, it is in general not possible to handle external magnetic fields or rf fields in accelerating cavities in the usual multipole expansion. They have to be modeled carefully in the neighborhood of the hollow beam.

For studies of the ring forming process, a particle-in-cell code TBCI-SF has been developed as an extension of TBCI.[7] This program solves Maxwell's equations and the Lorentz force equation simultaneously in order to model collective forces in the beam. The electromagnetic field is implemented on a two-dimensional, axisymmetric $r-z$ mesh. The particle momenta are modeled three-dimensionally. Arbitrarily static fields can be superimposed as calculated by PROFI[8] and resonant fields are precalculated by URMEL.[9] This code will soon be available for other applications.

Additionally, a fast code is desirable for on-line studies of the experiment and for feasibility studies of a TeV-collider based on the Wake Field Transformer principle. While this codes should handle external fields in a manner similar to TBCI-SF, high-energy approximations for the collective forces are possible. WAK-TRACK[10] has been developed for this purpose. This tracking program includes external static fields as calculated by PROFI, resonant cavity fields as calculated by URMEL, wake fields as calculated by TBCI, and calculates space-charge forces in a high energy approximation.

TBCI-SF

The program TBCI-SF is a particle-in-cell code. This method of solving Maxwell's equations and the Lorentz force equation self-consistently has been developed in plasma physics[11] and recently is being used in accelerator physics for high current beams.[12,13] After the initial conditions have been fixed, the following three steps are carried out in turn:
1. Calculate the current density at the mesh points which corresponds to the motion of the particles.
2. Advance fields in time using this current density as a driving term (equivalent to 1 step in TBCI).
3. Advance particle trajectories according to the Lorentz force.

Using the finite integration theory,[14] the fields and the current density are located in the mesh as shown in Figure 2.

The field evaluation with time is then calculated exactly as in TBCI, using the current density in the mesh as a driving term for the fields. Deeper discussions of this algorithm can be found elsewhere.[7,15]

The charge distribution in the bunch is described in TBCI-SF by macroparticles which represent a rigid charge distribution in a volume corresponding to one cell of the field mesh. These macroparticles are characterized by their position (r, z) and their rapidity $\vec{u} = (u_r, u_\phi, u_z) = \vec{p}/mc$. The position may be anywhere inside the region covered by the mesh, the rapidity is treated fully three-dimensionally. The Lorentz force equation can be written as

$$\frac{d}{dt}\vec{x} = \vec{v} \qquad (1)$$

$$\frac{d}{dt}\vec{u} = \frac{q}{mc}(\vec{E} + \vec{v} \times \vec{B}) \qquad (2)$$

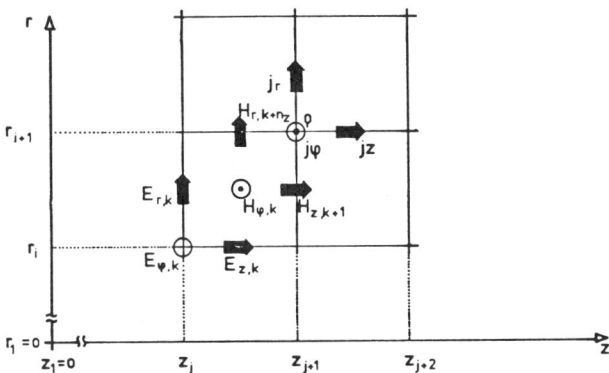

Figure 2. Allocation of field components.

Similar to Maxwell's equations, this system is solved by a leapfrog scheme. The velocity as well as the magnetic field must be time averaged when the velocity change is calculated. Furthermore, the second equation is implicit. In TBCI-SF it is replaced by an explicit algorithm: In the first step, the rapidity is advanced by half a time step using only the electric field. Then the rotation in the magnetic field is calculated and finally the second half step of acceleration is carried out.

In order to decrease noise amplitudes, pyramid-shaped particles are used. This allows a smooth approximation of any charge distribution at the cost of second order terms in the current density calculation.

A charge conserving scheme is used for the determination of current densities in the mesh. This is the same one which is used in ISIS[16] and was first described by Buneman.[17] Instead of multiplying the charge density by an average velocity, the current is calculated as a sum of charges which pass a cell wall during one time interval. The field at the particle position is calculated as a weighted mean of the fields at mesh points which are covered by the charge cloud represented by a macroparticle.

Only two of the four Maxwell's equations are needed to advance the fields in time. The others must be fulfilled implicitly. One of these equations, Gauss' Law, must be used to correct the fields if the current calculation is not charge conserving. In TBCI-SF, both are used to check the results every few time steps.

External fields created by sources different from the bunch current itself often cause problems in particle-in-cell codes. Sometimes static fields are not foreseen and rf cavity fields can only be produced by simulating the whole filling period, which is expensive in terms of CPU-time and may be inaccurate, or by using approximations like the "port approximation" in MASK.[12]

TBCI-SF allows external fields to be given as initial values of the electromagnetic field implemented on the mesh. Once they are set up, they do not need to be treated specially any longer. Static fields may be precalculated by PROFI. The PROFI mesh must be identical to the TBCI-SF mesh or to part of it in order to avoid numerical errors due to the interpolation. Resonant fields are precalculated by URMEL with the same restriction on the mesh. Both electric and magnetic fields are taken from URMEL results. Therefore, initial phase and amplitude of the rf field may be adjusted separately.

The CPU time needed for the simulation is a function of the number of macroparticles used n_M and of the number of meshpoints n_P:

$$t_{CPU} = \left[a\, \frac{n_M}{1000} + b\, \frac{n_P}{1000} \right] \cdot \frac{\text{number of steps}}{1000} \,. \tag{3}$$

Typically n_P and n_M are in the same order of magnitude. On an IBM 3084 Q the coefficients are $a \approx 13.5$ sec and $b \approx 160$ sec.

The output example (Figure 3) shows the bunching of the hollow beam in the buncher section of the DESY experiment. The cavity has initially been filled with a rf field calculated by URMEL. An overall solenoid field holds the particles on paths parallel to the z axis. The current snapshots in Figure 4 show more qualitatively the bunching efficiency.

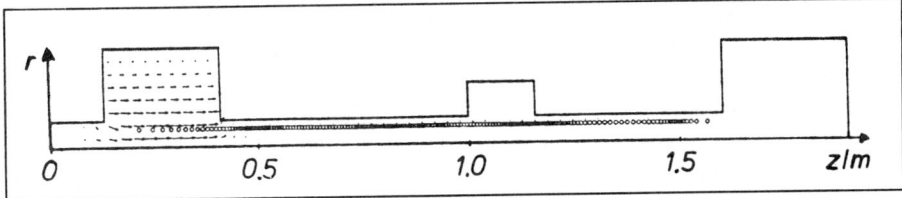

Figure 3. Example output of TBCI-SF: Bunches are being formed in a Cavity with drift.

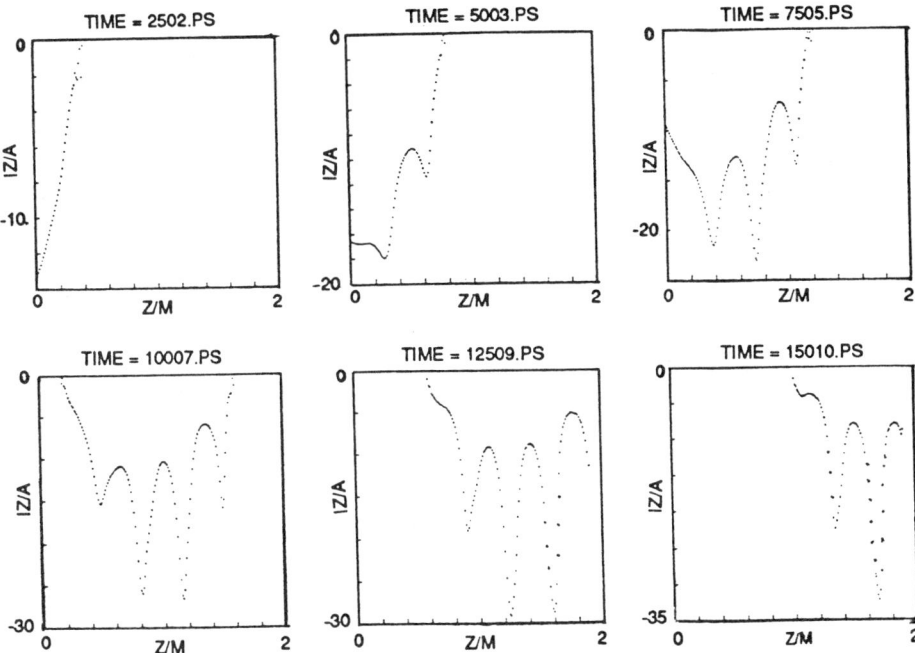

Figure 4. Current snapshots in the buncher region.

WAKTRACK

The computer code WAKTRACK is a tracking code: the particle trajectories are integrated element by element in a given structure. The independent variable is z, the axis of the linac. The time t, Lorentz factor γ, the transverse coordinates x and y and their derivatives dx/dz and dy/dz are integrated for each particle.

The element by element structure is overlaid by the field of solenoid coils which may surround the equipment. External fields may be calculated by PROFI (static fields) or URMEL (resonant fields). Collective forces of the other particles in the bunch are treated in a Green's functions approach. For some cases, the Green's function can be calculated analytically, but in most cases, a pseudo Green's function is calculated with TBCI.

In order to speed up the integration of the trajectories, an analytical solution of the Lorentz force equation for the case $\vec{E} \parallel \vec{B} \parallel \hat{z}$ is used. In the major part of the experiment these are the dominant fields. This analytical solution can be derived without any further approximations. The forces of the transverse fields B_x, B_y, E_x and E_y are treated in a kick approximation. In each step, the momentum is corrected according to these fields. The step length may still be fairly large (up to 3 cm) as long as the transverse fields are small compared to the longitudinal fields. The kick approximation is no longer appropriate if the transverse fields are of the same order of magnitude as the longitudinal fields. In this case, a Runge Kutta integration scheme is used.

WAKTRACK is used extensively as an operating tool for the experiment at DESY. It is fast enough (<1 min CPU on an IBM 3084 Q for a typical run) to be used on-line with the experimental work. The example run, shown in Figure 6, was used to maximize the charge passing through the transformer section in the stage 1 experiment. For more details on the experiment see reference 18. Figure 5 shows a technical drawing of the experimental setup which is schematically repeated between the top and the center part of Figure 6. For this optimization the most interesting part is the central one, where the trajectories of the hollow beam particles are plotted in an $r-z$ frame. Note that $r=0$ line is suppressed. The particles move on spiral paths around a magnetic field line. In the projection these spirals appear as oscillations. The lower plot shows that about 70 percent of the charge emitted by the gun reaches the transformer. In the top frame, the energy of the particles is plotted and their phase relative to a speed of light particle. In these plots, the acceleration can be observed as well as the bunch length.

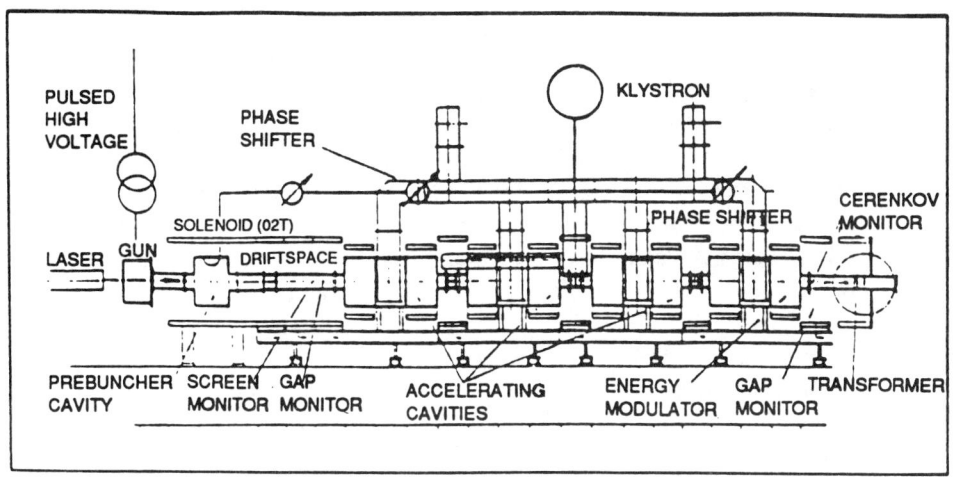

Figure 5. Overview of the Wake Field Experiment at DESY.

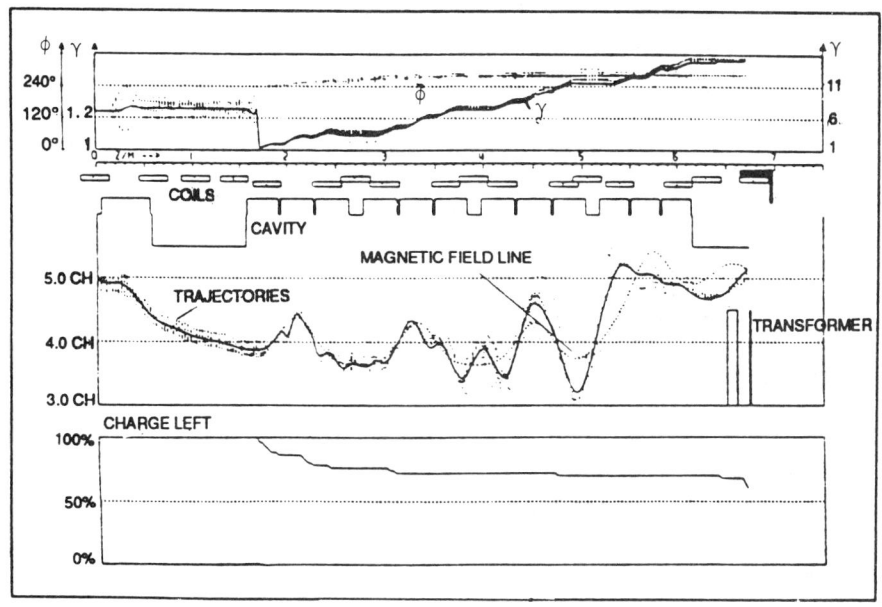

Figure 6. Example output of WAKTRACK.

REFERENCES

1. G.-A. Voss and Th. Weiland, DESY M-82-10, April 1982.
2. G.-A. Voss and Th. Weiland, Proc. ECFA-RAL Workshop, Oxford, September 1982, ECFA 83/68 and DESY 82-074, 1982.
3. W. Bialowons, H. Dehne, A. Febel, M. Leneke, H. Musfeldt, J. Rossbach, R. Rossmanith, G.-A. Voss, Th. Weiland, F. Willeke, Proc. 12th Intern. Conf. on High Energy Accelerators, Chicago and DESY M-83-27, August 1983.
4. W. Bailowons, H.D. Bremer, F.J. Decker, R. Klatt, H.C. Lewin, S. Ohsawa, G.-A. Voss, Th. Wiland, IEEE Trans. Nucl. Sc., Vol. NS 32-5, 2, pp. 3471-3475 and DESY M-85-08, 1985.
5. W. Bailowons, H.D. Bremer, F.J. Decker, M.V. Hartrott, H.C. Lewin, G.-A. Voss, Th. Weiland, P. Wilhelm, Xiao Chengde and K. Yokoya, Proc. Intern. Linear Accelerator Conference, SLAC (1986) and Symposium of Advanced Accelerator Concepts, Wisconsin (1986) and DESY M-86-07.
6. P. Schütt, Thesis, II. Institut für Experimentalphysik der Universität Hamburg (to be published, in German).
7. Th. Weiland, **NIM** 213, 13,21, 1983.
8. H. Euler, U. Hamm, A. Jacobus, J. Krüger, W. Müller, W.R. Novender, Th. Weiland, R. Wwinz, Archiv für Elektrotech. (AfE), Vol. 65 pp. 299 ff. (1982).
9. Th. Weiland, **NIM** 216, 329-348, (1983).
10. Th. Weiland and F. Willeke, Proc. 12th Intern. Conf. on High Energy Accelerators, Chicago, 1983, pp. 457-459 and following improved versions by G. Rodenz (on leave from Los Alamos), K. Yokoya (KEK), P. Schütt, Xiao Chengde (on leave from Univ. of Beijing) and R. Wanzenberg.
11. C.K. Birdsall and A. Langdon, *Plasma Physics via Computer Simulation*, McGraw-Hill, 1985.
12. S. Yu, SLAC/AP-34, September 1984.
13. G. Gisler, M.E. Jones, C.M. Snell, Bull. Am. Soc., **29**, (1984).
14. Th. Weiland, Proc. U.R.S.I. International Symposium in Electromagnetic Theory, Budapest, Hungary, August 1986, 537 ff. and DESY M-86-03 (April 1986).
15. Th. Weiland, CERN /ISR-TH/80-07, January 1980.
16. M.E. Jones, Los Alamos National Laboratory, private communication.
17. O. Buneman in *Relativistic Plasmas*, edited by O. Buneman and W. Pardo, p. 205, Benjamin, New York, 1968.
18. W. Bialowons, H.D. Bremer, F.J. Decker, H.C. Lewin, P. Schütt, G.-A. Voss, Th. Weiland, Xiao Chengde, Proc. Intern. Europhysics Conference on High Energy Physics, Uppsala, 1987.

MAGNETIC OPTICS DESIGN*
(A Designer's View of the Beam Transport Environment)

Edward A. Heighway
Los Alamos National Laboratory, MS-H808, Los Alamos, NM, 87545

ABSTRACT

This paper attempts to give a view of the beam transport and magnetic optics environment from the designer's perspective. It points to the many successes that have taken place in the last few years in the number and diversity of codes available and to the improvement in the physics content and capabilities of these codes. It also points to the lack of innovation in the way we access these codes and how we cope with their diversity; it pleads for an emphasis to be placed on the need for similar progress in the area of their human interface. As a lead to the community, a concept for a particular interface is presented.

INTRODUCTION

This Codes Workshop has produced several excellent papers describing the details of many powerful design codes. I will not, therefore, go into any details. I will not describe or laud any particular physics or mathematical approach. I will not state any preference for a given code or codes. Instead, I will attempt to describe the beam transport world from the designer's viewpoint. What is it in general terms that he or she needs? Is it a good environment, conducive to easily testing out new ideas, efficiently producing sound designs? If not, what can we do to improve that environment?

THE DESIGNER

Who is this elusive designer? What does he do, what does he need? Well, he's that fellow down the hall with the mounds of computer output on his desk (Fig. 1). He generally has project managers and engineers circling like vultures about him, waiting to pick his bones dry for the latest error analysis, the latest layout, because without that input, the building floor plan cannot be approved, the magnet construction drawings signed-off, the vendor briefing-package completed, et cetera, et cetera.

He is probably the same person who has exhausted the entire month's computing budget in the first ten days of the month and who will

---

*Work supported by Los Alamos National Laboratory Institutional Supporting Research, under the auspices of the United States Department of Energy.

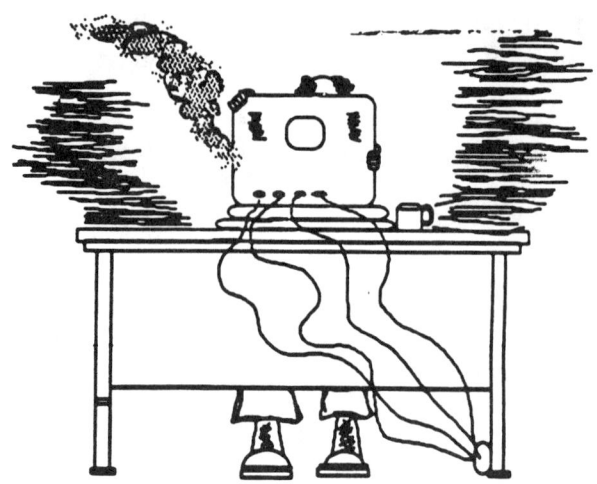

Figure 1. The Designer.

continue to do this every month no matter how many resources are allocated. So "code pushing" is an important part of his job, but on closer examination you will find that this is only a small part of the job and there are, and must be, much more important components.

The designer, more than anything else, is an expert. He has won that expertise through long experience, trial and error, and knowing what will and will not work as a starting point toward a particular solution. Much of that experience has been obtained at the keyboard (or if he is old enough, the keypunch) using the design tools to test ideas, to determine tradeoffs between alternate concepts. Just because you have run MARYLIE, read the manual, or even derived a few Lie algebraic functions does not qualify you as an expert. It may indeed be a starting point to adding MARYLIE to your designer's toolkit, but the expertise comes only with repeated use and total familiarity. Having a lathe in your basement does not make you a precision machinist.

THE DESIGN PROCESS

Let us examine the design process for a moment, just to get a better appreciation for the design environment. Often the first step is conceptual. Let us say that a particular beam dynamics problem requires an achromatic bending system, for example. What do we do first? Well we might remember that there is a standard solution using two dipoles and a central quadrupole and simply adapt that solution with the appropriate parameter changes to suit our conditions.

But then the physicist points out that, in this case, the system has also to be isochronous and our dipole-quadrupole-dipole solution will not meet specifications. We may or may not be able to find another remembered solution; if not, we have to be inventive or intuitive. For example, we might know from experience that a system with mirror symmetry about its midpoint is achromatic if there is no angular dispersion at the midpoint (or alternatively no positional dispersion) and isochronous if there is a momentum focus at the midpoint.

At this stage, we enter the trial and error scene. We start with several alternative configurations that are each likely to produce the beam properties required, optimize them a little to get close to a solution (if possible), and then look at geometry of the systems to see which is preferred. If we are fortunate to have a number of acceptable solutions, we can enter the tradeoff study phase.

Then we learn from our company's marketing department of some other constraints and have to add these to the global optimization of our solution. That is, the whole system might have to fit into an existing 10- by 10-foot room in a hospital basement, cost less than a hundred thousand dollars, and use less than 50 kW of prime power. Ouch! So now we must look at the parameter space (magnet size, complexity, power, etc.) for each of the potential solutions we have at hand to see if those parameters can be improved while still maintaining a good beam dynamics solution.

Once we have made our selection of the best of our options, we enter the last phase — the detailed design and specification stage. We must determine the exact specifications, the field strengths, the magnet locations, the pole-face rotations, etc; we must determine the tolerances allowed for each of these physical parameters. If the tolerance specifications are too stringent, the cost may become exorbitant; if the specifications are not stringent enough, the beam dynamics quality may be lost. The next step in the design process includes detailed layouts and cost analyses. With the successful completions of these tasks, we can commit to production.

These steps are not, of course, a unique prescription for design success, but at least they identify clearly that the design process is highly iterative and needs quick and easy concept evaluation. The tools at hand have to allow us to accomplish this.

THE DESIGNER'S TOOLS

Designers each have their favorite codes and have become highly skilled at using them. However, few have a solid repertoire of several codes; until recently, this was probably okay. There were several codes around and each was as good as the other, some perhaps adapted to a particular application like circular machines or linac design or space charge. But recently there have been breakthroughs, especially the improvements in modeling systems to high order.

TRANSPORT is now solidly third order and MARYLIE is no longer just a twinkle in the eye of Alex Dragt but a robust design code with recently added fitting capability and a solid stable of transport element descriptions to third order. MARYLIE has also added 2D space-charge capability to third order, through the strengths of its cousin CHARLIE. Alex and his team are forging ahead and are soon to release the fifth-order version of the code.

The Europeans have not been idle; Hermann Wollnik and his co-workers have their operational third-order code GIOS with a large list of elements and first-order 2D space charge. This is augmented by a new member of the Giessen family, COSY 5.0, which is now operational to fifth order.

Even though the ink on COSY 5.0 is hardly dry, a new COSY Infinity is in production from the Giessen group, using the COSY 5.0 input-output shell but with a differential algebra core and has the potential for operation to any order only limited by the number crunching power available.

TRACE 3-D is probably the code that has tried to make the human interface as easy as possible with its interactive operational mode, clear graphics, and powerful fitting and optimization capability. It continues to strengthen as a powerful design tool using first-order envelope equations that include a representation of 3D linear space charge.

RAY-TRACE, Stan Kowalski's code, continues to evolve and remains the workhorse of the spectrometer designer's toolkit, being used when the concept is solidly in place and detailed ray-tracing through mapped or analytic fields is required. The code now allows expression of the fields and integration of particle trajectories to fifth order.

Particularly exciting is the recent work of Martin Berz in developing the tool he calls differential algebra, which will probably revolutionize the core representation of the physics in our design codes.

Some have advocated "one code to rule them all" — a single code that can do everything (Fig. 2). I cannot resist presenting a slightly irrelevant but nonetheless delightful "Far Side, by Gary Larsen" image of a single-tool user at work (Fig. 3). Imagine how you might cope in your basement projects if you were restricted to a single tool.

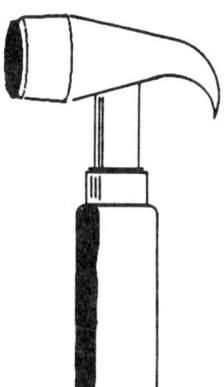

Fig. 2. The One Tool for Us All.

Fig. 3. The Tool User at Work.

[Permission to use Gary Larson's THE FAR SIDE cartoon granted (1988) by Chronicle Features, 870 Market St., San Francisco, CA 94102.]

I suspect, know in fact, that Dave Carey, Ken Crandall, Alex Dragt, Chris Iselin, Stan Kowalski, Hermann Wollnik and many others, will not stop adding to their codes new features that will allow their particular progeny to be more and more powerful and perhaps more and more alike. My personal belief is that it is unlikely that a single code will emerge to totally dominate the market. The innovators of this world will continue to think up new wrinkles, new angles, new applications that require special tools.

The trick is to make these tools easy to use, to produce a versatile toolbox (Fig. 4) that allows many tactics and provides a mechanism to be able easily to become skilled in the use of those tools.

The range of options in the tools we can choose has become very large and to the novitiate it is akin to a jungle (Fig. 5). We can use matrix algebra, Lie algebra, differential algebra, or integration methods. There was first order, then third order and now fifth order; the mind boggles as infinite order looms large. We can choose particle equations, envelope equations, linear space charge, 2D and 3D, third-order space charge, or distributions such as uniform, binomial, lorentzian, gaussian, 2D, 4D, and 6D. We have some sympathy for the new designer who just joined the team. No longer can we just give him a data file and say "Run this a few times."

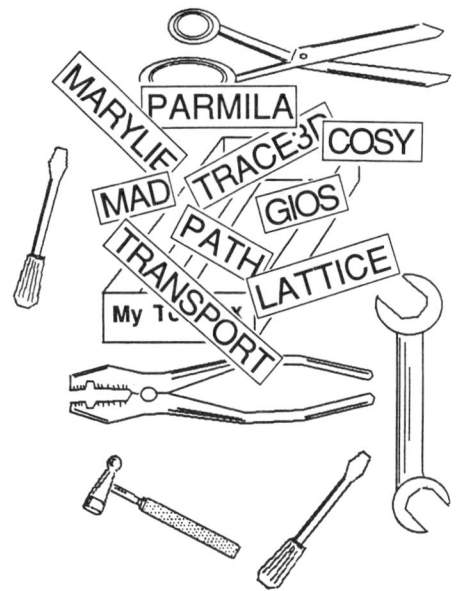

Fig. 4. The Designer's Toolkit.

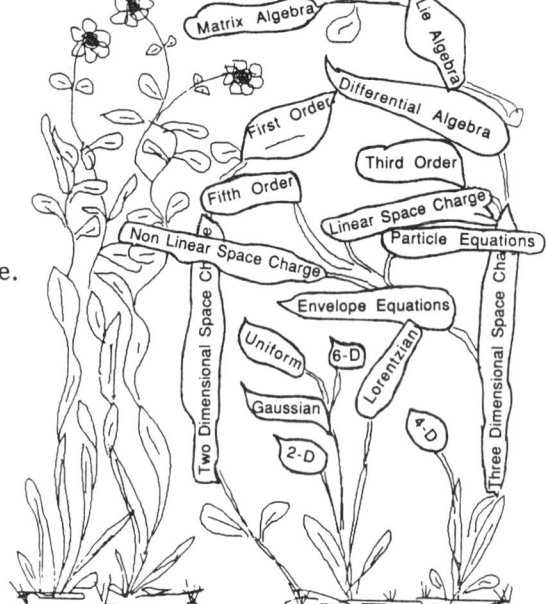

Fig. 5. It's a Jungle out there.

THE HUMAN INTERFACE

What does the interface to these codes look like? Let us take a peek at a few of the major players. Figure 6 shows us the data format for input to some major design tools in use around the world for beam transport

TRANSPORT:

5. 1.00 2.06 3.81 "Q1"

TRACE3D:

NT(1)=3, A(1,1)=4.2 , 48.5,

GIOS, COSY:

M   Q   0.52 1.33 1.00

MARYLIE:

Q1 QUAD
0.831   3.21   1.0   1.0

MAD:

Q1: QUADRUPOLE, L=0.75, K1=0.05

Fig. 6. Sample Data Files.

design. The list is not exhaustive but includes TRANSPORT, MARYLIE, GIOS (and COSY), TRACE 3-D and MAD; although the flavor in each is different, there is an unmistakable common thread: the input is a numerically coded sequential list describing beamline elements in a language that the user has to master and that has little or no immediate or obvious translation to a physical picture of what the system looks like or what the design problem is. The graphics of these codes is similarly diverse.

This diversity has two major drawbacks. Firstly, the new player on the block, the graduate student or new hire that we throw into the design team, is snowed under in an attempt to grapple with the several languages and will either spend considerable time translating decks from one code to another or, at worst, will restrict himself to one or two codes and give up on the rest. Secondly, the old players, the experienced designers, have become very set in their ways over the years and have invested much time to become skilled in the use of their favorite tools. It is hard for them to throw themselves wholeheartedly into use of a new code and a new set of beam element descriptions.

The interface to beam transport design has remained virtually unchanged for more than twenty years. While all around us the computer world has been in revolution, we have been concentrating on the physics and indeed making great strides. Yet now those strides themselves are difficult to exploit because the human interface has been neglected and fallen behind the physics progress. In an age where the menu/windowing-style interface has all but made instruction manuals obsolete, we in the beam transport design community are still struggling with an interface that has evolved extremely little from what was once a deck of punched cards.

A PLEA FOR INTERFACE DEVELOPMENT

Let us recap the questions that the design tools must answer and think of the human interface as they are asked. Does some concept work? Is it easy to try new concepts? What are the bounds for a particular type of solution? Are there tradeoffs between parameters in a given type of solution? Should the design be to first order, third order? How easy is it to choose between different beam transport models? Is it easy to specify the geometry? (See Fig. 7.) Is it easy to change it? Is it easy to see it? Does the beamline fit in the building? Is it easy to find out if it doesn't? (See Fig. 8.) Can the geometry be constrained to fit inside the building? How is the beam defined? Is it easy to choose between the several options? (See Fig. 9.) Does the beam fit in the beam pipe? (See Fig. 10.) Is the beam size big in the right places? Is it easy to find out? Can the beam be constrained to fit inside the pipe?

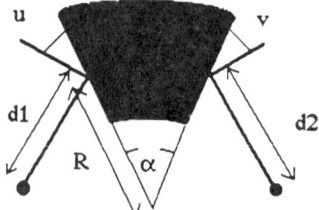

Fig. 7. Can we easily define the Geometry?

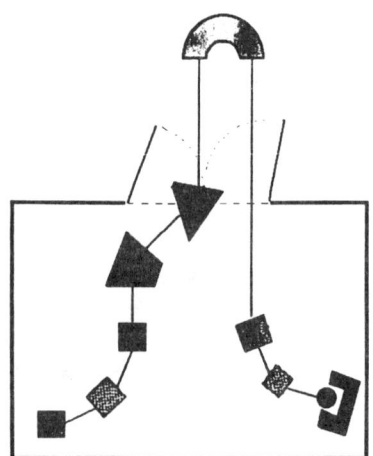

Fig. 8. Does it Fit in the Building?

The answers to these questions in today's environment is typically *no, maybe, sometimes, not easily*, etc., when really what we are after are answers like *yes, definitely, always, easily*, etc.

What now follows is not a description of an existing interface (although I wish it were) but a description of what I believe it could be. What I will describe is not perfect and complete but at least a considered

 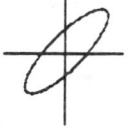

Fig. 9. How Hard is it to define the Beam?

Distributions
2-D, 4-D, 6-D
Uniform, Gaussian
etc.

Envelopes
σ matrices
Twiss parameters

Fig.10. Does the Beam Fit the Pipe?

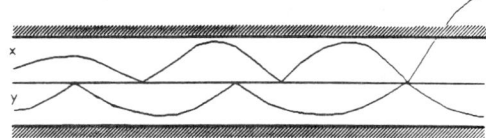

foundation for the specifications of a tool that the community certainly needs and one I hope the community will spend the time to invest in. Right or wrong, there has been little work on interface development in comparison with that on the mathematics and physics. Certainly our investment in the core physics and math has paid off and continues to pay off handsomely. I am sure an investment in the human interface will afford similar dividends.

AN INTERFACE PROPOSAL

Let us begin by envisaging our clean start-up interface screen as shown in Fig. 11. On the right, we see a graphic menu of selectable elements including (starting at the top) the beam description, a bending magnet, a quadrupole magnet, a sextupole, an octupole, a drift length, an output request, a fit request, and a constraint definition. It is envisioned that this list could be scrolled up or down to show elements beyond the limited set on the screen.

Command windows such as EDIT, RUN, DISPLAY, and FILE appear at the top of the screen and clicking on these windows reveals the respective blinds with the list of commands selectable below them as shown in Fig. 11. These will be illustrated in due course.

First, let us imagine how we might define a beamline problem. We click on the menu boxes with labels such as BEAM, DRIFT, BEND, DRIFT, QUAD, and CONSTRAINT. A chain of elements representing a sequence of beamline elements is then displayed in the body of the screen (Fig. 12). The interface software automatically labels the elements with a default name and number shown below the elements. These names are editable by clicking on the name in the usual way as shown for DRIFT2, where a vertical cursor is shown. Any element may be selected by clicking

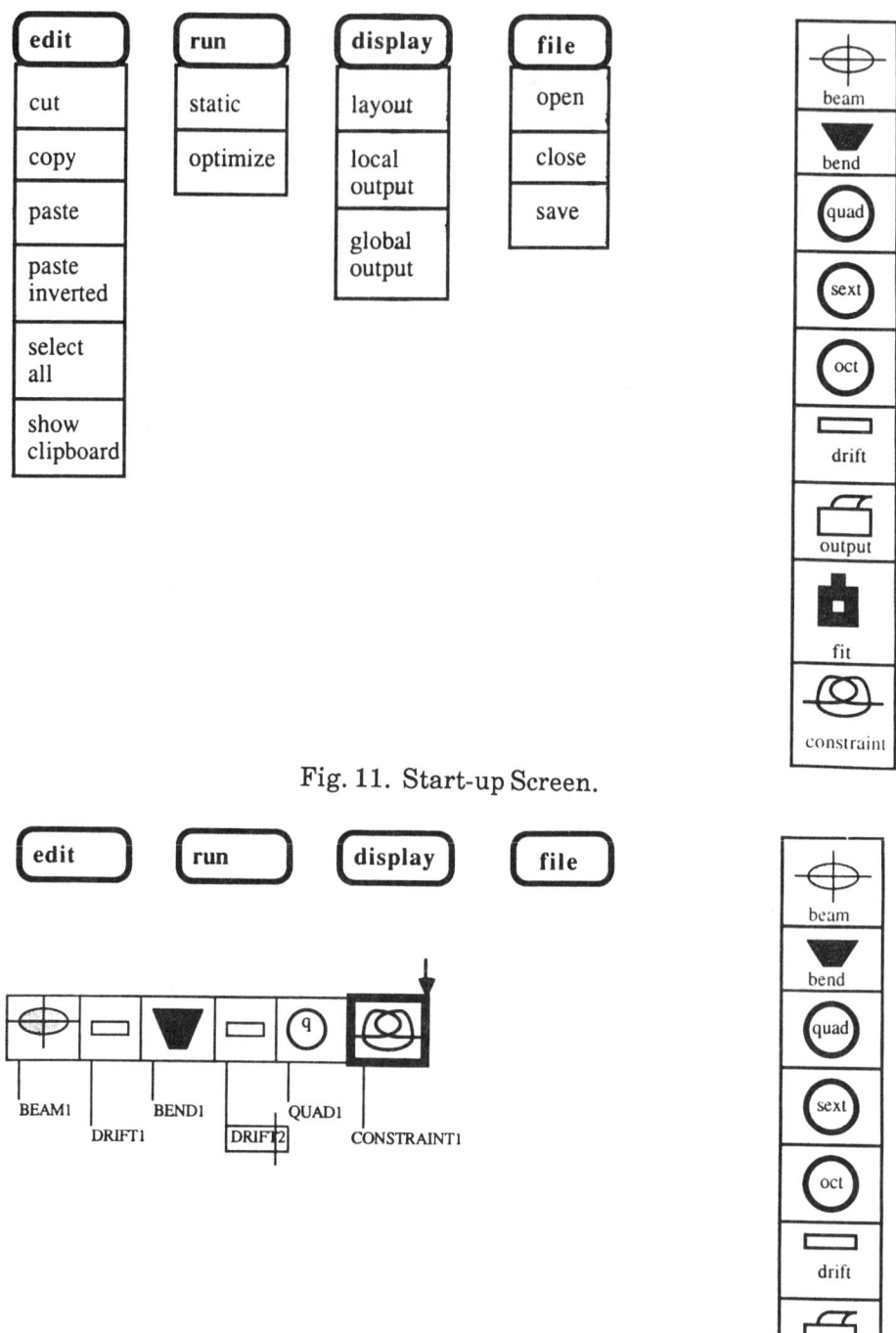

Fig. 11. Start-up Screen.

Fig. 12. Elements Selected to Form a Beamline Sequence.

on it, whereupon it becomes highlighted with a bold border as shown for the CONSTRAINT element. A backspace will delete a selected element. The arrow shows the position at which any further element selected from the menu will be added or inserted. The arrow can be moved by clicking on a new location in the element chain.

Now use the usual trick of shift and click to select the three elements BEND1, DRIFT2 and QUAD1 as shown in Fig. 13. A selection of the

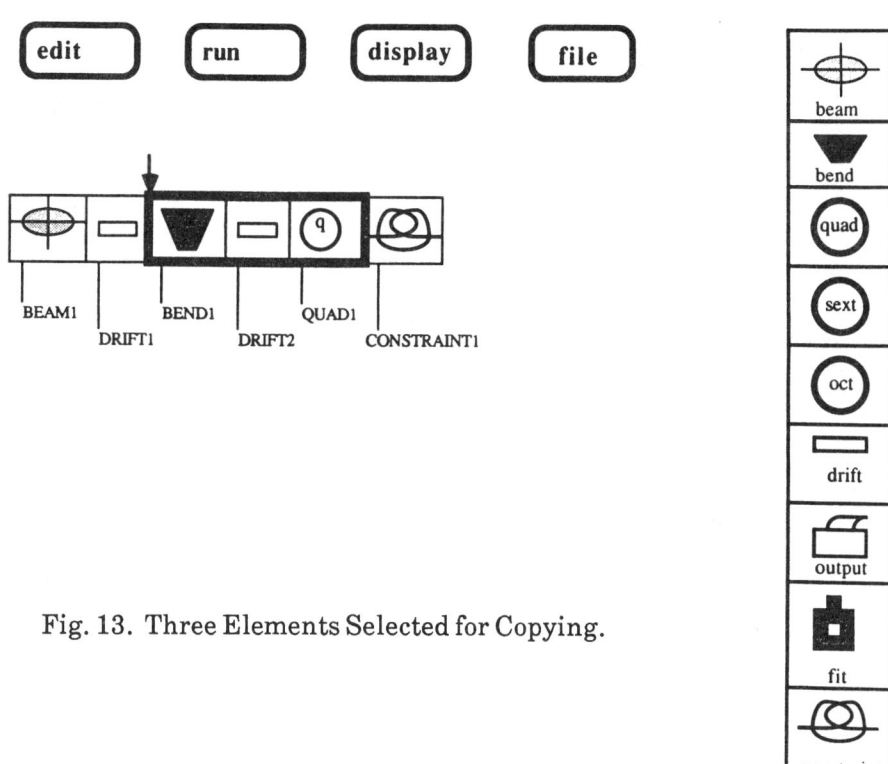

Fig. 13. Three Elements Selected for Copying.

COPY command from the edit blind copies these elements to temporary storage. A click on the end of the last element, CONSTRAINT1, moves the insertion pointer to the end of the element list (Fig. 14) and a selection of the PASTE INVERTED command from the edit blind inserts the three selected elements at the end of the element list in reverse order, resulting in the configuration of Fig. 15. These three new elements are renamed by editing their name windows (to allow them to be different from the originals); and, with the addition of a few more elements from the menu, we obtain the complete system shown in Fig. 16.

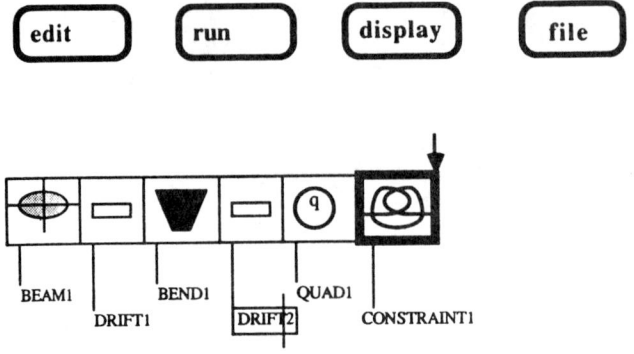

Fig. 14. Pointer Moved to End of Sequence.

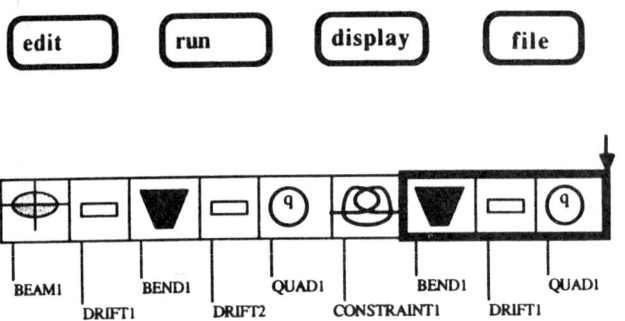

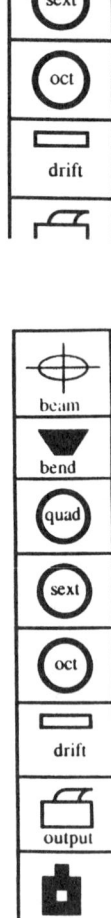

Fig. 15. Three Copied Elements Pasted Inverted at End of Sequence

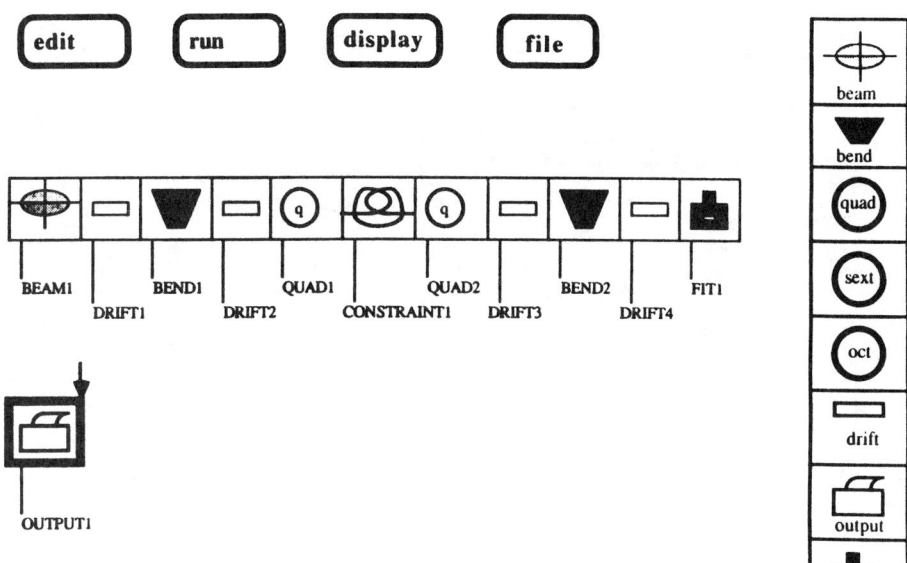

Fig. 16. Rest of Elements Added to Complete the Beamline.

At this point we have a logical beamline system identifying a sequence of particular types of elements. We have as yet no detailed description of the elements themselves, i.e., the BEND element, for example, is undefined as to bend field, field index, or length.

A double click on the BEND2 element now opens the element description menu for that element as shown in Fig. 17. The menu contains a parameter list for that particular element type; for a BEND, there are the usual radius, angle, field, edge angles, etc. The boxes can be filled by simply selecting a box with a click, typing into the space, and ending with a carriage return in the usual way. The element name is editable as well as the parameters. To return to the main menu one would simply click in the usual return box in the top left-hand corner of the BEND menu. An entry of an asterisk in any parameter box opens a special window for that parameter that allows the selected parameter to be a variable, as shown in Fig. 18. Here we define the starting value for the exit edge angle as well as minimum and maximum values acceptable for the parameter during any variation. There is a box that allows us temporarily to deselect or to reselect the variability of the parameter at any time. Clicking on the return boxes gets us back to the main menu.

Now we double click on the OUTPUT1 element to see the parameter menu of Fig. 19. We can see that the type of output wanted at that particular location in the beamline system can be defined. We can select a matrix output of some order, the generating functions of some order (applicable to MARYLIE for example). The beam description can also be

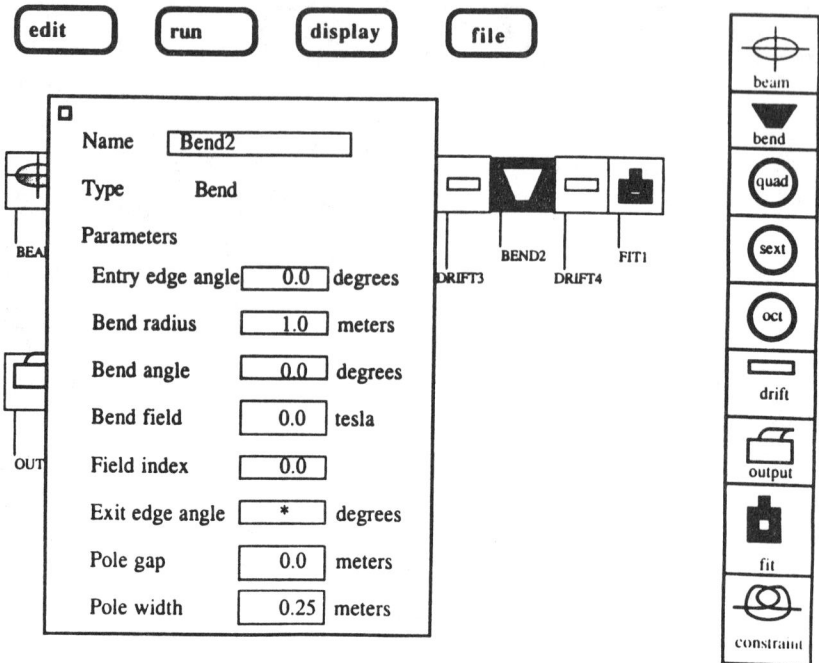

Fig. 17. Double Click on BEND Element Opens Menu.

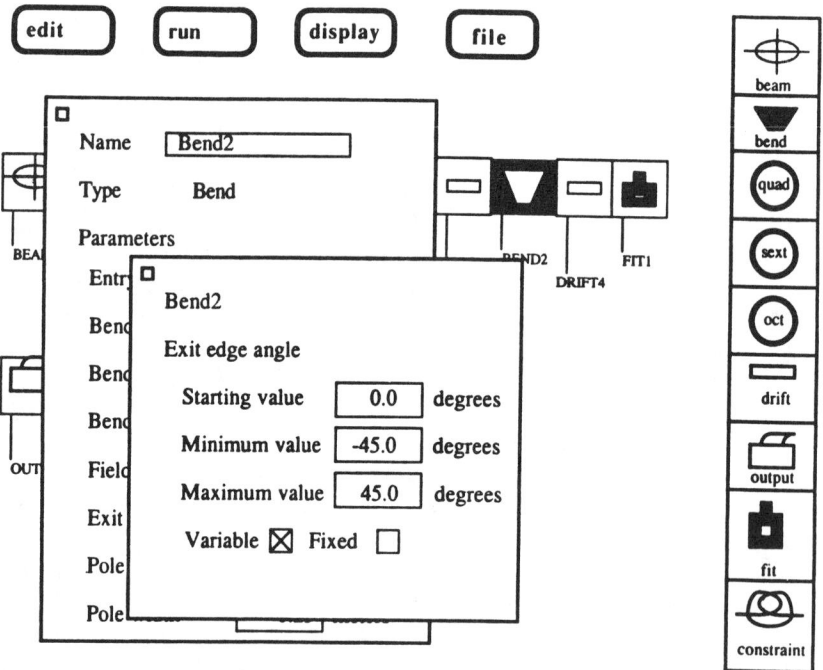

Fig. 18. Asterisk in Parameter Window Opens Menu for Variable.

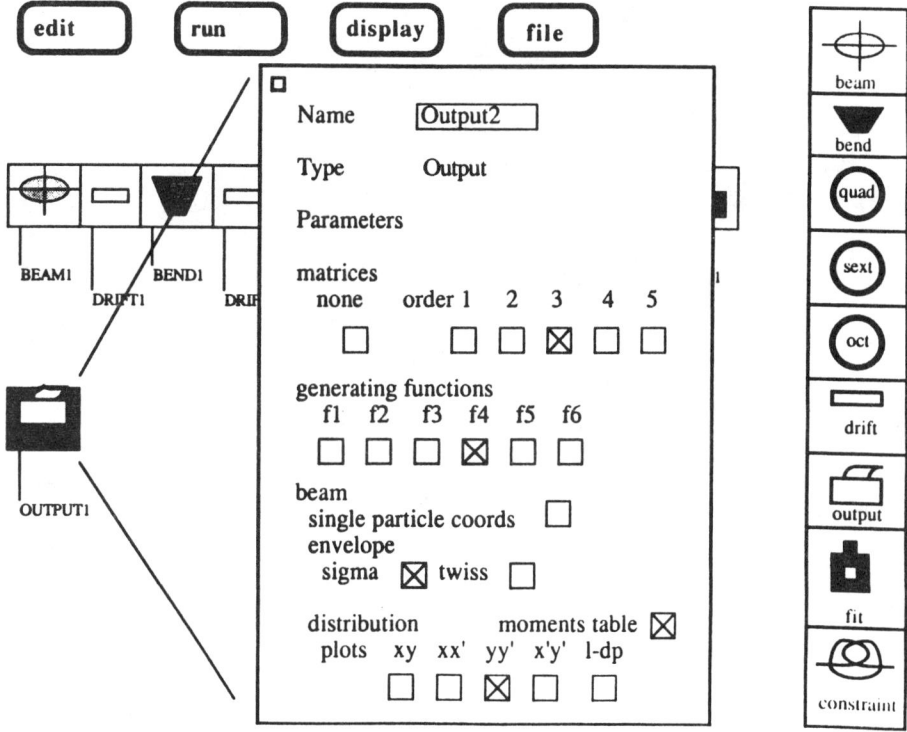

Fig. 19. Double Click on OUTPUT Element Opens Menu.

requested in the form of either particle coordinates or envelope description (twiss parameters or sigma matrices); if a distribution is available, we could elect to plot some phase-space projection. Other options can be imagined.

Double clicking open the FIT1 element shows the window menu of Fig. 20. At this location, we want the beamline or beam to have some particular properties. Let us imagine we select to fit matrix elements, then we open the menu window of Fig. 21. We simply fill in the matrix table by specifying the indices for the matrix element and the value requested. The table is scrollable (up and down) and can be edited in the usual way by click and drag selection, backspace erasure, etc. First-order elements will need two indices, second order needs three, etc. Illegal elements will be ignored.

The last element menu we will look at is for the CONSTRAINT1 element and is shown in Fig. 22. There may be other constraints imaginable, but the common ones are the floor coordinates and beam direction. Selection of either one would present a usual sub-menu allowing the coordinates or direction to be defined in a way similar to what we have already seen.

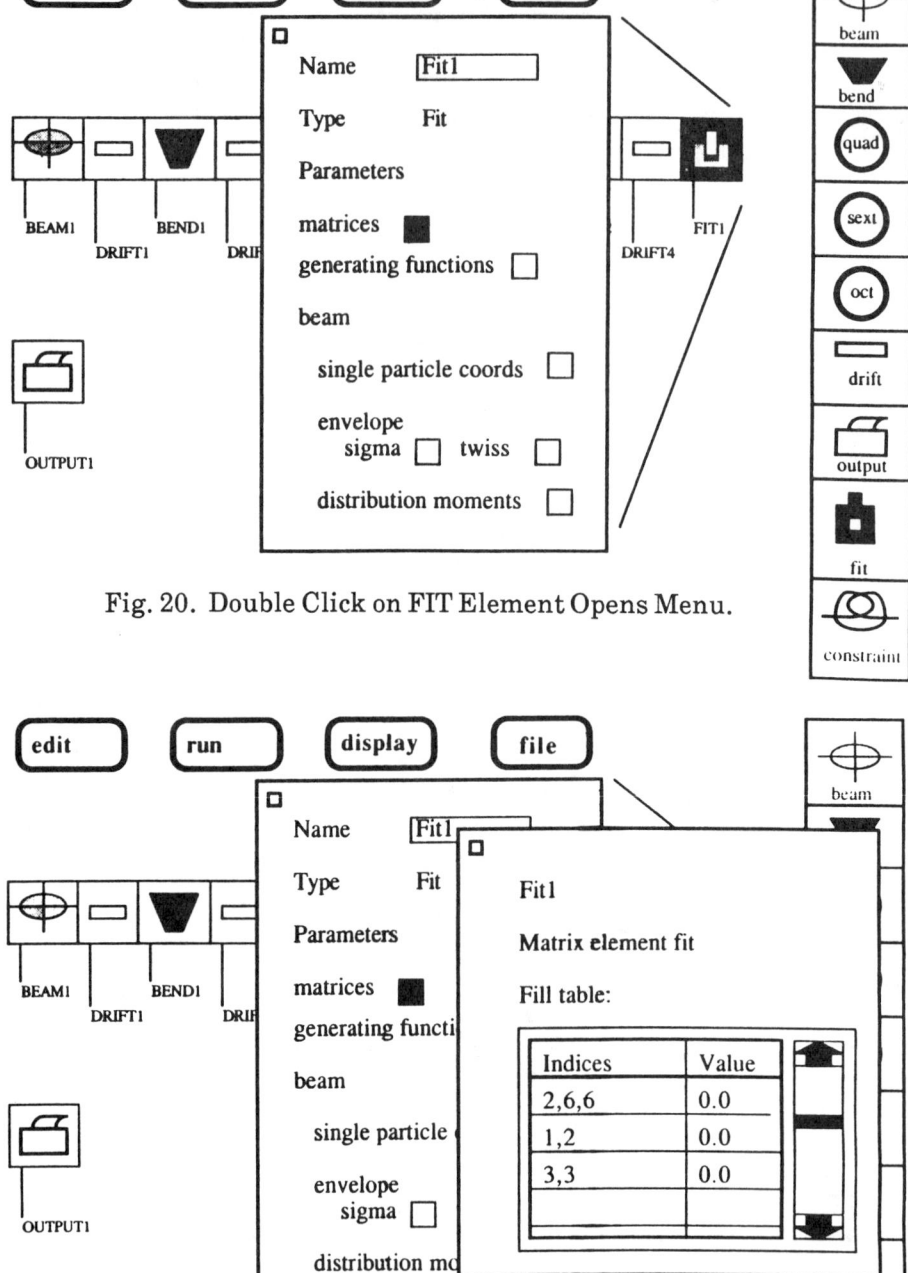

Fig. 20. Double Click on FIT Element Opens Menu.

Fig. 21. Click on MATRICES Box Opens Menu.

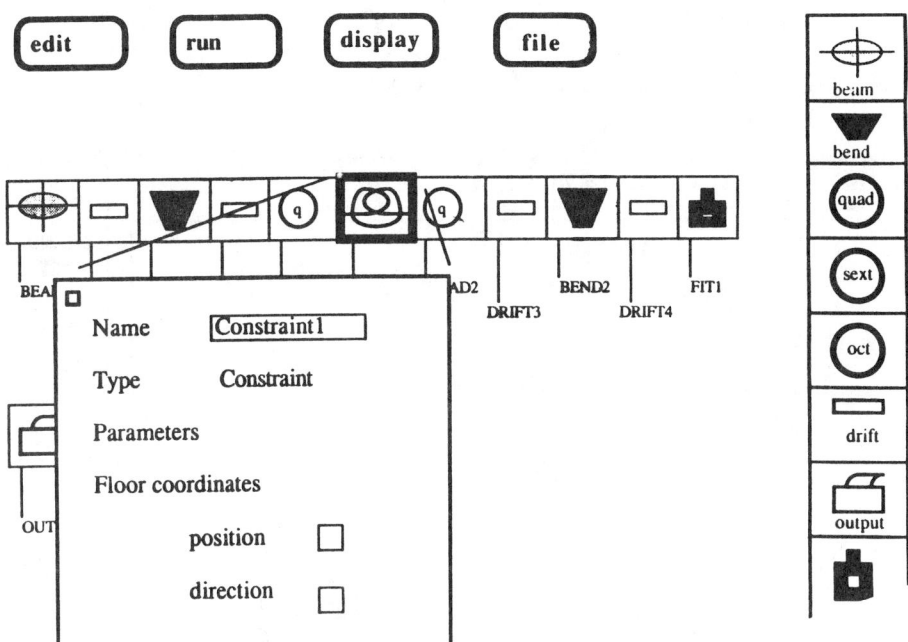

Fig. 22. Double Click on CONSTRAINT Element Opens Menu.

Now let us look at the main menu blind under the DISPLAY header and select the command LAYOUT. This option allows us to have a peek at the geometry of the system we have defined to make sure it is roughly the shape we expected. Figure 23 shows the resulting menu. The mini-window on the right allows us to select some subset of the system for either

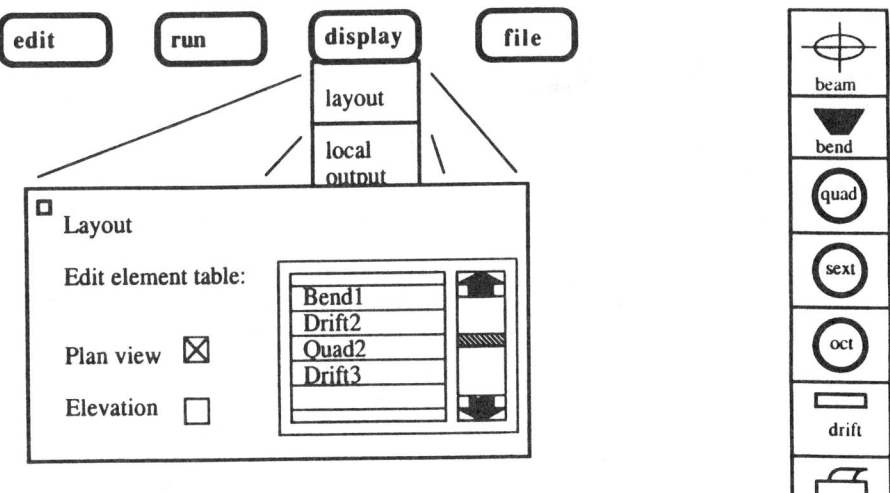

Fig. 23. Selection of LAYOUT Command from DISPLAY Blind.

PLAN or ELEVATION view. Selection of the plan view results in the scaled display of Fig. 24. We can elect to print the whole display or click and drag to select some portion for printing. Zoom in and out options can also be imagined.

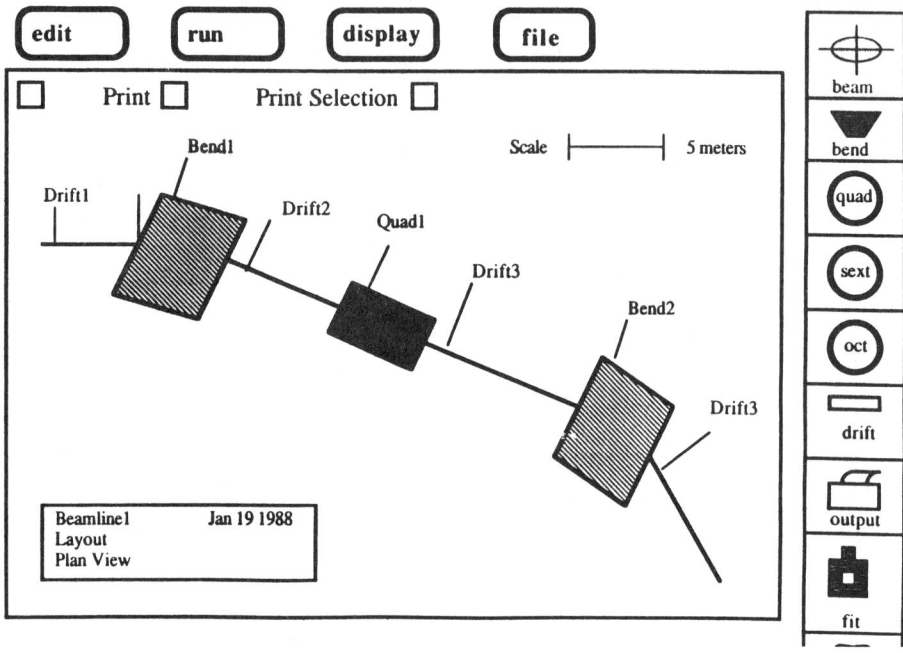

Fig. 24. Click on PLAN VIEW Box Displays Plan Layout of Selected Elements.

Selection of the LOCAL OUTPUT command, again from the DISPLAY blind, lets us have a look at what output is available along the beamline (it is assumed that a run has been made and that data is available for display). This display is shown in Fig. 25 and there may be text or graphics or nothing at all, depending on the OUTPUT elements that were requested. It may also be that our interface implementation for some particular code dumps some quantities to output or graphics by default. For example we might elect to always dump the first-order R and sigma matrices. A click on the tick, as instructed, reveals the OUTPUT available for that element (Fig. 26). The usual scrolling and printout of all or selected portions of the display are available. Figure 27 visualizes how a graphic output might look, complete with options to pan, zoom, and page through several graphics frames that might be available at that element.

Now we look at the global output command on the DISPLAY blind (Fig. 28), where selection of the envelope option results in the display of

199

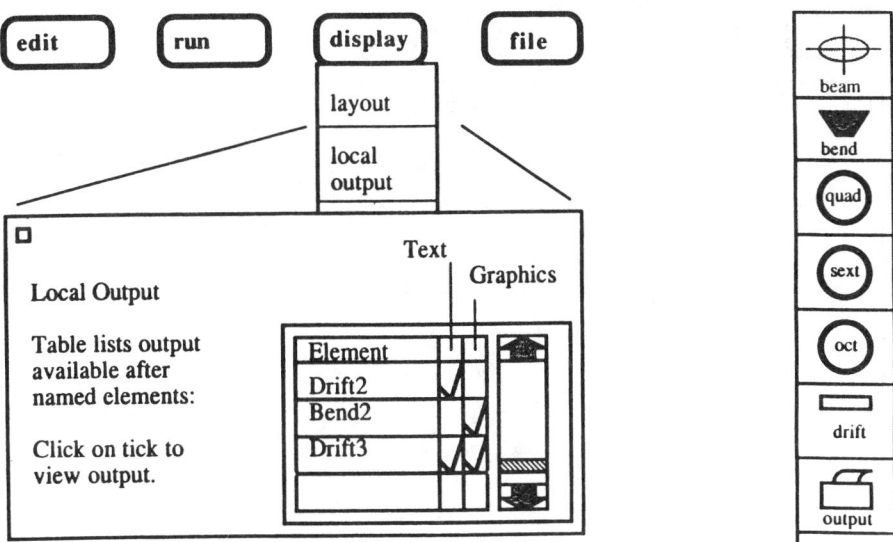

Fig. 25. Selection of LOCAL OUTPUT Command from DISPLAY Blind.

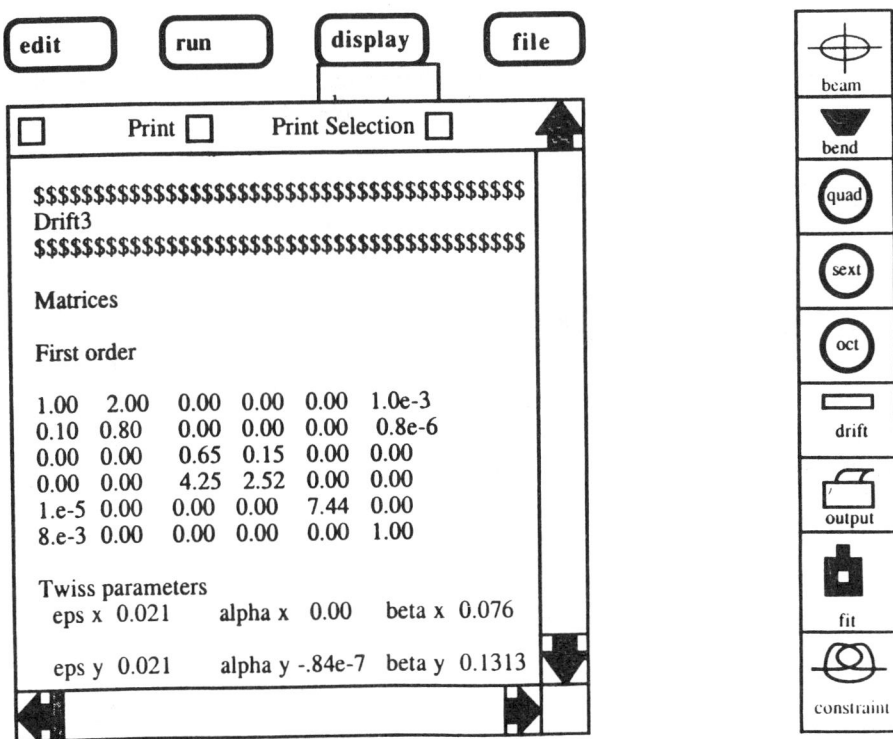

Fig. 26. Click on TEXT Box Displays Text Output for Selected Elements.

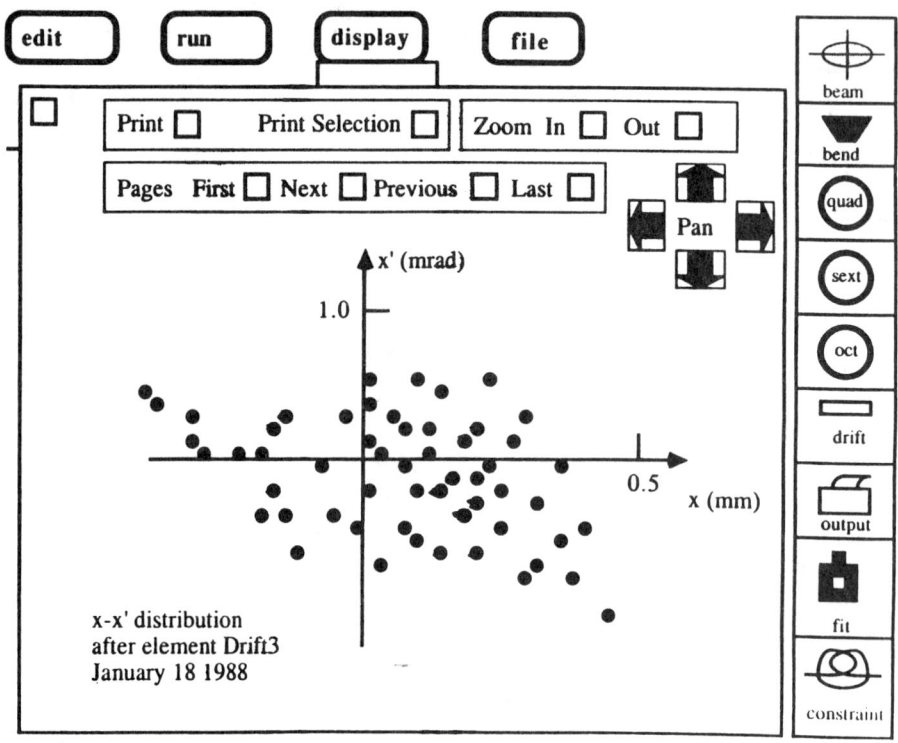

Fig. 27. Click on GRAPHICS Box Displays Graphics Available for the Selected Elements.

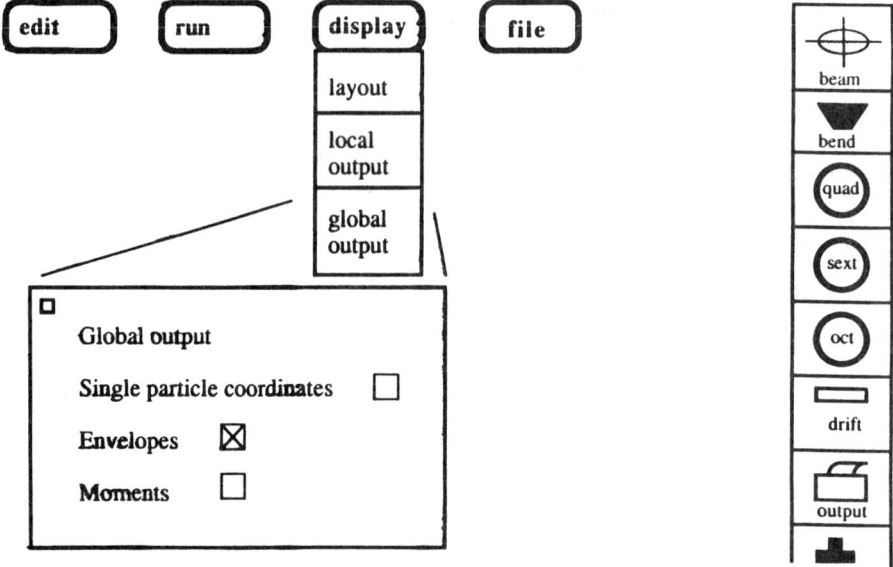

Fig. 28. Selection of GLOBAL OUTPUT Command from DISPLAY Blind.

Fig. 29. The vertical and horizontal envelopes and the apertures of the components in the beamline are shown.

As the last of our illustrations, we can look at the RUN blind, which allows two commands: one for a static or one for an optimization run (Figs. 30 and 31, respectively). Here we can select the code appropriate to the problem at hand. In the optimizing case, we can select to edit the variables or constraints or the fits before launching the job. We can also select the number of iterations and how often we print out information. Perhaps we could also select from a menu of available optimization schemes, if desirable.

The only blind we have not looked at is the FILE blind, and it should be easy by now to imagine what commands lie there. Here, as we would expect, we can select to store or retrieve data sets or output text and graphics files or quit the interface program itself.

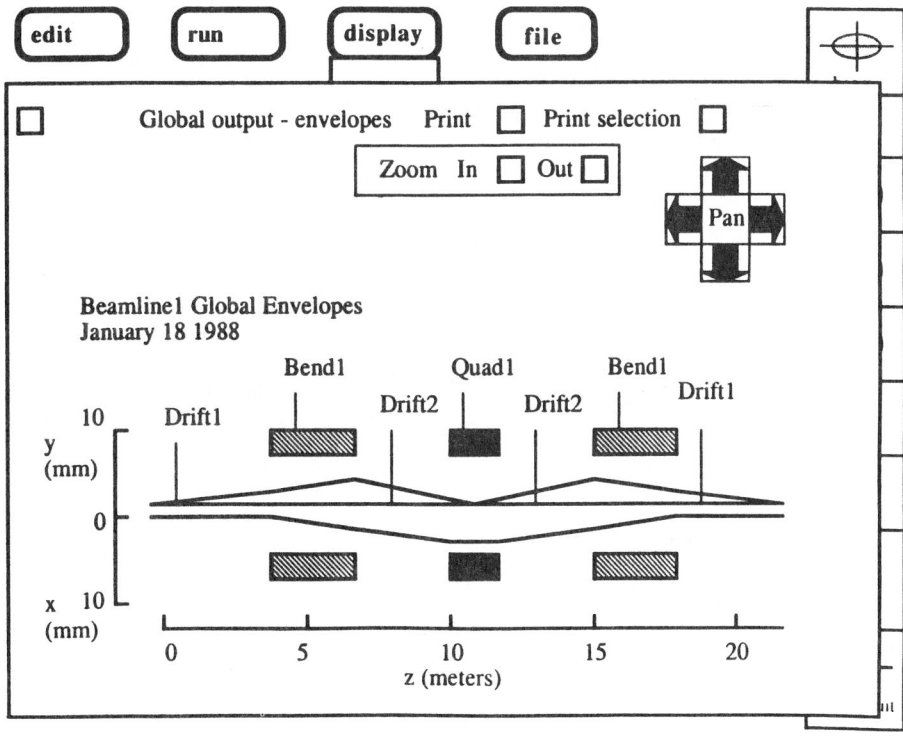

Fig. 29. Click on ENVELOPES Box Displays Envelope Plots for the Selected Elements.

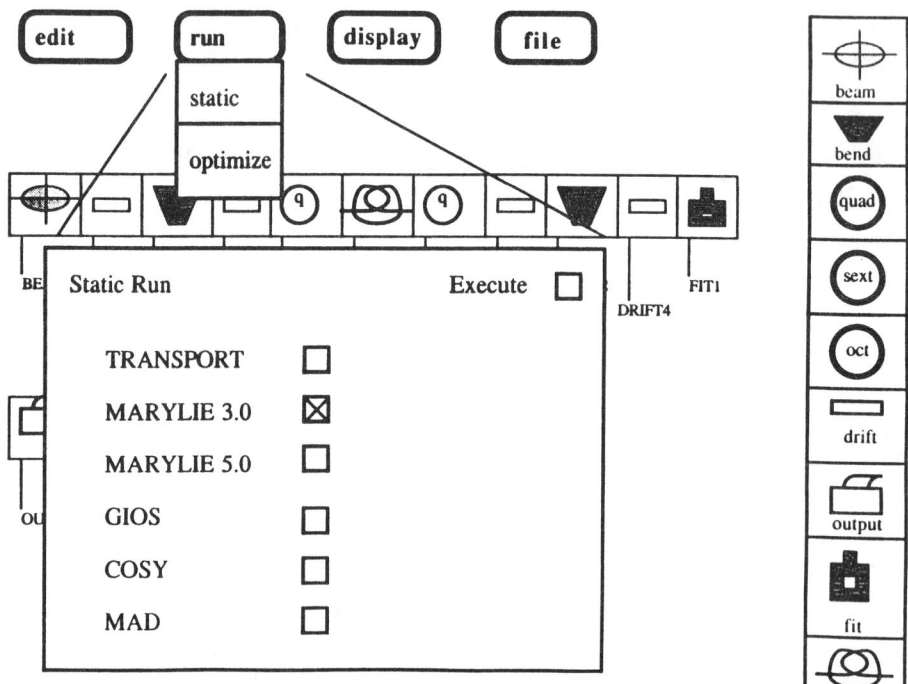

Fig. 30. Selection of STATIC Command from RUN Blind.

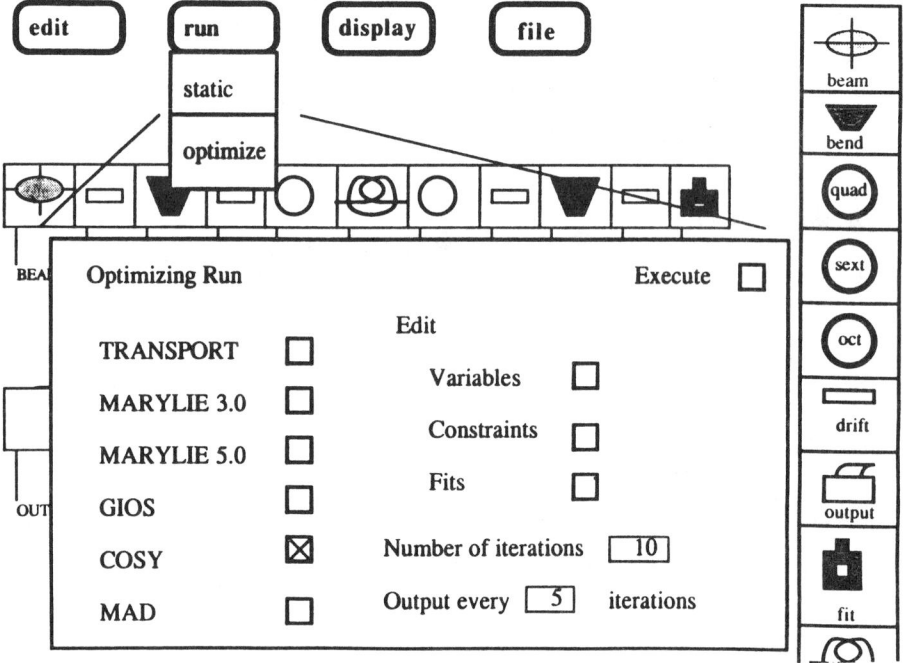

Fig. 31. Selection of OPTIMIZE Command from RUN Blind.

PARTING SHOT

At this point, I choose to end the guided tour. The system I have imagined is no doubt imperfect, but the flavor, I believe, is distinct and the goal, I hope, is clear.

Let me summarize what I wanted to say. Firstly, at this time, I think that the beam transport designer's world is richer and probably evolving faster than at any time since Karl Brown first put finger to keypunch. I personally find that very exciting. Secondly, this new-found wealth, while primarily in the representation of systems to high order, even infinite order, also extends to a potential for representation of even three-dimensional space charge to very high order. But this new wealth has its own problems. The codes are several and varied; and, although the laudable attempt to make the so-called MAD input format an international standard has rightly won many followers (and I believe TRANSPORT, MARYLIE and GIOS have various translators to MAD), the attempt only skims the surface of the iceberg. The revolution we see in the personal computer world today should be one that we in the beam transport design world should welcome and join.

Lastly, I have attempted to put together an example of what an interface should be. This was not meant to be definitive and is only a brief embodiment of a cry from the soul of an aging designer in this changing world. I hope it serves as a stimulant to others in the field, and if only a few groups take up the challenge, this paper will have been worthwhile.

This may not have been physics, but this too is reality!

BEAM DYNAMICS OF THE RF ELECTRON GUN OF THE BNL ACCELERATOR TEST FACILITY

K.T. McDonald
Princeton University, Princeton, NJ 08544

INTRODUCTION

The Brookhaven Accelerator Test Facility (ATF) is a 50-MeV electron linac designed to produce electron bunches which can be synchronized with the picosecond pulse of a 100-GWatt CO_2 laser. This facility will be used to study the acceleration of electrons by the laser-grating technique[1] as well as by an inverse free-electron laser,[2] and will also provide a picosecond source of x-rays via nonlinear Compton scattering.[3] Figure 1 is a block diagram of the linac and laser components. Here we report on the design of the electron gun which will provide rf bunches of up to 10^{10} electrons synchronized with the laser beam.

The gun is based on the design of Fraser et al.[4] who incorporate a photocathode in the end wall of an rf cavity which supports a strong standing-wave field. An ultrashort laser pulse ejects a bunch of electrons into the accelerating rf field. This concept offers many advantages:

- The electron bunch length is determined by the laser pulse width, which may be as narrow as a few picoseconds. This eliminates the need for a buncher section, and permits extremely narrow energy spreads for the accelerated beam.
- The electron bunch is synchronized with the laser pulse to picosecond accuracy.
- Very low emittance beams can be produced simply by decreasing the laser spot size at the photocathode.
- The rf cavity can support acceleration gradients of order 1 Mev/cm at the cathode, minimizing the space-charge growth of the emittance when the electrons are nonrelativistic.
- A multicavity gun can deliver a beam of several MeV.
- The gun can operate at the same rf frequency as the linac (2856 MHz for the ATF).
- Optimum design of the rf cavity minimizes emittance growth due to nonlinear components of the transverse electric and magnetic fields.

In the design of the gun for the ATF, we maximize the beam brightness by use of the highest possible accelerating field at the photocathode and the shortest laser pulse. The design parameters are summarized in Table I, and a section through the gun is shown in Figure 2.

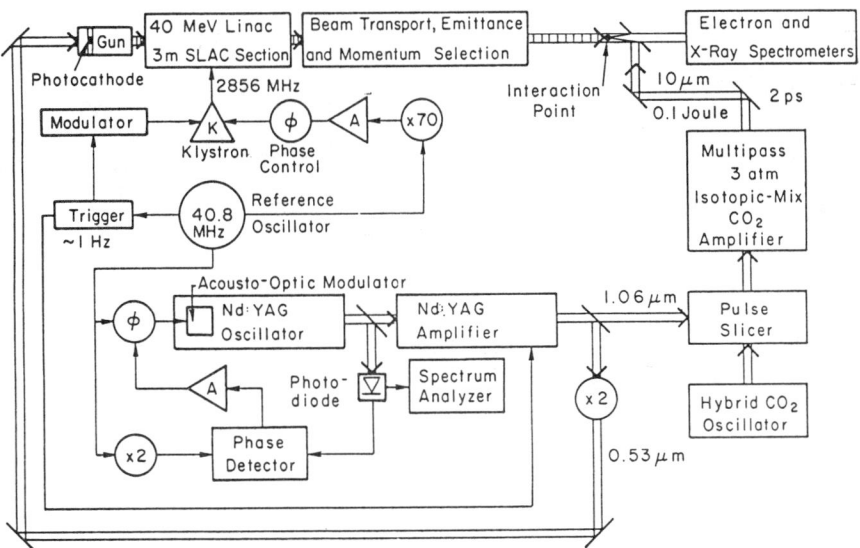

Figure 1. Block diagram of the Brookhaven Accelerator Test Facility.

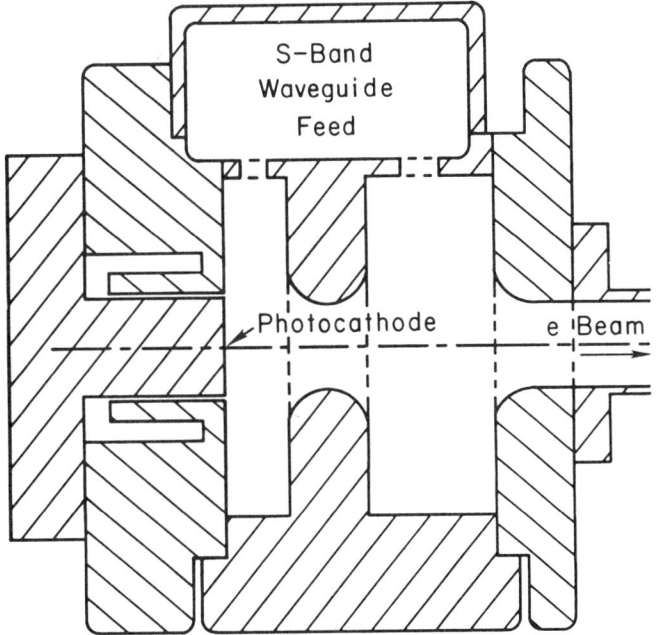

Figure 2. Section through the rf gun. Except for the waveguide feed the gun is axially symmetric. The $1\frac{1}{2}$-cells of the gun are 8 cm long.

Table I. Parameters of the ATF Gun.

RF Parameters	
RF Frequency (MHz)	2856
Cathode Cell length (cm)	2.625
Second Cell length (cm)	5.23
Cell diameter (cm)	8.31
Radius of aperture (cm)	1.0
Radius of nose (cm)	1.0
Field on cathode (MV/m)	100
Peak field on wall (MV/m)	106
RF Power (MWatt)	5.9
Cavity Q	12000
RF phase for laser pulse	67°
Final beam momentum (MeV/c)	4.7
Emittance Parameters	
Laser spot radius (σ in mm)	3
Laser pulse width (σ in psec)	2
Charge in bunch (nCoulomb)	1
ϵ_x† at cathode (mm-mrad)	3.5
$\Delta\epsilon_x$ due to self fields	6.2
$\Delta\epsilon_x$ due to rf fields	1.4
ϵ_x at exit	7.3
Beam energy spread (σ in keV)	17
Exit bunch length (σ in mm)	0.6
Exit bunch radius (σ in mm)	4.2
Exit beam angular divergence (σ in mrad)	28

† $\epsilon_x \equiv \dfrac{1}{mc}\sqrt{\langle x^2\rangle\langle p_x^2\rangle - \langle xp_x\rangle^2} = \sqrt{\langle x^2\rangle\langle \gamma^2\beta_x^2\rangle - \langle x\gamma\beta_x\rangle^2}.$

DESIGN CONSIDERATIONS

An ideal gun design would preserve the emittance of the electron bunch as it emerges from the cathode. In practice, two effects blow up this initial emittance. Electromagnetic interactions among the electrons (the space-charge effect) lead to a true increase in phase-space volume, while nonlinear radial dependence of the transverse components of external electromagnetic fields lead to an apparent growth of phase volume which cannot be later corrected with linear beam-optics.

The space-charge fields are only important relative to the strength of the accelerating field at the cathode. As the rf gun can have extremely large accelerating fields, only rather modest space-charge emittance growth will occur even for a relatively small bunch size. Once the electrons are relativistic, space-charge emittance growth largely ceases (as the electrons are far apart in the rest frame of the bunch). In an rf gun with a field strength of 1 MeV/cm, all space-charge growth occurs within 1 cm of the cathode. The field of the image of the bunch in the cathode plane serves to compress the bunch and reduces the space-charge growth by about 30 percent in the rf gun.

Thus, the primary design feature of the gun determined by consideration of space-charge emittance growth is that the accelerating field at the cathode be large. An optimal shape for the rf cavity of the gun is suggested by the criterion that the rf fields cause minimal nonlinear distortion of the phase space of the bunch.

From a Fourier analysis of a standing-wave rf field with circular symmetry, summarized in Appendix A, we find there is an ideal form for which the transverse electric and magnetic fields have linear radial dependence. A related form for static fields[5] was the basis for the design of the Los Alamos rf gun.[4] For the rf case it emerges that the length of a cell should be $\lambda/2$ where λ is the wavelength of the rf field, and that the fields in adjacent cells should be 180 deg. out of phase. This is the well-known result that a 'π-mode' structure causes the least distortion of the emittance of the beam.

The gun is made in a π-mode configuration by placing the cathode on a metal disk inserted at the midplane of one cell, so the first cavity is actually a half-cell, as shown in Figure 2. The ideal geometry for the rf case permits only this location for the cathode, unlike the static case which might favor a curved cathode surface.[5] The cathode cavity is followed by as many full cells as desired, subject to power limitations. An n-cell π-mode structure supports n modes whose frequencies may be rather close unless a side-coupling scheme is used. For the ATF we will use a $1\frac{1}{2}$-cell configuration for which the mode separation is about 2 MHz out of 2856.

A π-mode configuration alone does not assure linear dependence of the transverse fields. There is an ideal shape for the cavity walls, specified in Appendix A, which must extend to $r = \infty$. We find that a simple disk-and-washer construction provides sufficient approximation to the ideal shape so that rf-induced emittance growth will be less than that due to space-charge. A more complicated construction with rounded outer walls could reduce the power consumption, but would not improve the emittance performance.

The advantage of the π-mode configuration is maximized if the electron bunch crosses a cell boundary when the electric fields vanish. Then the deflection of the beam as the bunch nears an aperture is cancelled by the oppposite transverse fields encountered just beyond the entrance to the next cell. The laser pulse should strike the photocathode at an rf phase somewhat less than 90 deg. (where 90 deg. corresponds to maximum electric-field strength), so that the electrons leave the first cavity at a phase of 180 deg.. The beam should already be relativistic at the exit of the cathode cavity to insure a transit time of a half period in subsequent cells. Cell-by-cell compensation for the transverse distortions of phase space requires that each cell have the same peak field strength. The compensation is more exact the shorter the bunch.

There is no compensation for the transverse deflection at the exit of the last cell of the gun, and almost all of the rf-induced emittance distortion will occur here. Edge effects at the last aperture will increase the nonlinear components of the transverse fields and slightly detune the last cell. Computer simulations of the rf gun with the program SUPERFISH suggests that these edge effects are modest and the essential features of the ideal π-mode configuration can be achieved.

The transverse components of the vector potential of the rf field vanish on the cathode plane, so there is no contribution to the initial phase volume due to the electromagnetic component of the canonical momentum. In principle, if the laser field at the cathode is powerful enough, it also contributes to the canonical momentum of the photoelectrons. However, this would become important only at laser intensities well above the damage threshold for the photocathode.

PHOTOCATHODE

The photocathode should have a picosecond response time as well as good quantum efficiency, good mechanical stability, and low intrinsic emittance. In the initial operation of the ATF, we will use a Yttrium metal cathode to emphasize reliability; in a later phase, a Cs_3Sb cathode might be used.

The time response will be adequate if the photoelectrons can cross an optical absorption length in less than a picosecond. An electron of 1-eV energy has velocity $v/c \sim 10^{-3}$, so in 1 psec it can travel ~ 3000 Å. This is an order of magnitude greater than the absorption length of metals and of cesium-antimonide materials (but not that of GaAs and related materials).

Tests by members of the ATF group[6] indicate that the quantum efficiency for Yttrium metal illuminated by 4.65-eV light (as for a frequency-quadrupled Nd:YAG laser) is about 2×10^{-4}. The work function for Yttrium is about 3.1 eV, so the photoelectrons can emerge with up to 1.5 eV. If the electrons are emitted with an isotropic angular distribution at the cathode surface this relatively high energy leads to an initial emittance of 3.5 mm-mrad, as specified in Table I. A Cs_3Sb cathode has a work function of 2.1 eV and so could be illuminated with a frequency-doubled Nd:YAG laser. As the work function of copper is 4.6 eV, photoemission due to scattered laser light would be less severe in this case. The quantum efficiency of Cs_3Sb can be several percent, and the initial emittance would be about one third as large as for Yttrium.

The cathode material will be deposited on a removable plug of 1-cm radius in the end face of the gun, as shown in Figure 2. The plug will form one surface of an rf-choke joint, which avoids the need for direct electrical contact between the plug and the nearby wall of the gun. The cathode will be illuminated by a laser beam along the axis of the gun; the electron beam must be deflected away from the laser optics shortly outside the gun.

The vacuum requirements for stability of a surface over one day are of order 10^{-10} Torr; otherwise the surface becomes coated with residual gas molecules. This is likely to be more critical for a cesiated cathode where a fraction of a monolayer of cesium must reside on the surface for optimal performance. Vacuum pumping will be through the rf-coupling ports.

COUPLING OF THE RF POWER

Rf power at 2856 MHz will be supplied to the gun via an S-band waveguide which couples to the outer cylindrical wall of the gun, as sketched in Figure 2. In π-mode operation the magnetic field lines circulate in opposite senses in the two cavities of the gun and are matched to the sense of circulation in a TE_{01} mode of the waveguide. There will be essentially no coupling at all to the 0-mode of the gun.

Tests have been performed with an rf model of the gun (see also below) to determine the optimum configuration of the coupling holes. It appears that a $\frac{1}{2}$-in. slot will allow transmission of essentially 100 percent of the waveguide power into to gun when the waveguide contains a $\frac{1}{4}$-wave transformer just upstream of the gun for impedance matching.[7] A tunable short in the waveguide just after the gun will permit balancing of the power fed to the two gun cells.

The gun will be powered by the 6-MWatt "waste" from the traveling-wave linac section. Details of the coupling are also being modelled with the 3-d code MAFIA.[8]

SIMULATIONS OF GUN PERFORMANCE

The emittance of the gun beam has been calculated with a version of the program PARMELA, modified to include ejection of low-energy electrons from a photocathode by a laser pulse.

The photoelectrons are simulated with an energy spread of a fraction of an electron volt, and with isotropic directions. They are emitted randomly with a profile that is Gaussian in both radius and time, as for a laser pulse. The rf fields in the gun are taken from a Fourier analysis (see Appendix A) of the results of a SUPERFISH calculation. The effect of Coulomb interactions among the electrons is calculated with a point-by-point code (see Appendix B) rather than the standard code, PARMELA, as the latter seemed less suitable

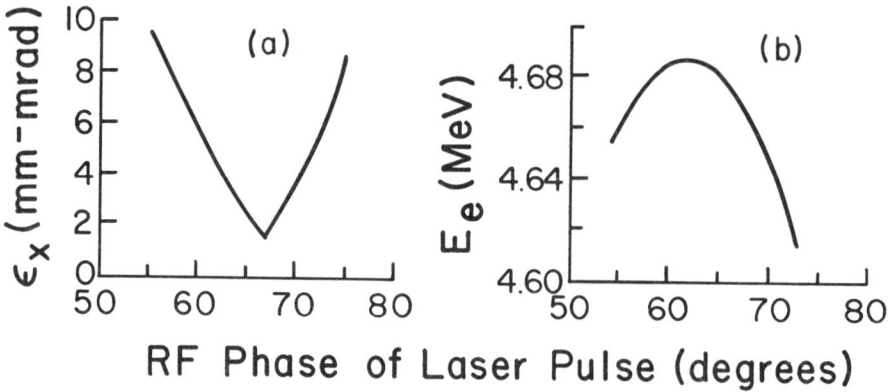

Figure 3. (a) Emittance growth due to nonlinear transverse rf fields, and (b) final beam energy as a function of the rf phase at which the laser pulse strikes the photocathode. The laser pulse has $\sigma_r = 3$ mm and $\sigma_t = 2$ psec.

for very short bunches. The important effect of image charges in the cathode plane is included. PARMELA assumes cylindrical symmetry for the rf fields but simulates a 3-d electron bunch with a small number of macroparticles. The simulation proceeds via a numerical integration of the equations of motion of the electron bunch, using the phase of the rf field as the time variable. We use the invariant emittance, ϵ_x, defined at the bottom of Table I to characterize the transverse phase space.

For a bunch emitted from a spot with $\sigma_r = 3$ mm with 1.5-eV initial energy, $\epsilon_x = 3.5$ mm-mrad at the cathode. This will be combined (effectively in quadrature) with emittance growth due to the rf fields near the apertures and that due to space charge.

We first explore the effect of the rf fields for a gun with peak field strength of 100 MV/m at the cathode. Figure 3a shows how the emittance growth due to the rf fields alone varies with the time at which the laser pulse strikes the photocathode. The initial emittance is set to zero and the space-charge forces are turned off. Time is measured in degrees, where 1 deg. $\sim$ 1 psec at 2856 MHz, and the electric field is at a maximum at a phase of 90 deg. The laser pulse length has a σ_t of 2 psec. The emittance growth is a rapid function of the initial phase and is optimized at 67 deg. where the contribution to ϵ_x is only 1.4 mm-mrad. For this initial phase the bunch leaves the gun at a phase of 360 degrees, as discussed above. Figure 3b shows how the beam energy varies slightly with initial phase.

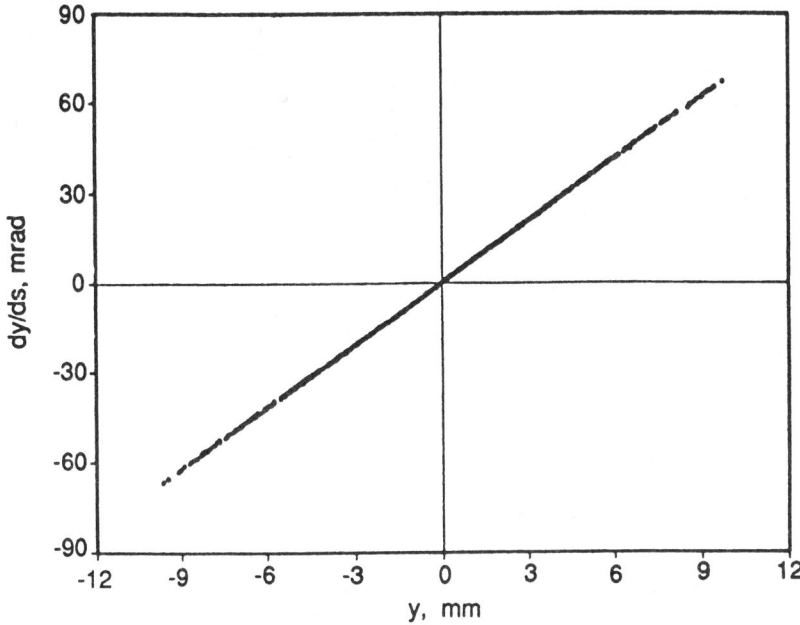

Figure 4. The transverse phase space at the exit of the rf gun for a bunch as given in Table I. $\sigma_z = 4.2$ mm, $\sigma_{z'} = 28$ mrad, and $\epsilon_z = 7.3$ mm-mrad.

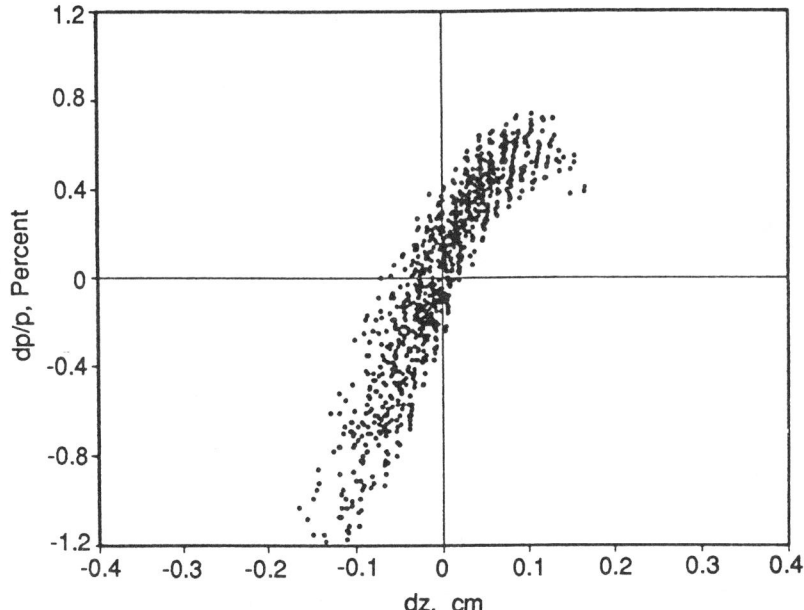

Figure 5. The longitudinal phase space at the exit of the gun. $\sigma_z = 0.6$ mm, and $\sigma_E = 17$ keV.

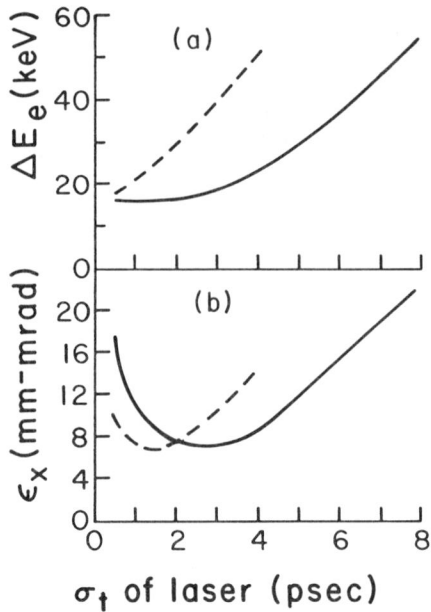

Figure 6. (a) The beam energy spread ΔE_e, and (b) the transverse emittance ϵ_x at the exit of the gun as a function of the laser pulse length. The solid (dashed) curves are for a peak rf field at the cathode of 100 (200) MV/m.

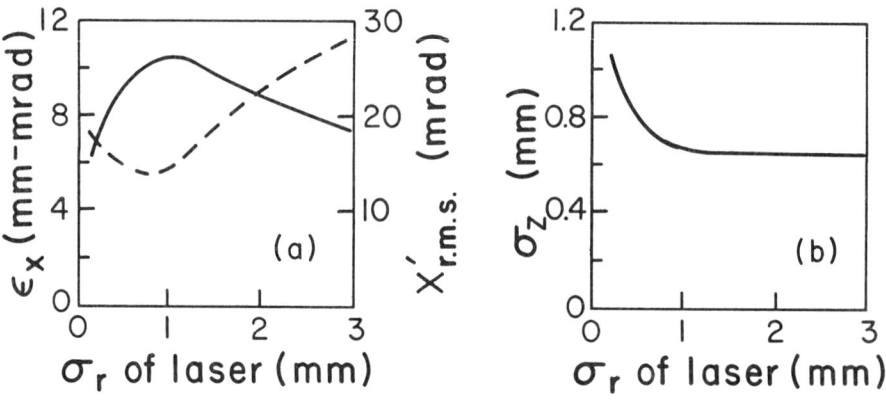

Figure 7. (a) The transverse emittance ϵ_x (solid curve), and the r.m.s. beam divergence (dashed curve) as a function of the radius of the laser spot size on the cathode. (b) The r.m.s. bunch length at the exit of the gun as a function of laser spot size. Other parameters are as in Table I.

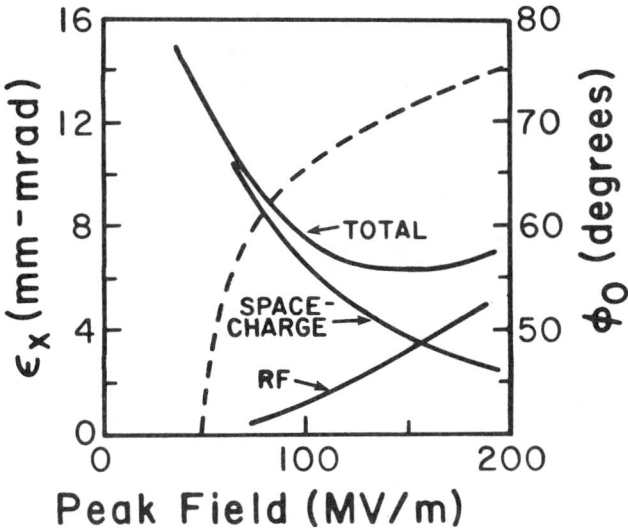

Figure 8. The transverse emittance (solid curves) as a function of the peak rf field on the cathode, and the optimum phase (dashed curve) for the laser pulse to strike the cathode. Other parameters are as in Table I.

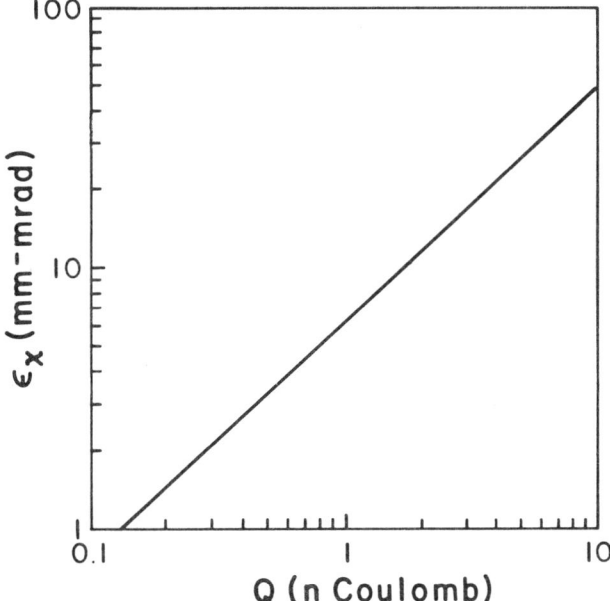

Figure 9. The transverse emittance ϵ_x as a function of the charge of the electron bunch, with other parameters as in Table I.

When the space-charge effects within the 1-nC bunch are restored, they add 6.2 mm-mrad to ϵ_x, for a total of 7.3 mm-mrad at the exit of the gun. Figures 4 and 5 show the transverse and longitudinal phase space at the exit of the gun. While the transverse phase-space area is small, it is highly elongated with an r.m.s angular divergence of 28 mrad and an r.m.s. radius of 4.2 mm. The large divergence requires some care in the subsequent beam transport to avoid additional emittance growth. The shape of the longitudinal phase space is perhaps not ideal; if the initial phase is increased, the shape becomes more diagonal at the expense of elongation in time and an increase in the transverse emittance due to the rf fields.

In Figure 6 we consider the effect of varying the laser pulse length. The bunch length at the gun exit is always proportional to the laser pulse, length, and greater peak currents can be obtained with shorter pulses. However, as seen in Figure f6a, the beam energy spread has a minimum value of about 16 keV r.m.s. due to longitudinal space-charge blowup. Figure 6b shows that there is an optimum pulse length for the transverse emittance; for very short pulses the space-charge forces become quite large, while for long pulses the rf fields cause considerable apparent emittance growth.

Figure 7a shows the dependence of ϵ_x on the laser spot size. On decreasing the radius from the nominal 3 mm, the transverse emittance at first rises and then falls. In all cases the beam nearly fills the exit port of the gun, and below a spot size of 0.15 mm at the cathode there is substantial scraping of the beam. The reduction in transverse emittance for small laser spot size is obtained at the expense of longitudinal blowup of the bunch, as shown in Figure 7b. For very small initial spot sizes, the bunch evolves into a nearly spherical shape at the exit of the gun. By moving the phase of the laser pulse slightly earlier as the initial spot size decreases, the energy spread of the beam can be kept less than 20 keV (σ).

The r.m.s. divergence of the beam on exiting the gun is shown as the dashed curve in Figure 7a. A minimum of about 13 mrad is achieved for an initial spot radius of 0.8 mm. If a lower beam divergence is more important than a short bunch length (= high peak current), there is an option to use an initial spot size of ~ 0.5 mm. The laser intensity required would approach 100 times that for a spot size of 3 mm and might lead to surface heating of a metallic cathode but should not be a problem for a Cs_3Sb cathode with quantum efficiency > 1 percent.

Figure 8 considers the effect of varying the peak field strength in the rf gun. At lower fields the initial phase must be earlier for the bunch to exit at 360 deg., as shown by the dashed curved. The optimal ϵ_x obtained for $\sigma_t = 2$ psec and $\sigma_r = 3$ mm is then shown by the solid curve. The individual contributions to ϵ_x from space-charge growth and nonlinear rf fields are also shown. Fields below 100 MV/m do not accelerate the bunch rapidly enough and the space-charge growth near the cathode becomes large. Above 150-MV/m field strength, the nonlinear rf fields cause the dominant emittance growth.

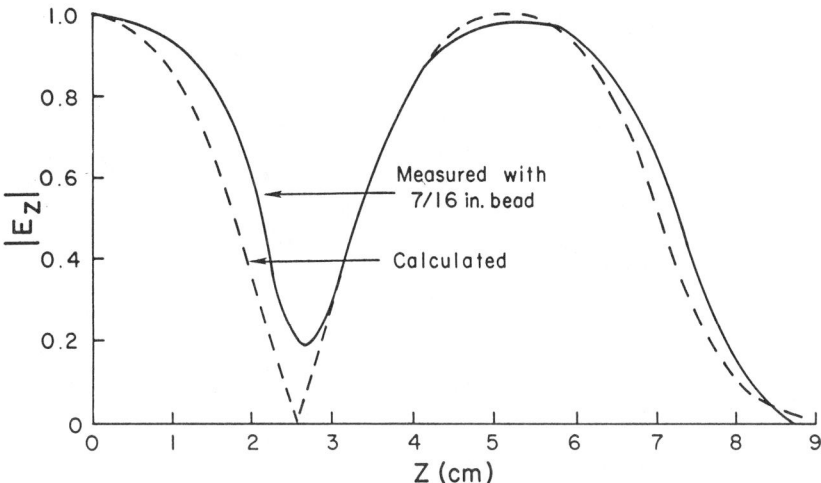

Figure 10. $|E_z|$ as a function of position z along the axis of the rf gun. The solid curve was measured and the dashed curve was calculated.

Figure 9 shows the emittance growth due to space-charge effects alone as a function of the total charge in the bunch. The trend is well described by $\epsilon_x \sim Q^{0.9}$. The beam brightness, which is proportional to $Q/\epsilon_x \epsilon_y$, varies with charge as $Q^{-0.8}$ in the space-charge-limited regime.

The design parameters of the rf gun seem well matched to the capabilities of this technique for producing low-emittance bunches of electrons with picosecond length. Higher fields in the gun combined with a shorter laser pulse would offer higher peak currents and greater beam energy with similar transverse emittance.

TESTS WITH A FULL-SCALE RF MODEL

A model of the gun has been built in brass to study the quality and external coupling of the rf fields in the gun. The cavity Q is measured to be 5500, compared to a calculation of 12000 for copper walls. The π-mode and the 0-mode are found to be 2 MHz apart when the gun has no external coupling, as was also calculated by SUPERFISH.

The field profile along the z-axis has been measured by observing the frequency perturbations caused by a $\frac{7}{16}$-inch-diameter teflon sphere. The results shown in Figure 10 agree well (after tuning of the two cells) with the SUPERFISH calculation. The bead size did not permit a clear demonstration of the zero-field point at the iris between the two cells.

Measurements of the power transmission coefficient from an S-band waveguide into the gun have been made, as mentioned above. An rf gun based on

APPENDIX A. FOURIER ANALYSIS OF THE RF FIELDS

We consider here a general form for standing-wave fields in a cavity (or extended structure) with cylindrical symmetry. From this form, we can anticipate that a series of cavities operated in π mode will impart the least apparent growth to the transverse emittance of the beam.

The unit cell to be analyzed extends over $-d \leq z \leq d$. We suppose that $z = 0$ is a symmetry plane, and hence an appropriate expansion for E_z which also satisfies the wave equation is

$$E_z(r,z,t) = \sum_{n=1}^{N} a_n I_0(k_n r) \cos(2n - n_0)\frac{\pi z}{2d} \sin(\omega t + \phi_0),$$

with

$$k_n^2 = \left((2n - n_0)\frac{\pi}{2d}\right)^2 - \left(\frac{\omega}{c}\right)^2.$$

The case $n_0 = 1$ corresponds to the condition that $E_z(r,d,t) = 0$, while $n_0 = 2$ corresponds to $\partial E_z(r,d,t)/\partial z = 0$.

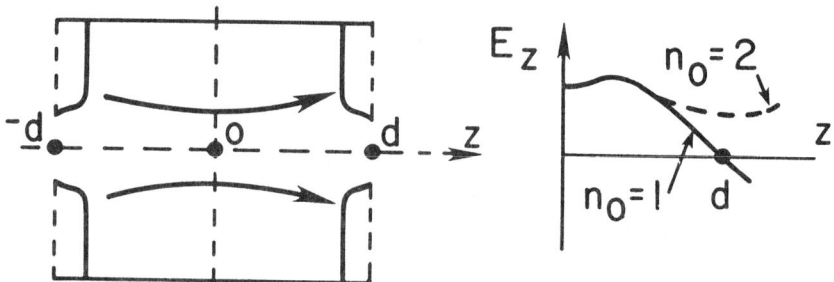

The radial electric field and azimuthal magnetic field then have the expansions:

$$E_r(r,z,t) = \frac{\pi r}{4d} \sum_{n=1}^{N} (2n - n_0) a_n \tilde{I}_1(k_n r) \sin(2n - n_0)\frac{\pi z}{2d} \sin(\omega t + \phi_0),$$

and

$$B_\theta(r,z,t) = \frac{\pi r}{\lambda} \sum_{n=1}^{N} a_n \tilde{I}_1(k_n r) \cos(2n - n_0)\frac{\pi z}{2d} \cos(\omega t + \phi_0),$$

where

$$\tilde{I}_1(z) \left(\equiv \frac{2I_1(z)}{z}\right) = 1 + \frac{(z/2)^2}{1!2!} + \frac{(z/2)^4}{2!3!} + \cdots$$

If E_r and B_θ are nonlinear in r, then the rf fields will cause a distortion of the phase space which cannot be corrected later with linear optics. From our expansion for E_r, we see that linear behavior can only be obtained when $k_n = 0$. The most relevant possibility is for $n = 1$ and, also, $n_0 = 1$, which leads to the requirement $d = \lambda/4$. The rf fields are then

$$E_z = E_0 \cos \frac{\pi z}{2d} \sin(\omega t + \phi_0),$$
$$E_r = \frac{\pi r}{4d} E_0 \sin \frac{\pi z}{2d} \sin(\omega t + \phi_0),$$
$$B_\theta = \frac{\pi r}{4d} E_0 \cos \frac{\pi z}{2d} \cos(\omega t + \phi_0).$$

The conditions that the full cell length be $\lambda/2$ and that E_z vanish at the cell boundaries are exactly those of operation of a π-mode structure. Note that the shaping of E_z to $\cos \pi z/2d$ near the z axis actually requires an aperture (beam port) in the cell wall.

Surfaces perpendicular to the electric field lines have the form

$$r^2 = a^2 - \left(\frac{4d}{\pi}\right)^2 \log\left(\sin \frac{\pi z}{2d}\right),$$

where a is the radius of the aperture at $z = d$. Near $z = d$ the curve is a hyperbola. As $z \to 0$, $r \to \infty$, so no finite cavity can support these fields. The figure below shows such a curve for the case $d = 2.625$ cm and $a = 1.0$ cm. The electrode shape for the ATF gun is also shown for comparison. For the ATF gun the Fourier coefficients of all higher harmonics are less than 15 percent of the fundamental.

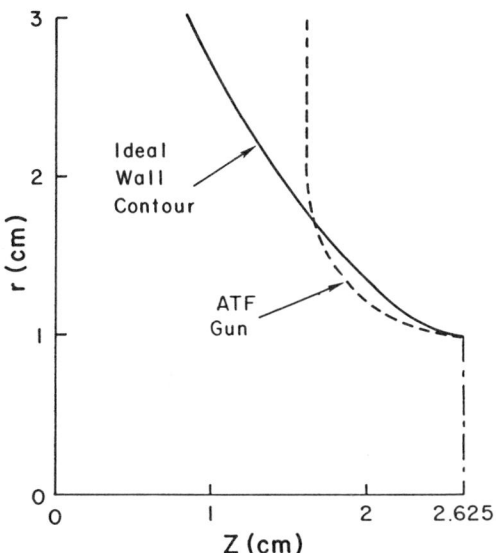

APPENDIX B. THE POINT-BY-POINT SPACE-CHARGE CALCULATION

A point-by-point space-charge calculation was written to replace the usual approximation found in program PARMELA.

In the new calculation the forces due to the electromagnetic fields at each simulated particle are evaluated based on the present positions and velocities of all other simulated particles. Options are available to calculate fields due to images in the cathode, and/or in a beam pipe. For n simulated particles n^2 fields must be evaluated at each step in the beam transport, so the space-charge calculation is somewhat time consuming.

The use of only the present positions and velocities in the calculation is not strictly correct, but a convenient approximation. Here we estimate the error due to this procedure.

We evaluate the fields due to another charge using

$$\mathbf{E} = \frac{Q\mathbf{r}}{\gamma^2 s^3} \quad \text{and} \quad \mathbf{B} = \boldsymbol{\beta} \times \mathbf{E},$$

where

$$s = r(1 - \beta^2 \sin^2 \theta)^{1/2},$$

$\mathbf{r}$ is the vector from the present position of charge Q to the observer, $\beta = v/c$ is the present velocity of the charge, and θ is the angle between $\mathbf{r}$ and $\boldsymbol{\beta}$. This expression would be exact if the charge had a uniform velocity.

In case the charge is accelerating, the electric field at the observer may be written

$$\mathbf{E} = \frac{Q\mathbf{r}_{\text{eff}}}{\gamma_{\text{ret}}^2 s_{\text{eff}}^3} + \frac{Q\mathbf{r}_{\text{ret}} \times (\mathbf{r}_{\text{eff}} \times \dot{\boldsymbol{\beta}}_{\text{ret}})}{cs_{\text{eff}}^3},$$

where the subscript $_{\text{ret}}$ means the quantity should be evaluated at the retarded coordinates of the charge, and the subscript $_{\text{eff}}$ means the quantity should be evaluated by extrapolating from the retarded time to the present assuming the retarded velocity remains constant.[9]

On comparing the exact and approximate expressions for $\mathbf{E}$, we see that the relative error in the approximation is

$$\frac{\delta E}{E} \approx \frac{\gamma^2 \dot{\beta} r}{c}.$$

Now the acceleration $\dot{\beta}$ is largely due to the applied rf field E_0, so we have

$$\frac{d\gamma\beta}{dt} = \gamma^3 \dot{\beta} = \frac{eE_0}{mc} = \frac{2\pi\eta c}{\lambda},$$

where $\eta = eE_0/m\omega c \sim 1$ for our rf fields, λ is the rf wavelength, and we suppose β is parallel to $\mathbf{E}_0$. Then

$$\frac{\delta E}{E} \approx \frac{2\pi\eta}{\gamma} \frac{r}{\lambda}.$$

Thus our use of the present parameters of the charge produces a significant error only when the distance to the charge is of the order of an rf wavelength. However, then the force due to the charge is so small as to be irrelevant.

At a frequency of 2856 MHz the wavelength, λ, is 10.5 cm, and then for $E_0 = 100$ MV/m we have $\eta \sim 3$. In a bunch of 5-psec full width, a typical value of r is 1 mm. Hence, we may expect our use of the present positions and velocities to yield an accuracy of 20 percent in the space-charge calculation.

The charge Q on each simulated particle is taken as the electron charge e multiplied by the ratio of the number of electrons in the bunch to the number of particles used in the simulation. As the number of simulated particles is typically less than 1000, the value of Q is quite large. Fluctuations in the simulation may then lead to overestimation of the space-charge force between nearby charges. We introduced a screening radius R of 1.75 times the average distance to the nearest charge, and scaled the simulated charge by Qr/R for pairs with separation $r < R$. This prescription proved stable against variation of the number of simulated particles in a series of lengthy test runs.

REFERENCES

[1] R.B. Palmer, A Laser-Driven Grating Linac, Part. Accel. **11**, 81 (1980).

[2] E. Courant et al., Proposal for the Design and Construction of an IFEL Accelerator Demonstration Stage, BNL (1987).

[3] R.C. Fernow et al., Proposal for an Experimental Study of Nonlinear Thomson Scattering, Princeton University preprint DOE/ER/3072-39 (1986).

[4] J.S. Fraser et al., Photocathodes in Accelerator Applications, *Proc. of the 1987 IEEE Particle Accelerator Conf.*, ed. by E.R. Lindstrom and L.S. Taylor (Washington, D.C., Mar. 16-19, 1987) pp. 1705-1709.

[5] M.E. Jones and W.K. Peter, Particle-in-Cell Simulations of the Lasertron, IEEE Trans. Nuc. Sci. **NS-32**, 1794 (1985).

[6] J. Fischer and T. Rao, private communication.

[7] K. Batchelor, private communication.

[8] H.G. Kirk, contribution to this conference.

[9] See for example, Panofsky and Phillips, *Classical Electricity and Magnetism.* 2nd ed. (Addison-Wesley, 1962), pp. 356-357.

OPTICS CODE DEVELOPMENT AT LOS ALAMOS [*]

C. Thomas Mottershead
Walter P. Lysenko
AT-3, MS H808, Los Alamos National Laboratory, Los Alamos, NM 87545

ABSTRACT

This paper is an overview of part of the beam optics code development effort in the Accelerator Technology Division at Los Alamos National Laboratory. The aim of this effort is to improve our capability to design advanced beam optics systems. The work reported is being carried out by a collaboration of permanent staff members, visiting consultants, and student research assistants. The main components of the effort are

- building a new framework of common supporting utilities and software tools to facilitate further development,

- research and development on basic computational techniques in classical mechanics and electrodynamics, and

- evaluation and comparison of existing beam optics codes, and support for their continuing development.

OPTICS CODE DEVELOPMENT

The Need for Code Development

At present, many independent beam optics codes exist, but none have all the desired capabilities. The existing codes are typically large, self-contained, batch-oriented programs that communicate via their own uniquely formatted input and output files. Expertise is necessary to use any of them, and modifications are difficult (often even for the author).

To design and operate advanced magnetic optics systems, we need to progress beyond this situation. We would like to do so by building on what has gone before, adding new capabilities while making maximum use of existing software to develop a portable, modular, expandable library of reusable simulation software suitable for both interactive and batch design processes, as well as model support in control systems.

The objective is to create a software development environment that makes it easy to merge, modify, and extend capabilities, with a common user interface to make expertise reusable. The interface should allow various control

---

[*] *Work supported and funded by the US Department of Defense, Army Strategic Defense Command, under the auspices of the US Department of Energy*

© 1988 American Institute of Physics

mechanisms (such as batch, interactive, or automatic) for running the codes and should use standardized optimization, fitting, and graphics utilities. The system must be capable of highly-accurate simulations, with the flexibility to allow trade-offs between speed and accuracy by using various levels of physics approximation. A key prerequisite is the definition of a simple and practical internal representation of a general beamline and of a general description of the beam.

Full implementation of these goals is a large undertaking, but plans can be laid so that small practical steps accumulate in that direction. The following sections describe these plans in more detail, along with implementation steps being taken or considered.

Internal Data Structure

The backbone of the system must be a simple, portable, internal data structure with well-defined units and variables. This probably means standard FORTRAN 77 common blocks, without computer specific extensions, and cannonical MKS particle coordinates as used by MARYLIE, COSY and MAD. The common blocks should contain a general machine description consisting of element sequence arrays, with the ability to repeat (nested) groups of elements, along with a simple, flexible, and efficient indexing scheme for accessing element parameter blocks. The common blocks should also contain a general beam description, allowing at least sets of particle coordinates or beam moments (σ-matrices) at selected locations. A set of database service routines could also be provided to make data access even more flexible and portable.

Beam Descriptions

There are three major possibilities for representing an accelerator beam. Our ideal system should allow all of them and provide means for transforming among them.

- Particle Sets: We need a standard particle set generator code to produce all the commonly used beam distributions in the forms needed by the optics codes. A start has been made, based on the BEAMGEN program from PATH, to collect every known algorithm for generating rays or particle sets. Our interim standard output is a simple ASCII file of the type used by MARYLIE, with one particle per record $(x, p_x, y, p_y, \tau, p_\tau)$. Generic phase-space scatter plot and analysis programs have been written to process this standard particle file.

- Beam Moments: The σ matrices as used by TRANSPORT may be considered second moments of the beam distribution. The higher moments of the distribution need to be considered for more accurate design, and a standard notation for them needs to be defined. The MARYLIE monomial sequence can be used as an interim standard storage format. Twiss parameters and higher moment invariants are related and can be used as input parameters for the particle set generators.

- Beam Distribution Functions: No standard format has been defined. Maximum entropy techniques may be useful for estimating distribution functions. Sums of Gaussians with adjustable amplitudes are being used to represent beam distributions for space-charge calculations.

Physics Levels

Design can be carried out to various levels of physics approximation, ranging from fast and simple to slow and accurate. The general data structure needs to be flexible enough to support all the commonly used formulations of accelerator physics. That is, while the physics modules themselves may use very different techniques, they should all be able to draw their machine parameters from, and store their results in, the common data structure. The design of this data structure must therefore allow for at least the following variations:

- Ray tracing can be done either by particle mapping or by trajectory integration. Mapping codes can use various map representations (Matrices, Lie Polynomials, or Differential Algebra) expanded to various orders.

- Magnetic fields can be represented by various expansions, nominal formulas, or empirical data tables. Fringe-field effects can also be computed in various approximations ranging from simple correction factors to full numerical integration of trajectories.

- Space-charge effects can also be computed to various levels of accuracy, ranging from simple approximations up to full particle-in-cell code calculations.

Reusing Software

To get the new system started, we need to make maximum use of existing software. One way to do this is to separate the input file parser from a major operational code and use it to initialize the standard commons. The tracking routines that propagate the beam through the machine would then be modified so that they can be driven by the new standard commons. New routines would then be written to display and modify the contents of the standard commons, at first by simple manual editing. A module would then be written to generate the operational code's input deck from the current contents of the standard common. As other operational codes receive this treatment, the set of programs to map their various input file formats to and from the standard commons will, in itself, constitute a useful translation facility. The initial operating capability of the new system would then be benchmarked on some familiar test cases. Work could then begin on new modules such as interactive graphic input, output, and control that would serve all the participating operational codes.

Control Mechanisms

It is desirable to have several options for controlling the basic physics routines that propagate a beam representation using the current machine parameters. An interactive system could allow manual adjustment of machine parameters, with prompt updates of diagnostic displays, such as ellipse plots. Prototypes of several such systems are under development [1,2]. A general purpose optimizer could also be used to adjust the machine parameters. Because design is generally aimed at achieving specified beam characteristics, the merit function guiding the optimizer is often calculated from the beam description. A batch process, using the same instructions as the interactive system, should be possible when the slow and accurate physics is being used. Finally, a set of common graphics modules should be able to display the results, no matter how they are obtained.

Optimizer Development

Optimizers and equation solvers are essential design tools. For our purposes, good ones need the following virtues: They should be FORTRAN callable subroutines, leaving the elaborate main physics codes in control. Most available optimizers are written the other way around, calling the function to be minimized as often as they please. They should be extremely frugal with function calls, because these can each be a very expensive model computation, and they should be as robust as possible, using the best available search algorithms to find the solution.

Significant effort has been directed toward the development of such optimizers and solvers. An efficient quasi-linear system solver was written using a multidimensional inverse interpolation algorithm (MDII)[3] and fully integrated into MARYLIE to provide a new parameter fitting capability. Available optimizer techniques and codes were surveyed, and a new optimizer (QMIN) was written addressing the above requirements. QMIN uses quadratic fitting of the Hessian to the best previous points to speed convergence. It performed very well on a suite of standard test problems [4]. Four of the most suitable optimizers (including QMIN) have been installed in a test version of MARYLIE for evaluation on real problems. An early version of QMIN is also operational in the McDonnell Douglas version of TRAVEL [5].

NEW CALCULATIONAL METHODS

This section surveys part of the basic research being done in new computational techniques in classical mechanics and electrodynamics.

Mechanics

A new, explicit, high-order multipoint symplectic Adams integrator for particle tracking has been written and tested for simple nonlinear Hamiltonians [6]. It does not require knowledge of higher derivatives of the forces.

A new technique is under development for generating high-order symplectic maps of machine elements by integrating the Hamiltonian-Jacobi equation using Differential Algebra methods [7,8].

The use of the moments of the distribution function as dynamical variables is under investigation. This includes the search for new moment invariants that are higher-order generalizations of the emittance [9].

Electrodynamics

The code CHARLIE uses Lie algebraic methods to study the effects of space charge [10]. CHARLIE is based on MARYLIE and, at present, can handle only 2-D space-charge problems with free-space boundary conditions.

A method of representing an arbitrary charge distribution by a superposition of regularly distributed Gaussian particles has been completed to fifth order for 3-D, using free-space boundaries. The corresponding field expansions are computed using Differential Algebra. This method has been tested statically but not yet in a dynamics code. It promises to be efficient and accurate for 3-D problems and can be made to handle 3-D conducting boundaries.

A formalism for computing the vector potential of Lambertson coils is under development [11]. The associated subroutines for use in tracking codes are being written.

WORK ON OPERATIONAL CODES

Improvements in Operational Codes

The operational codes in current use for beam optics design in the Accelerator Technology Division include MARYLIE, GIOS, COSY, MOTER, TRANSOPTR, TRANSPORT, TRACE-3D, and others. Modifications are often necessary in the course of the work. The following modifications have been carried out recently, either by or in collaboration with the authors of the codes.

The capabilities of MARYLIE [12] continue to grow. Recent additions include a general fitting procedure [3], a set of experimental optimization routines, and the ability to track moments. The ability to compute maps for strings of permanent magnet quadrupoles by numerical integration has also been added.

Some small improvements have been made to the linear space-charge routines in GIOS [13]. The new fifth-order code COSY [14] is now operational and can handle permanent magnet quadrupoles.

MOTER is an optimizing version of RAYTRACE. It was used extensively to design the Argonne Telescope experiment. Recent improvements include the ability to trace sextupole, octupole and decapole magnets.

TRANSOPTR [15,16] can now design beamlines containing dipole and quadrupole rf elements, which can be used for chromatic aberration correction. The transfer matrix formulas for these rf elements were calculated using the computer algebra system SMP, which also generated FORTRAN code for these elements. The computations were done to second order, although the automatic process could readily have generated higher-order elements.

Code Benchmarking

Benchmarking operational codes by comparing them with each other, and with experiments, is necessary to build confidence in the codes. The following benchmarking exercises were carried out recently.

The codes MARYLIE, MOTER, NIREC (numerical trajectory integration), and SYMPOPT (experimental symplectic integration code) were compared on a beam-expansion problem without space charge. Except for SYMPOPT, they agreed to within 0.5 microradian for this benchmark.

For a space-charge benchmark, the problem used was a drift space transporting a space-charge-dominated beam with 3-D free-space boundaries. On this problem, the codes TRACE-3D, TRANSOPTR, TRANSPORT, and PATH (nonlinear) agreed to better than 5%.

The Los Alamos/Argonne telescope was designed with MOTER and checked with MARYLIE and GIOS. The telescope worked [17] and produced measured aberration coefficients in good agreement with the code predictions when the correct fringe-field form was used.

ACKNOWLEDGMENTS

In addition to the regular staff members and students, two major groups of visiting consultants to the Accelerator Technology Division of Los Alamos National Laboratory have contributed to the work reported here. The group from the University of Maryland, lead by Alex Dragt and including Filipo Neri, Petra Wilhelm, and Rob Ryne, is resposible for the work on MARYLIE and CHARLIE. The group from the University of Giessen, W. Germany, lead by Hermann Wollnik and including Martin Berz and Bernd Hartmann, is responsible for the work on GIOS, COSY, and the application of Berz's Differential Algebra technique to space-charge problems.

References

[1] D. Flicker and L. Trease, *A Prototype for Integrating Modeling into a Control System: User Interface and Automatic Tuning Demonstration*, Los Alamos National Laboratory technical note AT-8:MOD:88-1, January 1988.

[2] R. R. Silbar, "An Interactive Interface to the Beam Optics Code TRANSPORT," this conference.

[3] L. Schweitzer and T. Mottershead, *A Multidimensional Inverse Interpolation Algorithm for Accelerator Design*, Los Alamos National Laboratory technical note AT-6: ATN-86-28, August 1986.

[4] H. K. Overley and C. T. Mottershead, *A Quadratic Form Fitting Algorithm for Unconstrained Minimization*, Los Alamos National Laboratory technical note, AT-3:TN-87-26, August 1987.

[5] P. Meads and R. Schmitt, McDonnell Douglas Corp., private communication, January 1988.

[6] P. J. Channell and C. Scovel, "Symplectic Integration of Hamiltonian Systems," to be submitted to *Nonlinearity* for publication.

[7] G. Pusch, Group AT-3, Los Alamos National Laboratory, private communication, 1988.

[8] M. Berz, "Differential Algebraic Description of Beam Dynamics to Very High Orders," Lawrence Berkeley Laboratory SSC Central Design Group report 152 January 1988.

[9] W. P. Lysenko, "Moment Invariants for Particle Beams," this conference.

[10] R. D. Ryne, "Lie Algebraic Treatment of Space Charge," Ph.D. thesis, University of Maryland Dept. of Physics and Astronomy (1987).

[11] E. A. Wadlinger, *Fringe Fields of Current Dominated Multipole Magnets - Theory*, Los Alamos National Laboratory technical note AT-3:88-1, February 1988.

[12] A. J. Dragt, L. M. Healy, F. Neri, R. D. Ryne, D. R. Douglas, and E. Forest, "MARYLIE 3.0, A Program for Nonlinear Analysis of Accelerator and Beamline Lattices," *IEEE Trans. Nucl. Sci.*, 32, 2311, 1985.

[13] H. Wollnik, J. Brezina, and M. Berz, "Gios-Beamtrace - A Program Package to Determine Optical Properties of Intense Ion Beams," *Nucl. Instrum. and Methods* A258, 408, (1987).

[14] M. Berz, "Generalization of Nonlinear Particle Optics to Aberrations of Fifth Order," Ph.D. thesis, Justus Liebig Univ., Giessen, W. Germany (1986).

[15] W. P. Lysenko, *Feasibility Study of E. Heighway's Scheme for Correcting Chromatic Aberrations in Beam-Expanding Telescopes*, Los Alamos National Laboratory technical note AT-6:ATN-87-9, February 1987.

[16] W. P. Lysenko, *Second Order Matrix Elements for RF Deflector*, Los Alamos National Laboratory technical note AT-6:ATN-87-11, April 1987.

[17] T. Dombeck, "The Argonne Telescope," to be presented to the European Accelerator Conference, Rome, 1988.

NONLINEAR BEAM DYNAMICS IN A FUNNEL
FOR COMBINING TWO INTENSE ION BEAMS*

J.H. Whealton, R.J. Raridon, K.E. Rothe,
W. R. Becraft,[†] and T.L. Owens,[‡]
Oak Ridge National Laboratory, P.O. Box 2009, Oak Ridge, Tennessee 37831-8071
USA

1. ABSTRACT AND INTRODUCTION

The concept of funnels was introduced over the last few years with an endeavor to increase the beam intensity by combining two beams in the following fashion:

The beam is, in each case, produced by an rf accelerator and thereby composed of bunches. The beam bunches are made to occupy relatively small fractions of the longitudinal phase in these cases. The bunches from each of the two beams are made to interlace and enter an rf deflector which produces the interlacing of the beams into one beam with twice as many bunches occupying twice the phase. The funnel itself, in one embodiment called the magnetic funnel, is composed of many transport elements with strong transverse focusing produced by quadrupole permanent magnetic fields. An occasional rf rebuncher is introduced to recompress the beam longitudinally so the beam occupies the appropriate small fraction of velocity space in the parallel direction. Crucial elements of the funnel are the beam dynamics in the rf rebuncher and in the deflector. Beam dynamics in either case must be assessed using an analysis which we are going to describe below.

Several components in a magnetic[1] funnel have been examined by dint of a full three-dimensional solution to the time-dependent Vlasov-Poisson equations with all image charges included.[2] Specifically, the rms emittance growth of subsystems is examined in detail. For the systems considered, a significant parallel emittance growth occurs. Details of the cause and ephemerality of this emittance growth are studied. These systems[1] were originally designed using the 2-1/2-D PARMILA-type analysis[3] which does not account for image charges and neglects azimuthal nonlinear space charge forces. Designs based on PARMILA are referenced for the subject evaluation. A highly resolved, accurate assessment of rms emittance growth has not been obtained with such analysis. However, for the subject analysis at least the precision is significant as will be shown. First, we will consider the simple rebuncher; second, the rf deflector; and third, we will introduce a nonlinear longitudinal emittance reducing "optical" element. The rms parallel emittance growth is due to the longitudinal velocity kick being

---

*Research sponsored by the Office of Fusion Energy, U.S. Department of Energy, under contract DE-AC05-84OR21400 with Martin Marietta Energy Systems, Inc.
[†]Consultant, Grumman Space Systems, P.O. Box 3056, Oak Ridge, TN 37831.
[‡]Now at Fermilab

dependent on a transverse dimension as well as the longitudinal dimension forming an oblique surface of zero volume, for example, in the three-dimensional phase; z, V_z, and x. A plane surface parallel to x in this space has zero rms emittance. A curved surface or a plane not parallel to x in this space has finite rms emittance under the conventional definition:

$$\varepsilon_{z, rms} = \frac{\beta \gamma}{n} \left\{ \sum_j (z_j^2) \sum_j (z'_j{}^2) - \left(\sum_j (z_j z'_j) \right)^2 \right\}^{1/2}.$$

This is considered conventional, because of the protocol suggested by Los Alamos National Laboratory (LANL), and practical, because the occupation in phase space of an uncontrolled surface, even for a moment, eventually makes that surface uncontrollable. To effect control would require two identical particles seeing a different force even when they are in the same position at the same time. Control must be exerted as soon as possible upon the deviant. Having the longitudinal velocity depend differently on both transverse positions does not help matters as this is what happens in an rf deflector as presently considered.

2. NONLINEAR BEAM DYNAMICS IN AN RF REBUNCHER (2-D)

The dynamic systems which we will consider are:

$$\nabla^2 \phi(\mathbf{x},t) = \int f_-(\mathbf{v}_-, \mathbf{x}, t) d\mathbf{v} - exp[-\phi(\mathbf{x},t)/T_e] \qquad (1)$$

$$\frac{\partial f_-(\mathbf{v}_-,\mathbf{x},t)}{\partial t} + (v_- \bullet \nabla_x) f_-(\mathbf{v}_-,\mathbf{x},t) + [\nabla \phi(\mathbf{x},t) \bullet \nabla_{v_-}] f_-(\mathbf{v}_-, \mathbf{x},t) = f_o(\mathbf{v}_-, \mathbf{x},t) \qquad (2)$$

The subject analysis can best be understood with reference to Figure 1, which shows the path of the calculation. First, the Poisson equation is considered. For this first pass, the source terms are set equal to zero and a Laplace equation is solved by SOR, finite difference, and boundary interpolation within a cell, using a Gauss-Seidel implicit method.[4] Considering the attributes A1, resource utilization, and A2, accuracy, iteration reduces memory requirements (A1), and boundary interpolation contributes to the accuracy per cell (A2). Generally, individual convergence of the solutions is not warranted on each pass (contributes to A1), since the iteration procedure lends itself to incomplete convergence of the intermediate solutions. As noted before, the finite difference method compared with the finite element method has in our experience reduced A1 by a factor of 20 for the Poisson solution (Reference 5 vs Reference 6) for the same accuracy. Boundary conditions for arbitrarily shaped metal surfaces can be specified as time-dependent Dirichlet or ramped Dirichlet conditions (contributes to A2). Neumann boundary conditions can also be specified.

Second, the Vlasov equation is solved for an arbitrary initial condition using the solution to the Laplace equation above for a time step Δt. The technique is described in References 7 and 6 where significant advances in A1 and A2 are reported. Reference 7 speeds up the Vlasov solver by a factor of 10 (contributes to A1) from that in References 8 or 9 while at the same time improving the accuracy by over a factor of 10 (contributes to A2). Reference 6 decreases resource utilization (A1) over Reference 5 by a factor of 400 with the same accuracy. The trivial relationship between the coordinates inside an element and the global elements for the uniform Cartesian grid used in this algorithm allows a factor of 20 (of the 400) savings in the Vlasov solver (A1) over that employed in the irregular elements of Reference 5. As mentioned in Reference 7, the Vlasov solver is made self-regulating in accuracy; trajectory refinement is undertaken only in those places that need it (A2).

Third, charge deposition is done in three dimensions by interpolation over the grid and is "exact" in the sense that as the three-dimensional grid is made more fine and the number of trajectories is increased, a result as accurate as desired can be obtained (A2). Notice that nowhere is any paraxial-like assumption made, and the fields "to all orders" are directly calculated (attribute A3, nonlinear effects). Therefore, aberrations (to all orders) are also directly computed. Other nonlinear optics effects (A3) computed include space charge "to all orders" caused by nonuniform beam density and/or boundaries. (Boundaries cause nonlinear space charge forces also because they alter the delicate dependence of ϕ on r required to keep it linear.)

Fourth, the beam charge and the exponential plasma term (A3) are taken as inhomogeneous terms to the Laplace equation solved in step 1 above. Now the two inhomogeneous terms are, in many cases, large, of opposite sign, extremely nonlinear, and three dimensional. This is the cause of numerical difficulties that were first surmounted (in two-dimensional steady state) in Reference 8. The technique used, accelerated under-relaxation, improved the prior art[10] by a factor of 1000 (A1) in the beam perveance of interest and by a greater factor for higher perveance. Another factor of 10 (A1) increase in speed was achieved, while at the same time the accuracy was increased by more than a factor of 100 (A2) in Reference 9. This technique was extended to three dimensions in References 5 and 6. Essentially the best technique we have found is to use an unconverged Newton SOR outside its established range of validity.[11]

Fifth, the time is moved back by Δt, the ions are moved back to their phase space positions a time Δt ago, and the Vlasov equation is resolved with the new fields computed from the Poisson equation solution of step 4. The trajectories are different from those computed in step 2 because of the presence of the space-charge terms (steps 3 and 4).

Sixth, since the trajectories of step 5 are different from those of step 2, steps 3, 4, and 5 are repeated (Vlasov-Poisson iteration) until no change obtains. This completes the convergence procedure (A2), and it is time to proceed to the next time step. However, one should note the implication of the iteration consisting of steps 5 and 6.

Seventh, the time is advanced by Δt and steps 2 through 6 are repeated. This performs the beam evolution through the device under consideration.

The attributes A1 through A3 provide orbit accuracies of up to 10^{-8} radians in speedy calculations with significant nonlinearities. Six items contributing to a decrease in resource utilization (A1) total about 2×10^9 in the product of memory saved and CPU time (however, the accounting procedure leading to this figure is somewhat ambiguous). Five items contributing to increased accuracy (A2) make an improvement of about 10^6 for a significantly nonlinear problem.

We turn now to a preliminary examination of a 425 MHz rebuncher. A rebuncher is generally a cylindrically symmetric affair as illustrated in Figure 2. An emittanceless beam bunch of constant density (waterbag) is shown entering the fringe fields of an rf rebuncher. The fields are near the maximum and the nonlinear forces are evident in the accompanying phase space occupation diagrams.

When the center of the bunch is in the center of the rebuncher, the rf fields are at a null and only the space charge and image charge fields are present. At the end of the rebuncher, the longitudinal phase space occupation is supposed to have a negative slope. Nonlinear shear aberrations are also shown. Longitudinal emittance as a function of time is shown in Figure 3 for both zero and 100 mA beam current. The double lumped structure represents partial canceling of aberrational shear forces on both sides of the null field (denoted as $\phi=0$ on Figure 3). The partial cancellation occurs because of the extreme fringe fields such as shown in Figure 2(a). The radial center of the bunch is molested much more than the edges. As the bunch proceeds, the edge catches up, partially mitigating the aberrations. Immediately after the null, the edge gets more action than the center (overshoots) and the emittance climbs again. Near the trailing edge of the fringe fields, the center catches up again. The net result, at least in the case of zero beam current is that most of the aberrations cancel (90 percent). The reason the rms emittance, at zero current, does not completely round out is that some ion relative motion occurs during traversal of the rebuncher. However, in the high current case in Figure 3, there is a noticeable space charge/image charge component which is superimposed on the above described shear aberrational phenomena.

The effect of waterbag bunch shape on longitudinal emittance is shown in Figure 4 (for a somewhat different beam radius). The emittance growth for the hard (square) beam (S0) is significantly greater than the softer (elliptical) beams (S1 and S2); however, there is not much difference between S1 an S2 over the region considered. Numerical noise is 11 orders of magnitude lower than emittance values of interest.

A parallel normalized rms emittance growth on the order of 0.003 πcm. mr. is expected on the "simple" rebuncher. This is in contrast to the negligible growth predicted by PARMILA. Since there are several rebunchers, the total emittance growth due to rebunchers is expected to be higher than 0.003. The complex double rebunchers, necessary near the rf deflector, will probably add more than this because of the possible degradation of mode purity; the smoothness of the phase space distribution will possibly half this. For the whole funnel the estimated total parallel emittance growth due to the rebunchers is greater than 0.005.

3. NONLINEAR BEAM DYNAMICS IN AN RF DEFLECTOR (2-D)

Visualization of emittance growth in the rf deflector is aided by reference to a three-dimensional isometric which is shown in Figure 5. For openers, we are going to consider a two-dimension variant of the deflector by considering strictly slot geometry. This produces some errors which will be checked later.

Three bunches in an rf deflector are shown at a particular instant of time in the lower part of Figure 6, labeled B1, B2 and B3. The dashed lines are electric potentials at that instant of time. For this figure, the beam space charge is zero. The potential contours are at linear increments: a coarse increment for extreme potentials, and a finer linear increment near the center of the potential range where the beam bunch is located. Also shown in Figure 6 are the transverse phase space occupation (upper left-hand side) and longitudinal phase space occupation (middle upper side and expanded on right-hand upper side).

Fringe fields due to the boundary conditions are clearly shown in Figure 6. The part of the beam bunch nearest the deflector gets kicks in both the transverse and longitudinal directions. The bunches begin with zero emittance. Bunch number 1, as indicated on the $x-x'$ phase space diagram, is steered approximately into place (aberrations are noticeable); bunch number 2 is approximately 1/3 steered into place also with much aberrations; bunch number 3 shows very little change at this time (as indicated on both the phase space occupation diagrams shown). Longitudinal emittance of bunch number 2 is relatively large as indicated by the B2 $z-z'$ phase diagram while the first bunch parallel emittance is decreasing since the aberrational shear fields in the second gap, G2, are canceling out some of the rms emittance produced by the shear fields in the first gap, G1. At a later time, as illustrated in Figure 8 taken near a null in the rf driving frequency, bunch number 1 has completed its emittance reduction and B3 is at a maximum in emittance because almost no cancellation has taken place. For 50 mA average beam current, the corresponding cases are shown in Figure 8 and 9, respectively. Space charge fields interact with the fringe fields in the case of bunch numbers 1 and 3, respectively, in Figure 8, whereas the main steering force is bent visibly for bunch number 2. The longitudinal emittance is even qualitatively different. The null field illustrated in Figure 7 for zero current is now dominated by interacting bunches in Figure 9.

Now we will consider the effect of a 425 MHz deflector on a 50 mA beam composed of bunches occupying 28 degrees out of 360 degrees longitudinal phase and 2.5 mm transverse width (see Figure 10). The longitudinal emittance growth of this bunch is shown in Figure 11. Both 50 mA and 0 mA are shown in Figure 11. To get an idea of the properties of such a deflector, we consider first the time reversed deflector in slot geometry. This is done to expedite the calculation. Time reversal is not seriously inhibiting. Similar results apply either way when correctly interpreted. Slot geometry probably shows the emittance growth to within a factor of two in either the parallel or transverse direction. An azimuthal nonlinear space charge issue is neglected by such a representation. The main point of the computation is to get a clear idea of the space charge, image charge and applied field aberration issues as quickly as possible, so we can focus on the relevant causes.

In Figure 10(a), a bunch is entering (leaving) the fringe fields of the deflector. The space charge fields and image charge fields are clearly interacting with the fringe fields. In Figure 10(b) and 10(c), the time is near an rf null [as is Figure 10(f)] and the space charge fields dominate the applied rf fields. In Figure 10(d)-(e), the applied fields dominate, but are clearly perturbed by the space charge fields.

The shear fields are partially canceled by having the field reversed in the gap. This is one of the advantages of having the field reversed in the gap. Another advantage is that the nonlinear shear fields will on the average be smaller in the gap since the rf fields are nearer zero. A major disadvantage is that the steering is mitigated as a result (see Figure 12 for $T_D=1\beta\lambda/2$). As the bunch enters the gap, the steering is initially in the wrong direction; exiting the second gap also produces wrong direction steering. Therefore, the intended steering has to compensate for this. The steering will be less than intended. In this instance, where the thickness of the deflector is $\beta\lambda/2$, the actual steering is 1.36 degrees instead of the 1.50 degrees as planned. Incidentally, the computations agree exactly with the simple calculations (it was only tested to one part in 1000) for the steering in a configuration with no fringe fields. Such a deficit in steering may appear at first sight to be an issue; however, to produce higher steering, say 10 percent higher, requires higher fields in the deflection section which impact reliability and higher heat loading since the rf power to the device is proportional to E^2, so the rf power/heat loading must go up to 20 percent. There may be a heat dissipation problem even without this inconvenience.

The effect of changing the deflector thickness on the longitudinal emittance growth and steering angle are shown for four values of L: $\beta\lambda/2$, 1.2 $\beta\lambda/2$, 0.7 $\beta\lambda/2$, and 0.8 $\beta\lambda/2$. Emittance growth is shown in Figure 13 and steering is shown in Figure 12. A thickness somewhat smaller than $\beta\lambda/2$ appears desirable since the steering increases a few percent with no reverse fringes to counter and the emittance decreases.

Also shown in Figure 13, is the situation without deflector fields (therefore no field aberrations). The parallel emittance grows, even in this case, about the same as for the thin plate $T_D \sim 0.7\ \beta\lambda/2$ case, indicating that virtually all of this emittance growth is due to nonlinear space charge and image charges.

An important feature of Figure 11 is a qualitative difference between the emittance growth in the high current case and the low current case. The low current case, when not molested by fringe fields, shows no rms emittance growth. However, the high current case seems to suggest a prevailing emittance growth with time, or distance traversed, on which the shear aberrational fields are only a perturbation. This prevailing emittance growth is caused by nonlinear space charge forces and nonlinear image charge forces and is denoted by the line L2.

Results for a 100 mA deflector whose pulse width is 4 mm and longitudinal phase occupation is 28.30 is shown in Figure 14. The convergence of the solution as a function of axial resolution is shown in Figure 15 for an order of magnitude variation.

4. THREE-DIMENSIONAL CALCULATIONS ON RF DEFLECTOR

The heart of the deflector in full three-dimension is shown in Figure 16. Figure 16(b) is rescaled and the three axes are not the same. The beam is chosen to be a square bunch and is seen drifting approximately in the middle of the device in Figure 16. Because of the scale of the figure the fact that the rf deflector plates are much longer than the space between them is not evident from the drawing. However, the spacing was indicated previously in the two-dimensional section; for example, Figure 10 actually reflects the situation. In Figure 17 is shown again unscaled coordinates of the rf deflector in full three-dimension showing the entrance aperture, the rf parallel plate deflector itself. In this case, in part A of this figure, potential contours on the symmetry axis occur at an instant of time when the pulse happens to be in the middle of the deflector and these fields, at a maximum, are shown. The fringe fields can easily be seen. The contours of the fringe fields are compressed over the contours of the fields in the normal deflection region so as to show the extent of the fringe fields. In part B, the same situation is shown except the 100 mA pulse is being deflected. The pulse happens to be at this time in the middle of the rf deflector plates. On the symmetry axis one can see an indentation. A depression of this equal potential contour is due to this pulse. This is similar to the findings in the two dimensional situation.

The situation in Figure 18 is rather similar to Figure 17, except that here is exhibited explicitly the fringe fields near the aperture. This is the situation shown for both 0 mA and 100 mA current in Figure 18(a) and 18(b), respectively. The issue is these fringe fields coming from the parallel plate deflector and entering and impinging upon a circular hole at the entrance of the exit aperture. The effect of these fringe fields is a principle result of these three-dimensional calculations.

The parallel emittance growth for these three-dimensional deflectors is shown as a function of time for a particular pulse in Figure 19. Here we see the characteristic emittance growth at the beginning for either 0 mA or 100 mA current. We see a gradual but slower emittance growth in the region of the rf deflector itself for the case of the 100 mA beam and no growth for the case of the 0 current beam. The various bunches given to the beam at the fringe field upon exiting the deflecting region are shown in both the 100 mA and the 0 mA cases by the twitches thereof, and finally, the post emittance variation is shown. There is no increase of emittance above a certain level for the 0 current beam, which is expected since no nominal space charge forces are present and nothing else would be existing in this region to increase the emittance of the beam. However, as was shown in other cases previously, for the 100 mA beam there is the characteristic nonlinear space charge induced longitudinal emittance growth.

In this case, as in so many cases, the initial distribution of the beam has zero emittance, which corresponds to a shoebox waterbag model. If the emittance was larger to begin with, then the emittance growth fortunately would be presumably less than is shown here. All the emittance growth that we see here is obtained in the funnel itself for a 0 emittance. If the beam, for example, has an initial emittance of 0.005 and a current of 100 mA, then instead of the final emittance being 0.0085 as is shown here at the final time, the final emittance would probably be approximately 0.01, reflecting the quadratics zoning of the emittance growth with the initial emittance of the beam.

The transverse component of the emittance transfers both to the beam direction and transverse to the surface of the deflector electrodes is shown in Figure 19 as a function of time. Again we have the usual twitches on both as the beam bunch passes both ends of the deflection plate and we have in both cases a significant emittance contributed by non-cancellation of these fringe fields from one end to the other.

The emittance growth calculation here for the rf deflector are approximately 0.008 ε_z, or ε_x, and slightly less for ε_y, These values are significantly in excess of the PARMILA projection (Reference 3).

5. NEW PARALLEL RMS EMITTANCE REDUCING LENS

For a typical rebuncher, the parallel emittance growth for a hard (square in r-z space) and soft (elliptical) beam as a function of distance traveled by the bunch is shown by the upper curves in Figures 20 and 21. Suitable phasing of the rebuncher, as determined by the subject analysis, with respect to the bunch produces a deliberate smashing of the beam to counteract the natural nonlinear space charge/image charge forces. The successful use of this technique is indicated by the bottom curves of Figures 20-21, showing significant improvements in parallel emittance growth (factor of ~2). Further development of this, and similar techniques, have the prospect of enhancing beam quality significantly.

A look at Figure 22, which is the same as Figure 20 for the symmetry rebuncher but is continued for twice the time or distance, will show that the peak of the parallel emittance as a function of the time for a bunch increases after the region of the rebuncher starting at the point labeled A. The parallel emittance continues to increase in the drift rather strongly to point B, continues to increase, after a while starts to level off, reaching a maximum at point C, then levels off for a significant time, finally dips a little bit (approximately 10 percent at point D), and then starts increasing. An examination of the beam properties in these four cases will show why the emittance has this property. Figure 23 is a simple rebuncher with a beam pulse. This corresponds to point A of the curve in Figure 22. The rebuncher is joined by a transport channel with no fields connected but a Dirichlet metal boundary condition as indicated. The beam simply drifts along this channel and the emittance is examined. The top diagram in Figure 21 is the transverse emittance where the transverse philosophy is plotted versus the transverse distance. There is a finite emittance which can be seen. The bottom right hand diagram is the longitudinal emittance with the longitudinal velocity as the vertical axis and the longitudinal position of the beam as the horizontal axis. The points approximately fall along a straight line with relatively low emittance (point A in Figure 22). Slight aberrations occur on the ends. The velocity is characteristic of a rebuncher, which means the beam is compressing. Figure 24 is after the beam has drifted a significant distance, as shown. This corresponds to point B on Figure 22. The transverse emittance has increased significantly and the parallel has obviously increased. Enormous aberrations, basically from nonlinear space charge forces between points A and B, can be seen on the longitudinal emittance diagram on the lower right hand side. This is consistent with the curve in point B of Figure 22. As the beam drifts still further, as indicated in Figure 25 (corresponding to point C of Figure 22), it covers an enormous distance. The longitudinal emittance is very high and very different from that presented in Figure 24. Here the beam is essentially crossing over from a rebunch case to a debunched case. Figure 26, corresponding to point D of Figure 22, shows the beam transported still further. The beam is now debunching. Space charge forces have caused the beam to expand instead of compress, and the emittance growth is slowing down for the moment, but will increase again as the beam simply expands in a very nonlinear fashion. This kind of growth is a significant concern for intense beams drifting in long, unneutralized transport systems where there is a requirement for very low longitudinal emittance growth.

6. CONCLUSIONS

One embodiment of a funnel will probably contribute to a parallel emittance growth, end to end, of about 0.015 ᴨcm mr, assuming perfect alignment according to calculations performed thus far. The transverse emittance growth is expected to be slightly less, but still on the order of 0.010ᴨcm mr. The value for the parallel emittance may be halved with the deployment of the lens discussed in Section 5.

References

1. K. Bongardt et al., Proc. Symp. Accel. Aspects of Heavy Ion Fusion, GSI-82-8, 224 (1982), also GSI-84-11, 389 (1984); F.W.Guy, AIP Conf. Proc. 139, 207 (1985); F.W. Guy and T.P. Wangler, ibid, p. 185.
2. "Ab Initio Calculation of Nonlinear Transient Beam Dynamics for Ion LINAC and Injectors," J.H. Whealton, R.J. Raridon, M.A. Bell, K.E. Rothe, B.D. Murphy, and P.M. Ryan, Workshop on Space Charge Effects in Beam Dynamics, ORNL (1986); "Ion Beam Dynamics in High Intensity RFQ LINAC," J.H. Whealton, R.J. Raridon, M.A. Bell, K.E. Rothe, B.D. Murphy, Twenty-Eighth Annual Meeting of the Division of Plasma Physics of the American Physical Society, Baltimore, Maryland, November 3-7, 1986; 1987 Particle Accelerator Conference, Washington, D.C. March 16-19, 1988, Bull. APS 32, 189 (1987); "3D, Full Implicitly, Ion Time Scale, Self-Consistent Plasma Edge Calculations Near ICRH Antenna/Faraday Shield," J.H. Whealton, P.M. Ryan, and R.J. Raridon, ICRF/Edge Physics Workshop, Boulder, Colorado, March 30, 1988.
3. T.L. Owens, private communications (1987). Some of the PARMILA calculations were done by K. Crandall of AccSys Inc.
4. G. D. Smith, Numerical Solution of Partial Difference Equations: Finite Difference Methods, 2nd ed. (Oxford Univ. Press, Oxford, 1978).
5. J. W. Wooten, J. H. Whealton, D. H. McCollough, R. W. McGaffey, J. E. Akin, and L. J. Drooks, J. Comput. Phys. **43**, 95 (1981); K. Ota, N. Inone, H. Hikei, J. Marikawoo, S. Ishida, and T. Uchida, J. Appl. Phys. **23**, 1241 (1984);J. W. Wooten, J. H. Whealton, and D. H. McCollough, J. Appl. Phys. **52**, 6418 (1981).
6. J. H. Whealton, R. W. McGaffey, and P. S. Meszaros, J. Comput. Phys. **63**, 20 (1986).
7. J. H. Whealton, J. Comput. Phys. **40**, 491 (1981); J. H. Whealton, Nucl. Instrum. Methods **189**, 55 (1981); J. H. Whealton, IEEE Trans. Nucl. Sci. **NS28**, 1358 (1981).
8. J. H. Whealton, E. J. Jaeger, and J. C. Whitson, J. Comput. Phys. **27**, 32 (1978).
9. J. C. Whitson, J. Smith, and J. H. Whealton, J. Comput. Phys. **28**, 408 (1978); J. H. Whealton and J. C. Whitson, Part. Accel. **10**, 235 (1980).
10. E. F. Jaeger and J. C. Whitson, ORNL/TM-4990, 1975.
11. J. M. Ortega and W. D. Rheinboldt, Iterative Solution of Nonlinear Equations in Several Variables (Academic Press, N.Y., 1970).

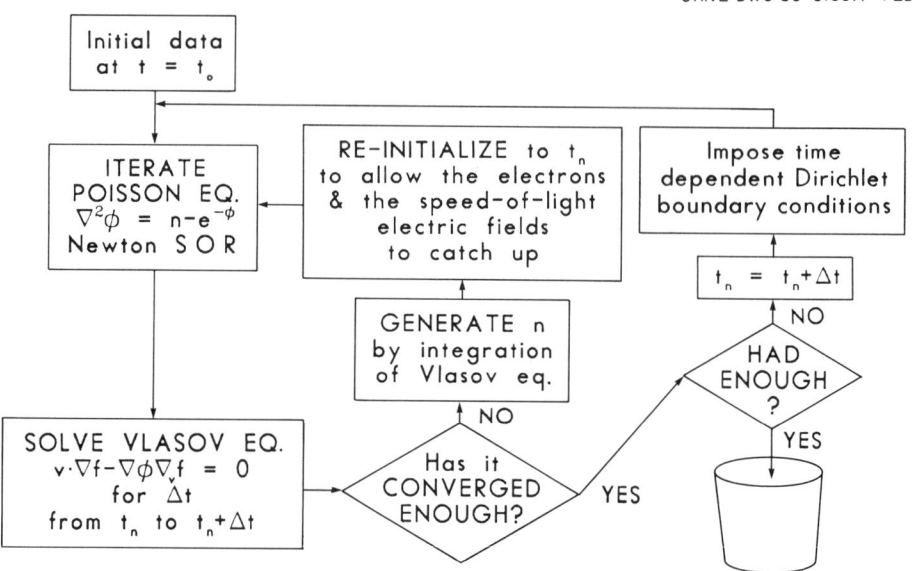

Fig. 1. Path of calculation for 3-D, time-dependent Vlasov-Poisson analysis.

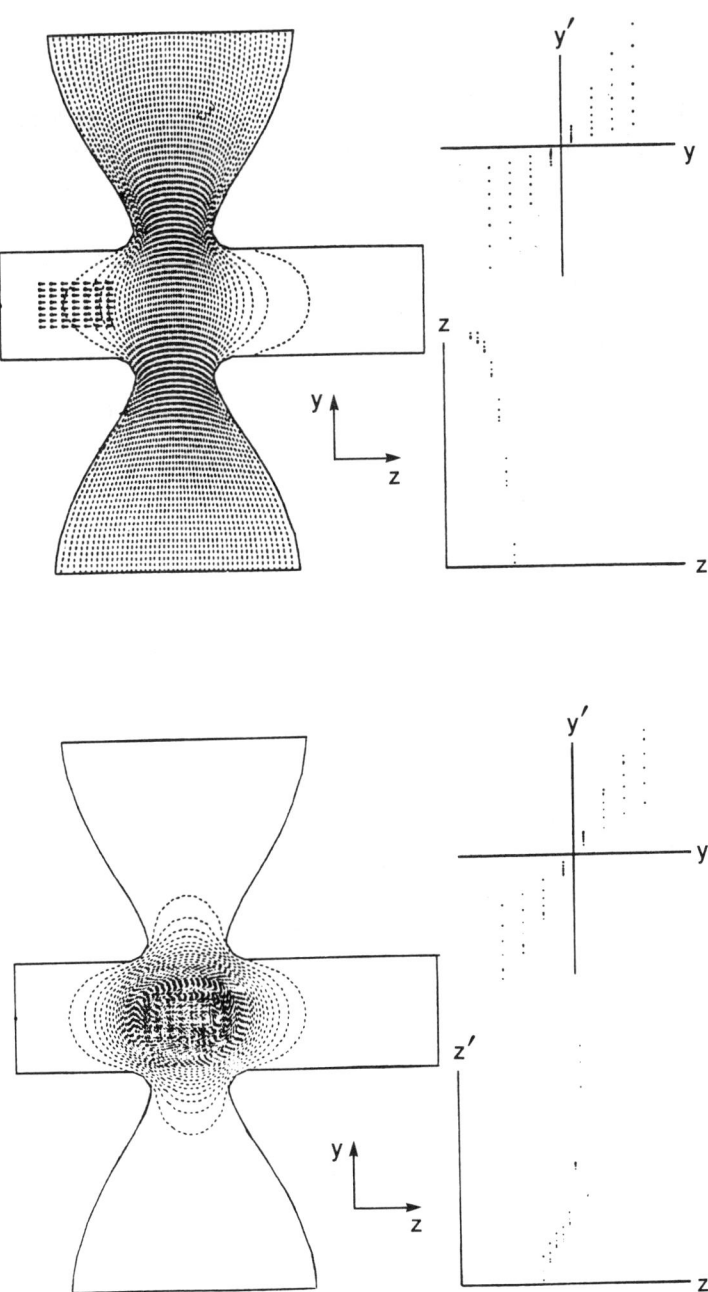

Fig. 2. Rebuncher at two different times.

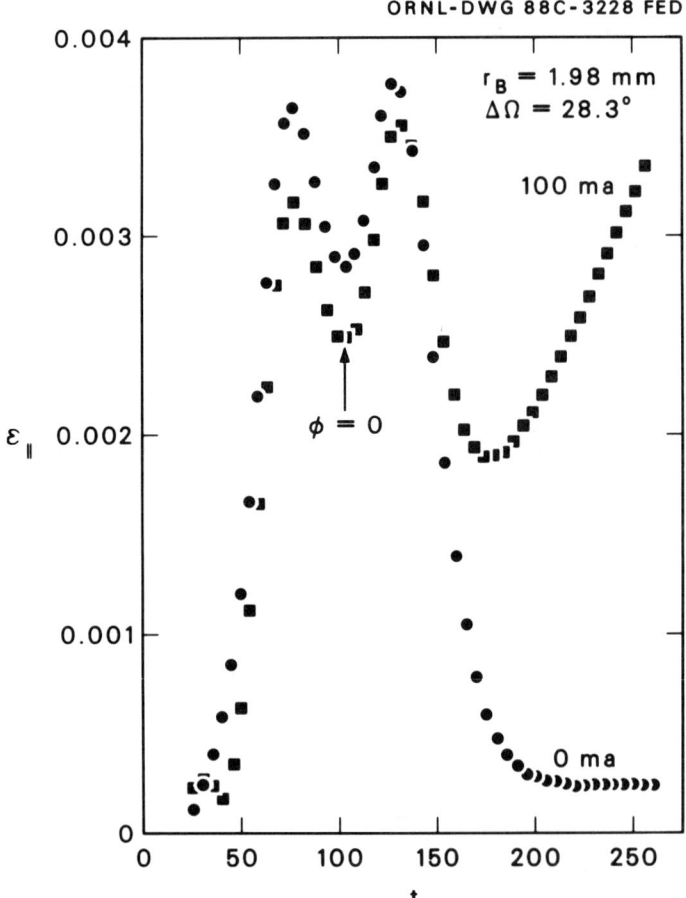

Fig. 3. Longitudinal emittance as a function of time.

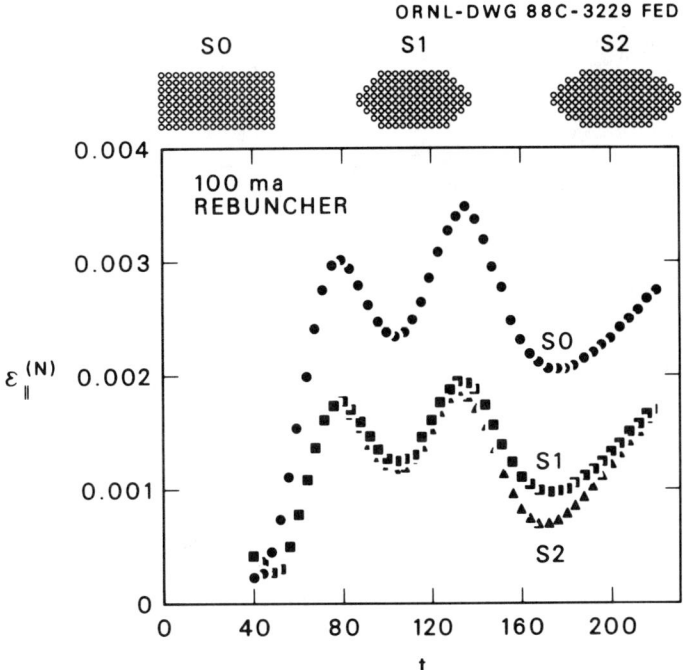

Fig. 4. Effect of beam shape on longitudinal emittance.

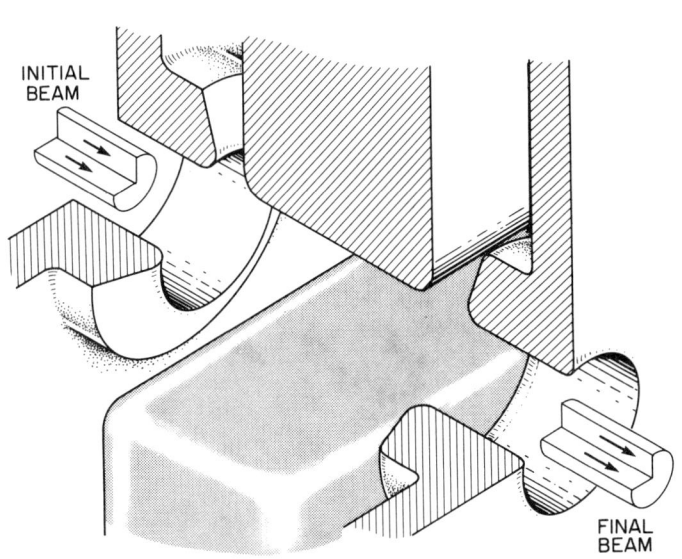

Fig. 5. Isometric of an rf deflector.

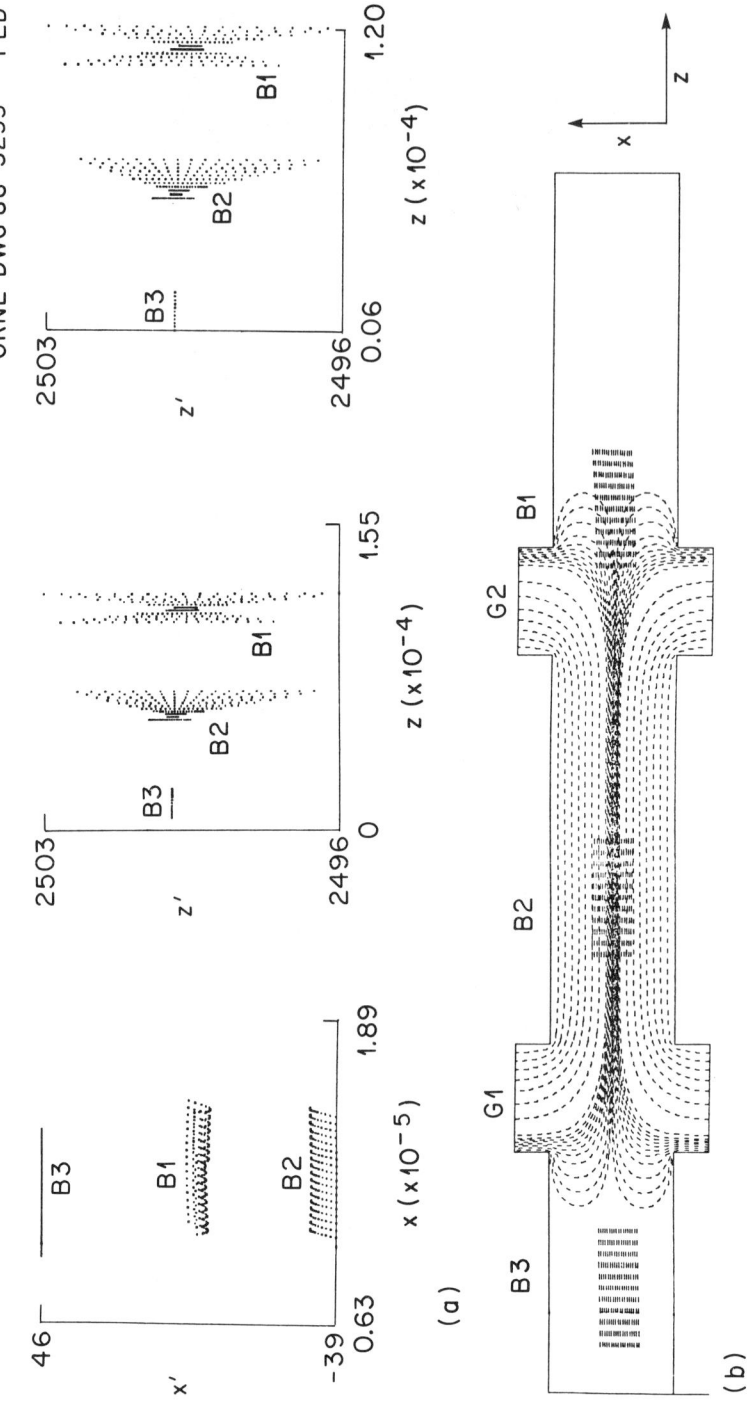

Fig. 6. Phase space occupation and deflector.

243

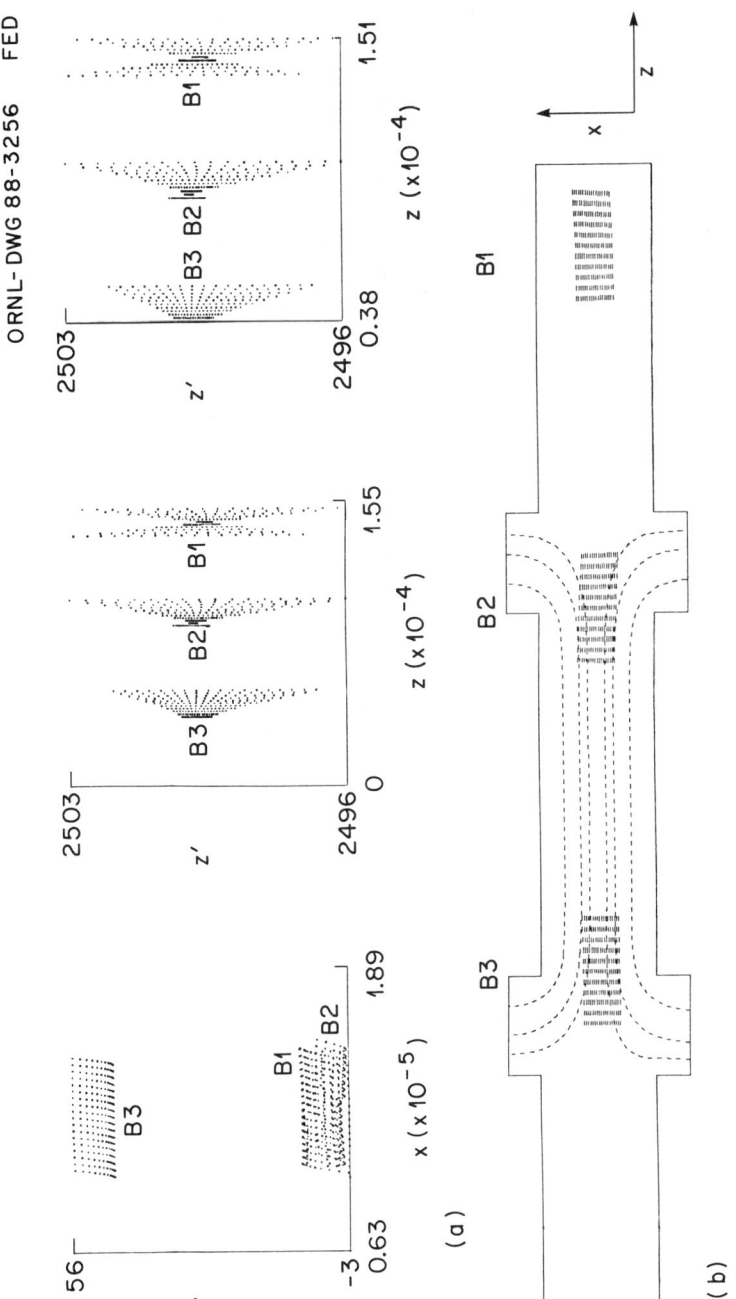

Fig. 7. Phase space occupation and deflector near a null.

244

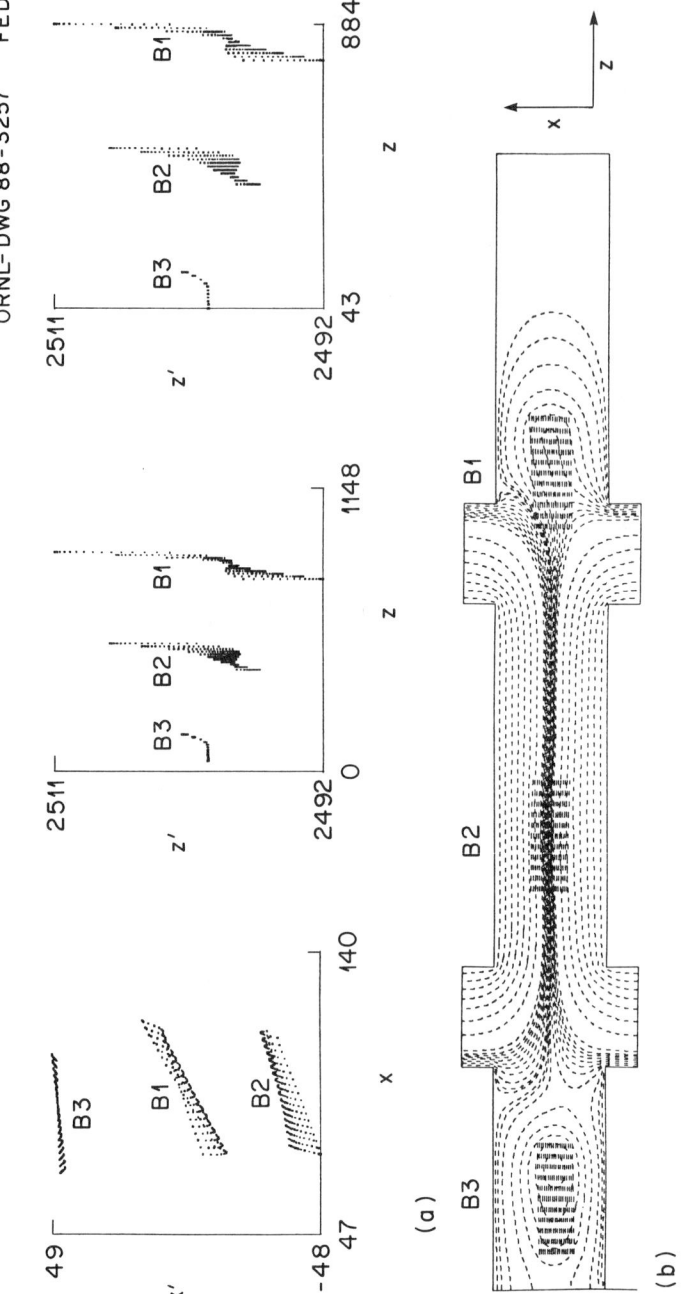

Fig. 8. Phase space occupation and deflector for $I_B = 50$ mA.

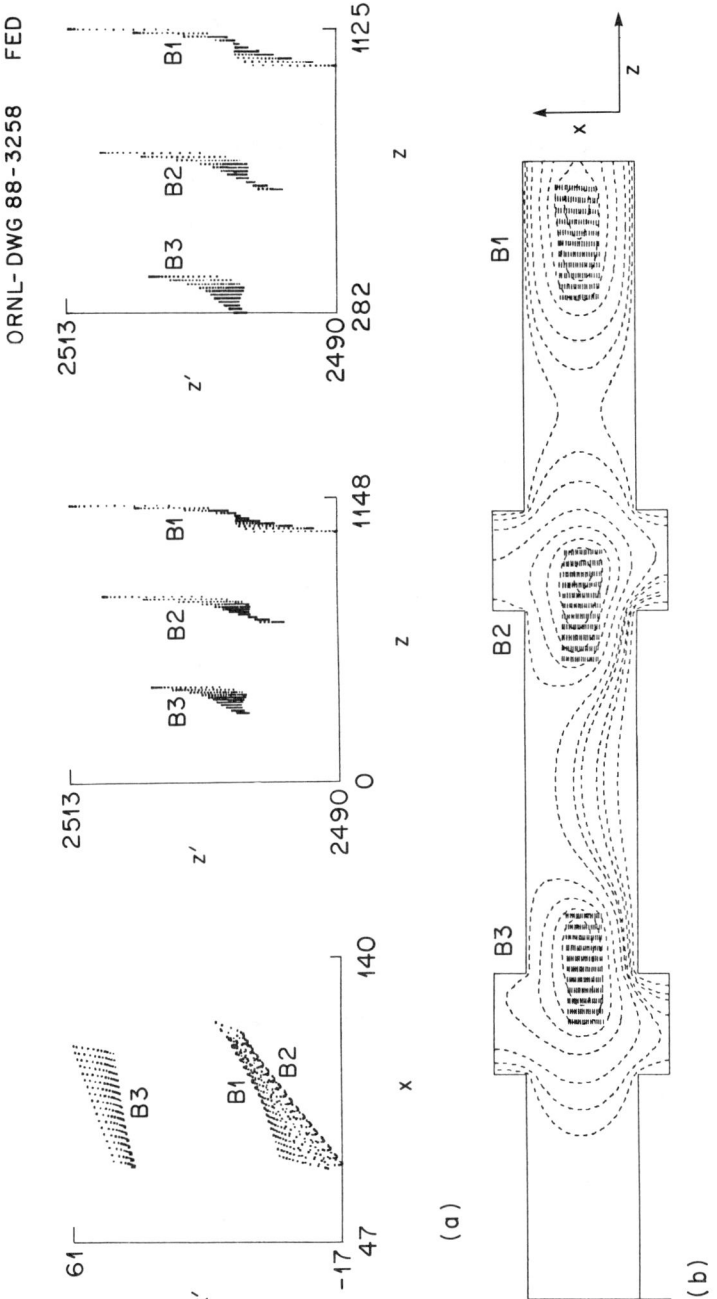

Fig. 9. Phase space occupation and deflector near a null for $I_B = 50$ mA.

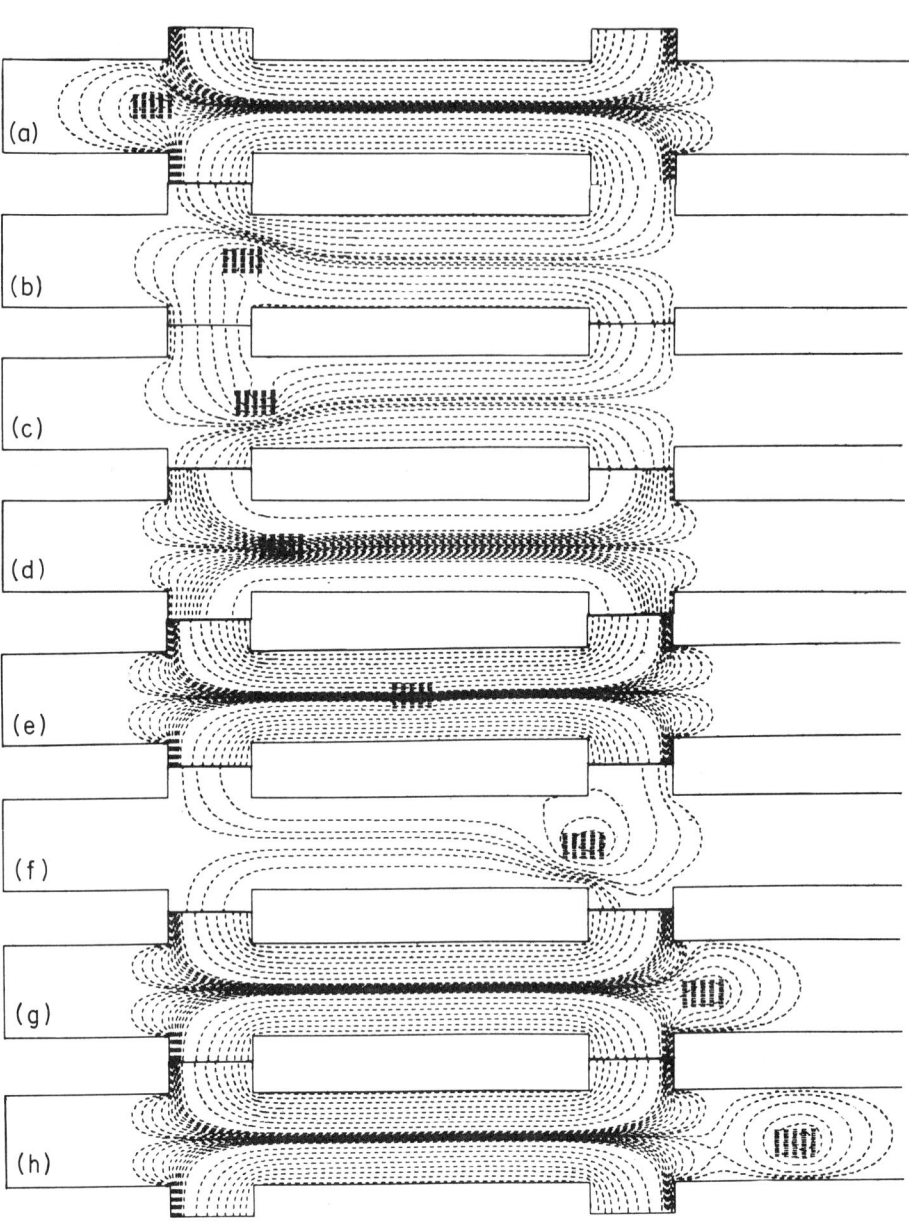

Fig. 10. Bunches occupying 28 degrees out of 360 degrees longitudinal emittance.

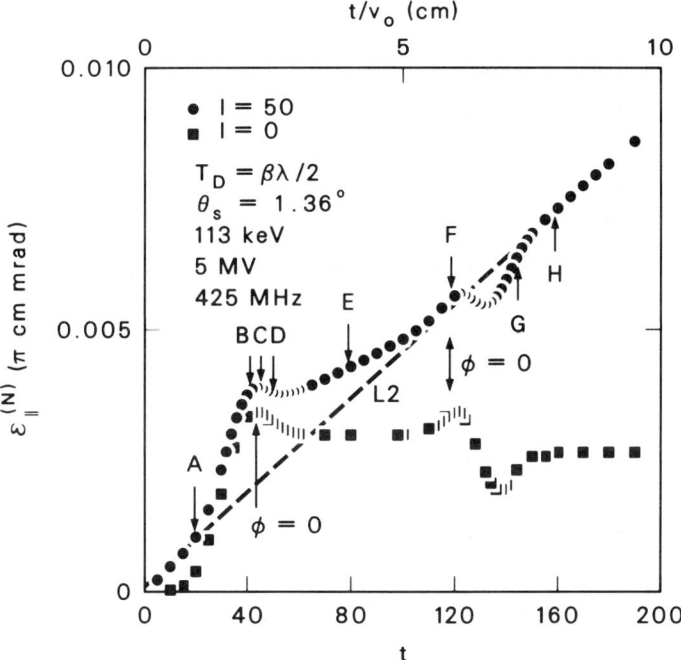

Fig. 11. Longitudinal emittance growth of bunches in Figure 10.

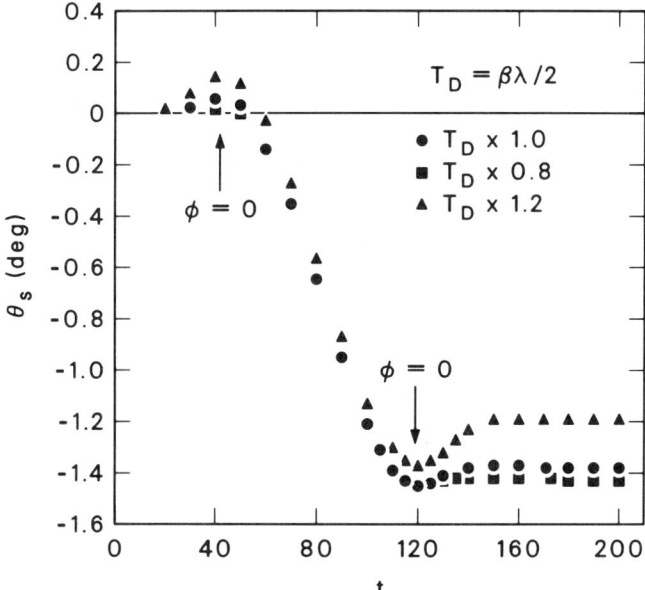

Fig. 12. Rf deflector steering showing mitigation by fringe fields.

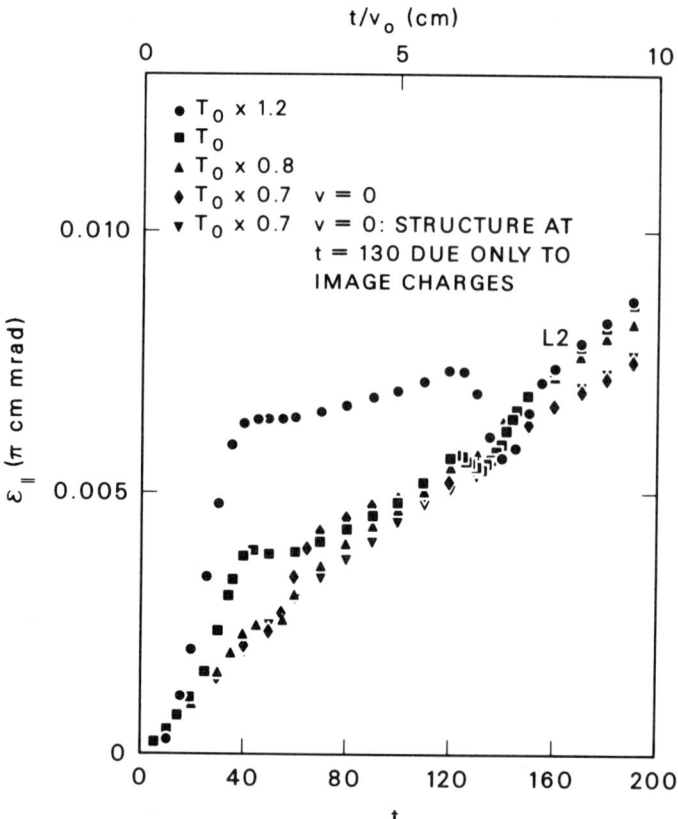

Fig. 13. Effect of deflector thickness on longitudinal emittance growth.

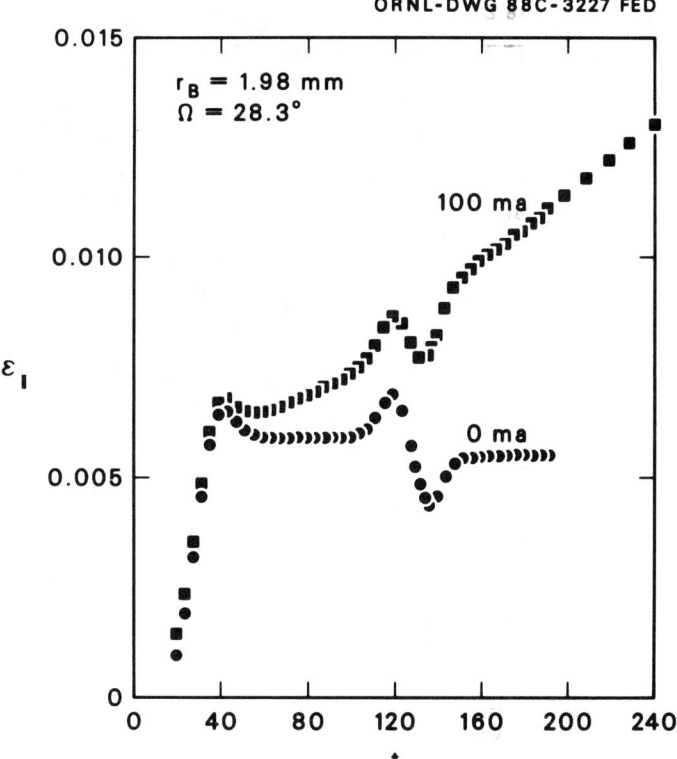

Fig. 14. 100 ma deflector.

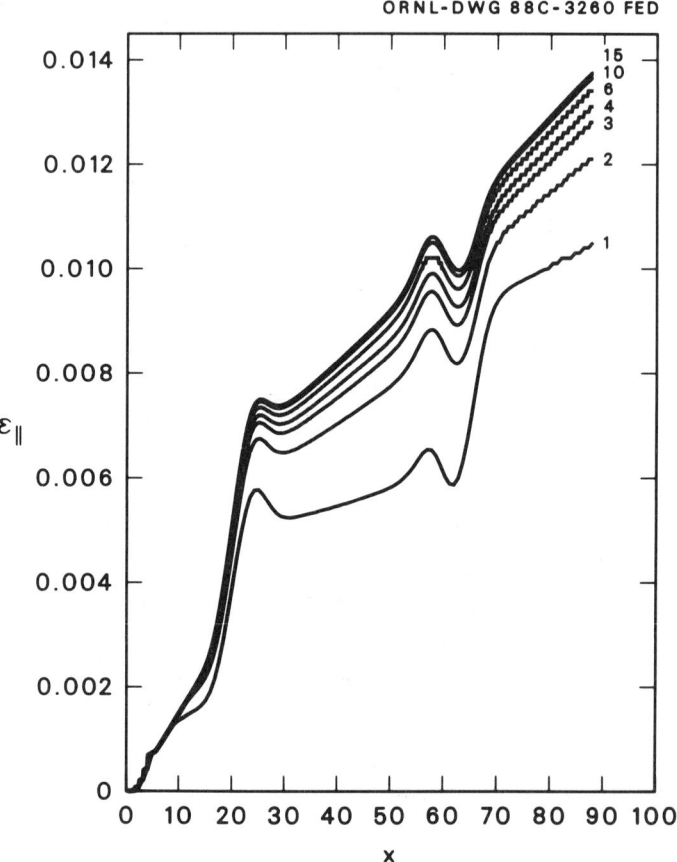

Fig. 15. Convergence as a function of axial resolution.

ORNL-DWG 88-3261 FED

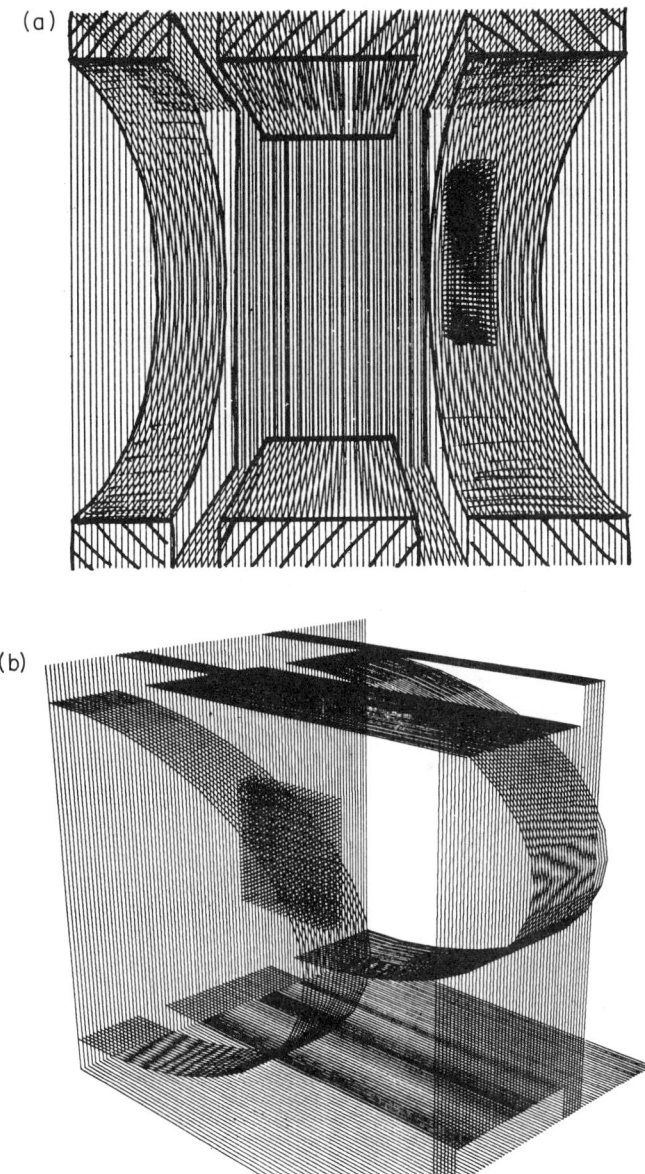

Fig. 16. A deflector in two views.

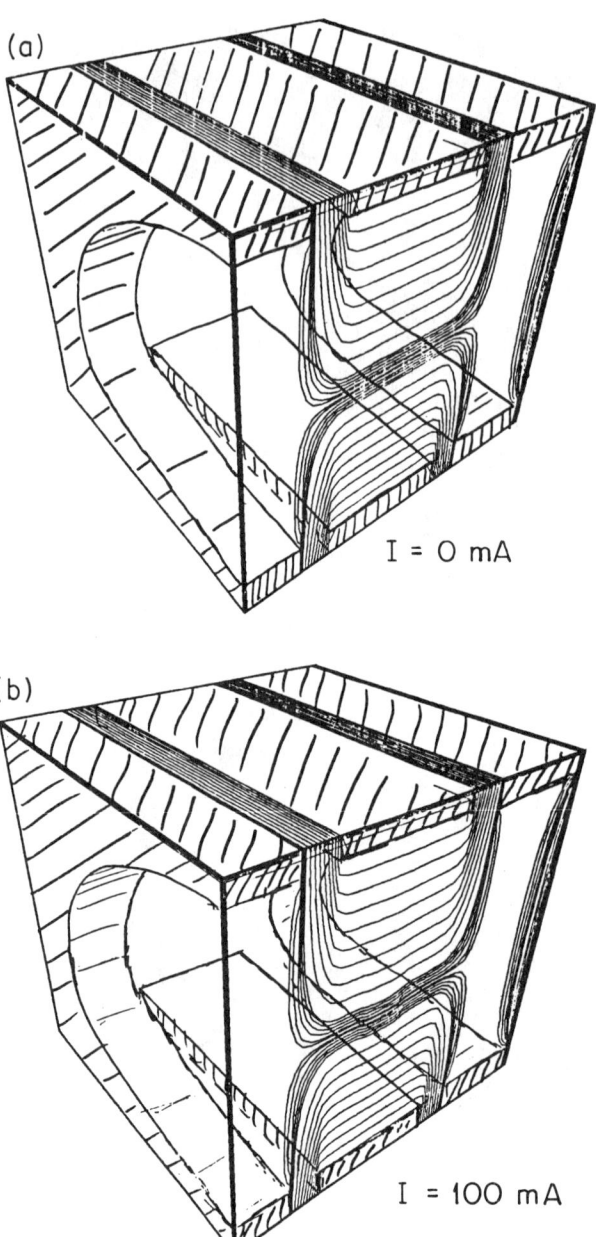

Fig. 17. Rf deflector potentials in 3 dimensions.

253

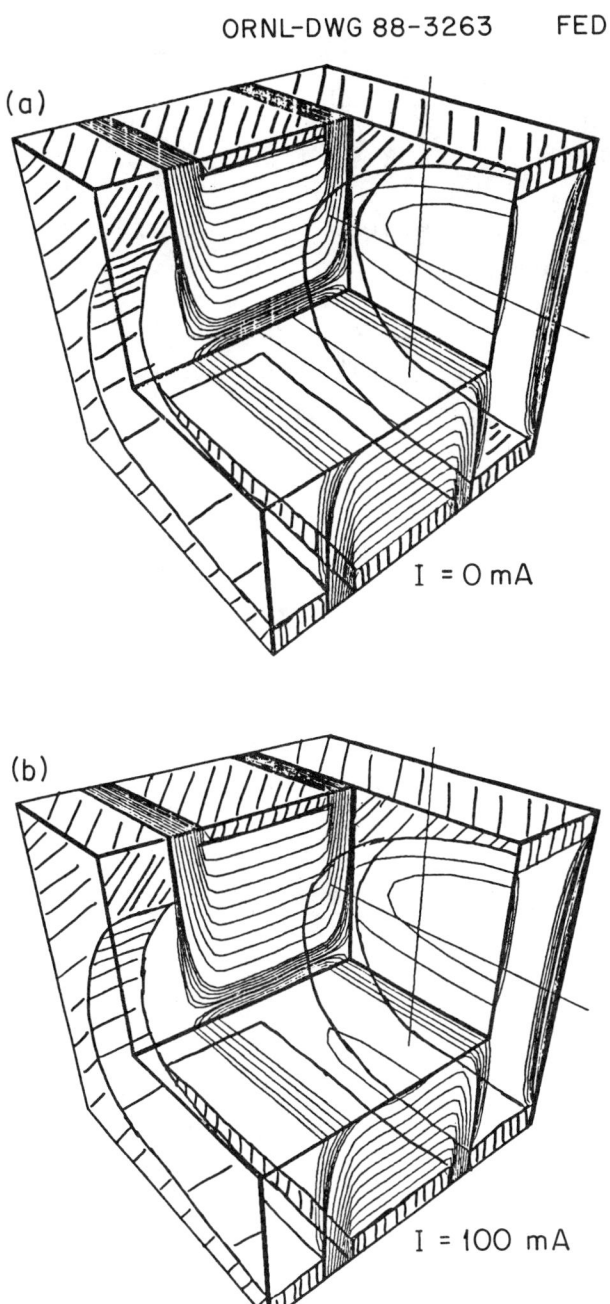

Fig. 18. Rf deflector with fringe fields.

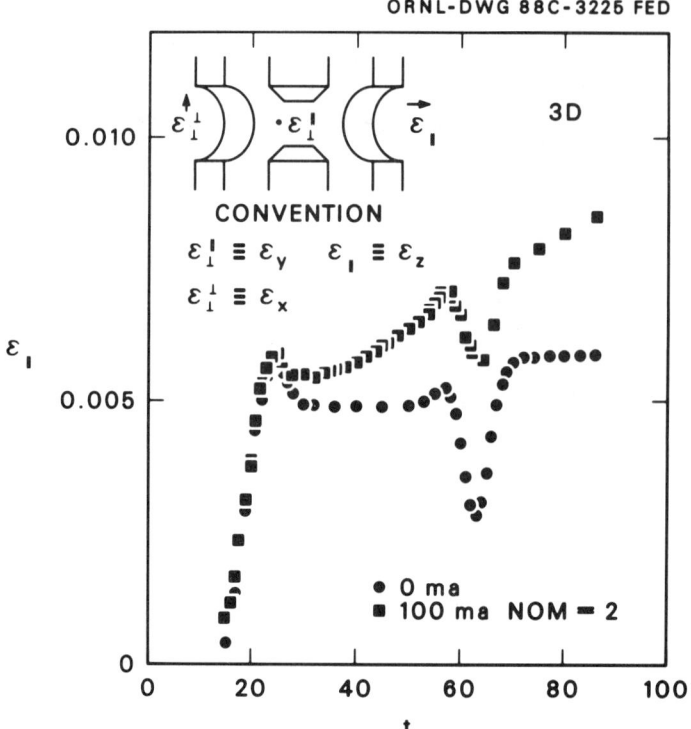

Fig. 19(a). Emittance growth as a function of time.

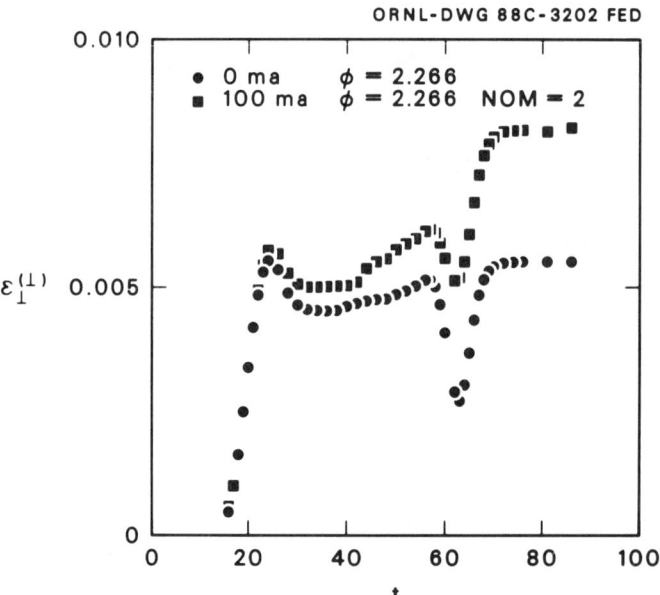

Fig. 19(b). Emittance growth as a function of time.

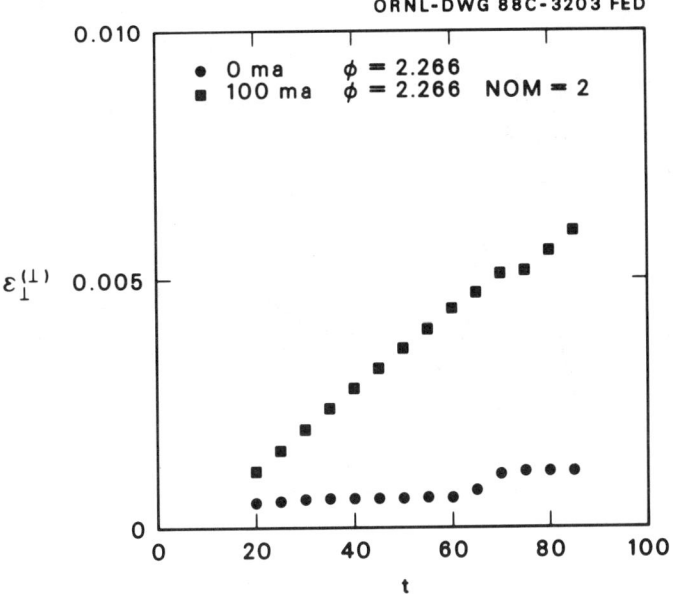

Fig. 19(c). Emittance growth as a function of time.

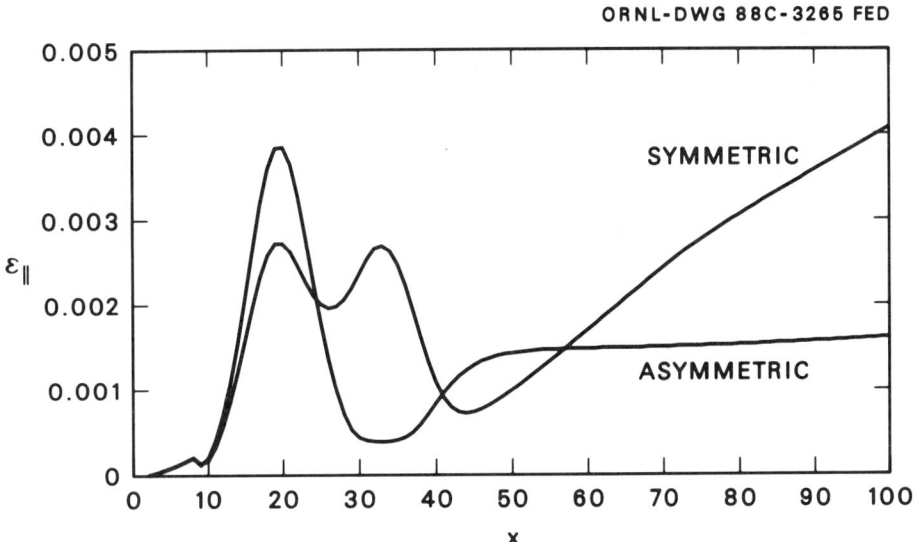

Fig. 20. Parallel emittance growth vs distance traveled for a soft beam.

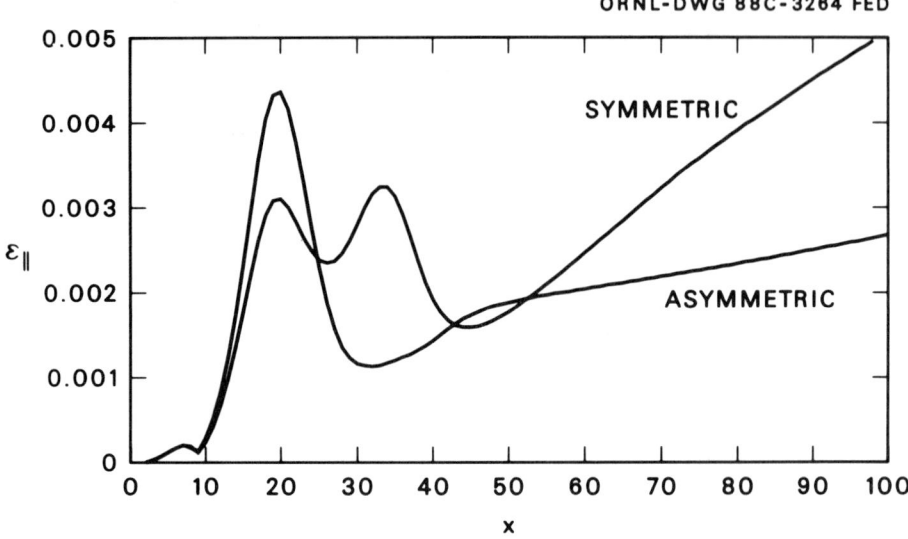

Fig. 21. Parallel emittance growth vs distance traveled for a hard beam.

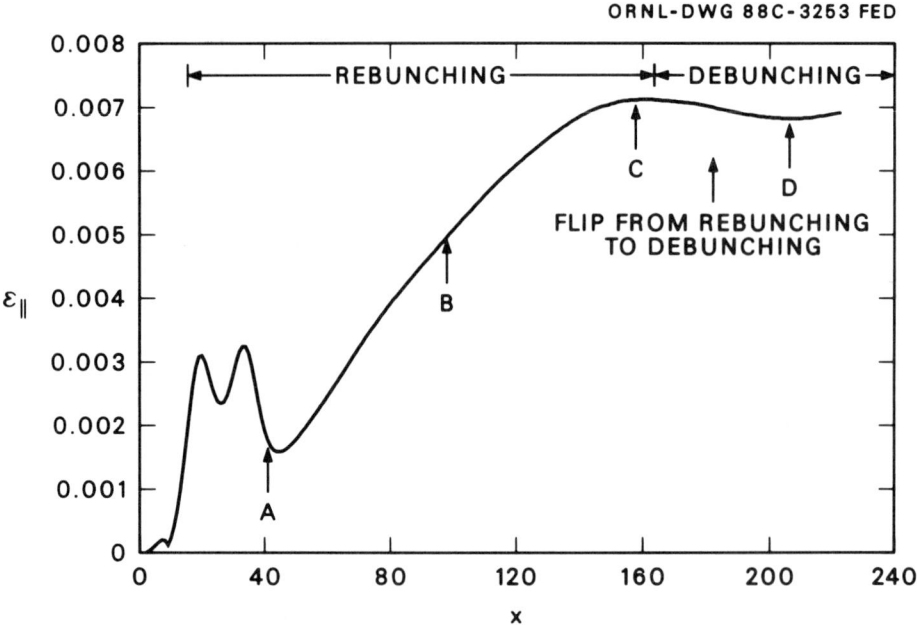

Fig. 22. Hard beam parallel emittance growth for a longer distance.

ORNL-DWG 88-3266 FED

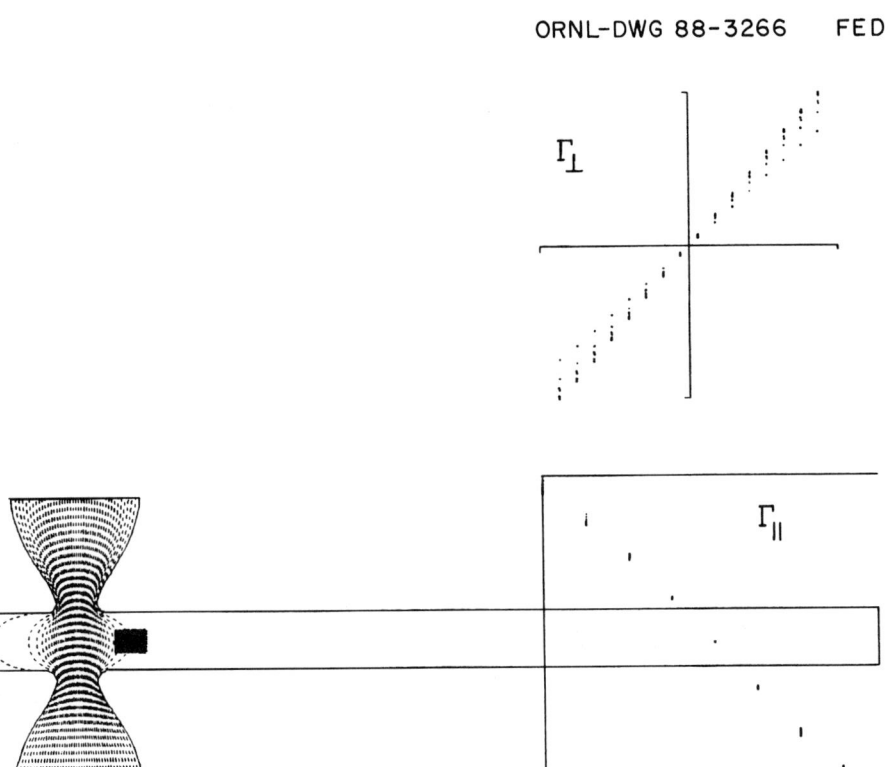

Fig. 23. Simple rebuncher with a beam pulse.

258

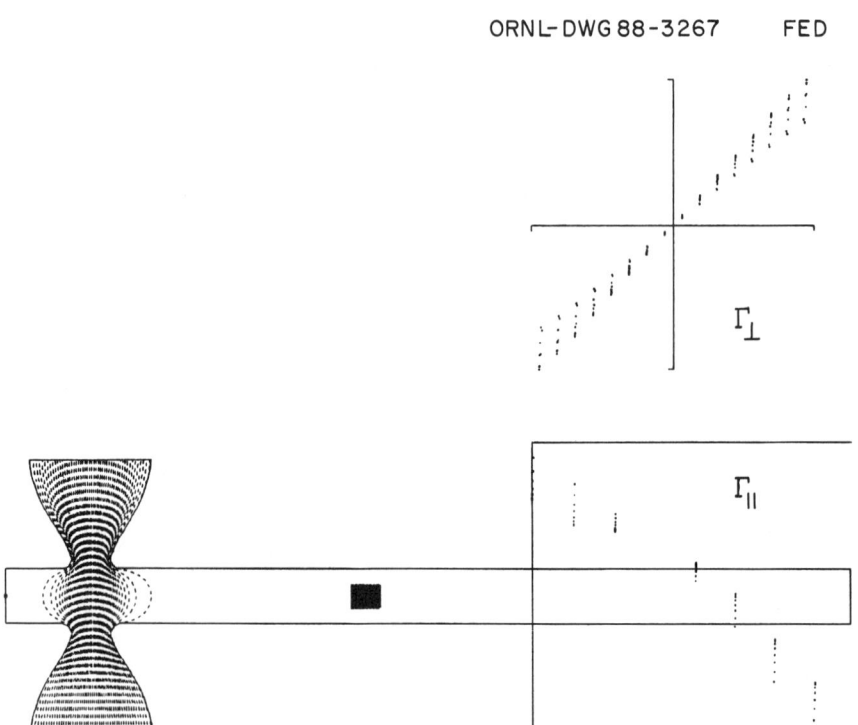

Fig. 24. Same as Figure 23 but at a later time.

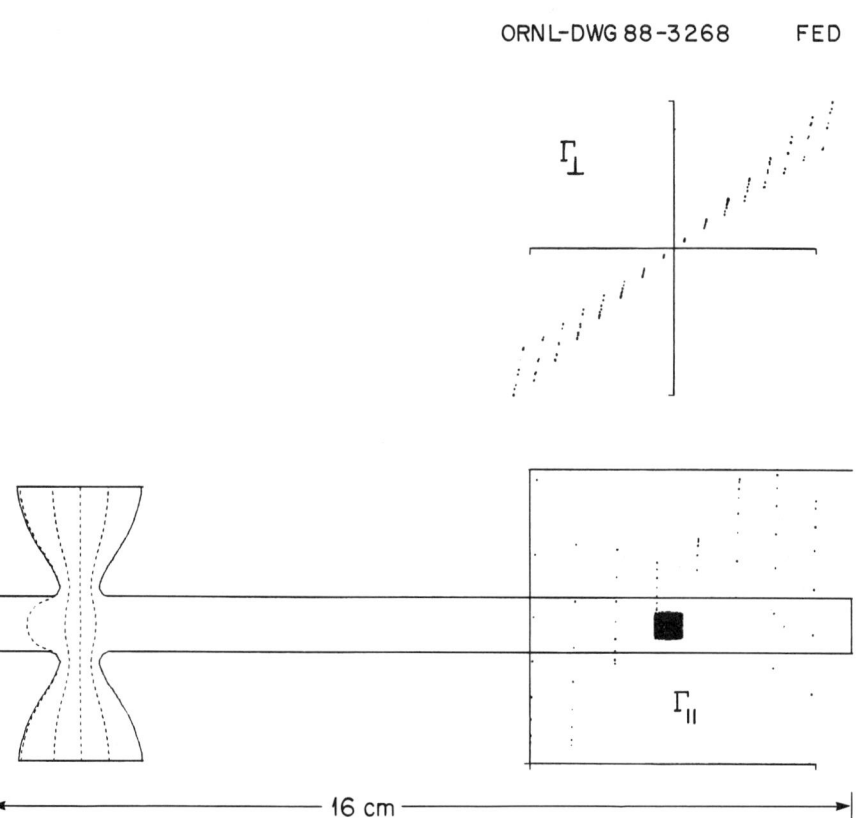

Fig. 25. Yet later.

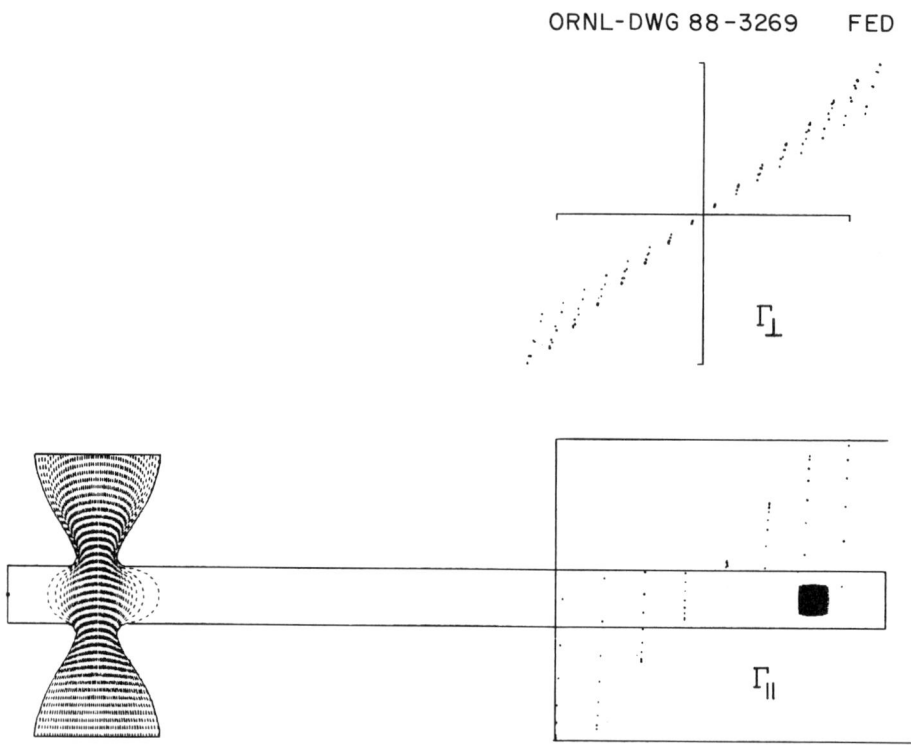

Fig. 26. Yet later: the beam debunching.

LIE ALGEBRAIC METHODS FOR CHARGED PARTICLE OPTICS

Alex J. Dragt
Center for Theoretical Physics, University of Maryland, College Park, MD 20742

ABSTRACT

New Lie algebraic methods have been developed for characterizing charged-particle optical systems. These methods are applicable to accelerator design, charged-particle beam transport, electron microscopes, and also light optics. They exploit the fact that the equations of motion for trajectories in these systems can be written in Hamiltonian form. The new methods represent the action of each separate element of a compound optical system, including all departures from paraxial optics, by a certain operator. The operators for the various elements can then be concatenated, following well defined rules, to obtain a resultant operator that characterizes the entire system. Finally, this resultant operator can be analyzed using various Lie algebraic and group-theoretical tools in order to determine the performance of the system.

The use of Lie algebraic methods has several benefits. First, the calculation of high-order nonlinear effects is facilitated. Second, Lie algebraic methods provide an optimal method for characterizing optical systems. Consequently, their use is expected to minimize computer storage, maximize computational speed, and facilitate communication, insight, and improved design. Finally, Lie algebraic methods and associated group theoretical concepts both suggest entirely new ways of looking at problems and provide a powerful analytic and computational calculus for analyzing and describing the behavior of optical systems. In particular, it should be possible eventually to analyze and control nonlinear behavior with the same facility and completeness that we now handle linear behavior.

BIBLIOGRAPHY

A. Dragt and J. Finn, "Lie Series and Invariant Functions for Analytic Symplectic Maps," J. Math. Phys. **17**, p. 2215 (1976).

A. Dragt, "A Method of Transfer Maps for Linear and Nonlinear Beam Elements," IEEE Trans. Nucl. Sci. **NS-26**, No. 3, p. 3601 (1979).

A. Dragt, "Transfer Map Approach to the Beam-Beam Interaction, *Nonlinear Dynamics and the Beam-Beam Interaction,* M. Month and J. Herrera, eds., AIP Conf. Proc. No. 57, p. 143 (1979).

A. Dragt and O. Jakubowicz, "Analysis of the Beam-Beam Interaction Using Transfer Maps," Proceedings of SLAC Symposium on the Beam-Beam Interaction, SLAC-PUB-2624, Stanford University, Stanford, CA (1980).

A. Dragt and D. Douglas, "Charged Particle Beam Transport Using Lie Algebraic Methods," IEEE Trans. Nucl. Sci. **NS-28**, p. 2522 (1981).

D. Douglas, "Lie Algebraic Methods for Particle Accelerator Theory," Ph.D. Thesis, Department of Physics, University of Maryland (1982).

A. Dragt, "A Lie Algebraic Theory of Geometrical Optics and Optical Aberrations," J. Opt. Sci. Am. **72**, p. 372 (1982).

A. Dragt, "Lectures on Nonlinear Orbit Dynamics," *Physics of High Energy Particle Accelerators,* AIP Conf. Proc. No. 87, R.A. Carrigan et al., ed. (1982).

A. Dragt and D. Douglas, "Lie Algebraic Method for Charged Particle Beam Transport and Particle Tracking," Proceedings of the Brookhaven Conference on Accelerator Orbit and Particle Tracking Programs, Brookhaven National Laboratory Report BNL-31761 (1982).

A. Dragt and D. Douglas, "Lie Algebraic Methods for Particle Tracking Calculations," Proceedings of the 12th International Conference on High-Energy Accelerators, F.T Cole and R. Donaldson, eds., Fermilab (1983).

A. Dragt and D. Douglas "MARYLIE, the Maryland Lie Algebraic Beam Transport and Particle Tracking Program," IEEE Trans. Nucl. Sci. **NS-30**, p. 2442 (1983).

A. Dragt and E. Forest, "Computation of Nonlinear Behavior of Hamiltonian Systems Using Lie Algebraic Methods," J. Math. Phys. **24**, p. 2734 (1983).

A. Dragt and D. Douglas, "Particle Tracking Using Lie Algebraic Methods," *Computing in Accelerator Design and Operation,* W. Busse and R. Zelazny, eds., Lectures Notes in Physics **215**, Springer-Verlag (1984).

E. Forest, "Lie Algebraic Methods for Charged Particle Beams and Light Optics," Ph.D. Thesis, Department of Physics, University of Maryland (1984).

J. Milutinovic, "Comparison of MARYLIE 3.0 with Numerical Integration," Department of Physics Technical Report, University of Maryland (1984).

D. Douglas et al., "A Method to Render Second Order Beam Optics Programs Symplectic," IEEE Trans. Nucl. Sci. **NS-32**, p. 2279 (1985).

A. Dragt, et al., "MARYLIE 3.0 - A Program for Nonliner Analysis of Accelerator and Beamline Lattices," IEEE Trans. Nucl. Sci. **NS-32**, p. 2311 (1985).

A. Dragt, "Nonlinear Lattice Functions," *Proceedings of the 1984 Summer Study on the Design and Utilization of the Superconducting Super Collider,* Snowmass, Colorado, R. Donaldson and J. Morfin, eds., Am. Phys. Soc. (1985).

E. Forest, "Normal Form Algorithm on Nonlinear Symplectic Maps," Report SSC-29, Lawrence Berkeley Laboratory (1985).

E. Forest, "Equivalence of Michelotti's Normal Form and the Map Normal Form as Used by the MARYLIE Code," Report SSC-30, Lawrence Berkeley Laboratory (1985).

A. Dragt, "A Comparison of Third-Order Geometric Aberrations and Second- and Third-Order Chromatic Aberrations for Quadrupole Systems as Computed by MARYLIE, GIOS, and TRANSPORT," Department of Physics Technical Report, University of Maryland (1986).

A. Dragt and E. Forest, "Lie Algebraic Theory of Charged Particle Optics and Electron Microscopes," *Advances in Electronics and Electron Physics,* **67**, P. Hawkes, ed., Academic Press (1986).

A. Dragt, E. Forest, and K. Wolf, "Foundations of a Lie Algebraic Theory of Geometrical Optics," *Lie Methods in Optics,* J.S. Mondragon and K.B. Wolf, eds., Springer-Verlag (1986).

A. Dragt and R. Ryne, "Use of MARYLIE to Compute Second- and Third-Order Aberrations," Department of Physics Technical Report, University of Maryland (1986).

E. Forest, "Analytical Computation of the Smear," Report SSC-95, Lawrence Berkeley Laboratory (1986).

L. Healy, "Lie Algebraic Methods for Treating Lattice Parameter Errors in Particle Accelerators," Ph.D. Thesis, Department of Physics, University of Maryland (1986).

F. Neri, "Notes on Symplectification through Fifth Order," Department of Physics Technical Report, University of Maryland (1986).

A. Dragt, "Elementary and Advanced Lie Algebraic Methods with Applications to Accelerator Design, Electron Microscopes, and Light Optics," Nucl. Instrum. Meth. Phys. Res. **A258**, p. 339 (1987).

A. Dragt and L. Healy, "Lie Algebraic Methods for Treating Lattice Parameter Errors in Accelerators," Proceedings of the 1987 IEEE Particle Accelerator Conference, p. 1060 (1987).

A. Dragt and H. Moser, "Influence of Strongly Curved Large-Bore Superconducting Bending Magnets on the Optics of Storage Rings," Nucl. Instrum. Meth. Phys. Res. **B24/25**, p. 877 (1987).

A. Dragt et al., "MARYLIE 3.0, a Program for Charged Particle Beam Transport Based on Lie Algebraic Methods," Department of Physics Technical Report, University of Maryland (1987).

A. Dragt and R. Ryne, "Lie Algebraic Treatment of Space Charge," Proceedings of the 1987 IEEE Particle Accelerator Conference **2**, p. 1063 (1987).

A. Dragt and R. Ryne, "Numerical Computation of Transfer Maps Using Lie Algebraic Methods," Proceedings of the 1987 IEEE Particle Accelerator Conference **2**, p. 1081 (1987).

E. Forest, "Hamiltonian-Free Perturbation Theory: The Concept of Phase Advance," Report SSC-111, Lawrence Berkeley Laboratory (1987).

B. Leemann and E. Forest, "Brief Description of the Tracking Codes FASTRAC and THINTRAC," Report SSC-133, Lawrence Berkeley Laboratory (1987).

E. Forest, "Canonical Integrators as Tracking Codes (or How to Integrate Perturbation Theory with Tracking)," Report SSC-138, Lawrence Berkeley Laboratory (1987).

E. Forest, "Lie Algebraic Maps and Invariants Produced by Tracking Codes," Particle Accelerators **22**, p. 15 (1987).

E. Forest et al., "Study of the Aberrations of Periodic Arc Using the Lie Algebraic Techniques," Nucl. Instrum. & Meth. Phys. Res. **A258**, p. 355 (1987).

R. Moses et al., "Scaling Laws for Aberrations in Magnetic Quadrupole Lens Systems," Proceedings of the 1987 IEEE Particle Accelerator Conference, **3**, p. 1764 (1987).

F. Neri, "Lie Algebras and Canonical Integration," Department of Physics Technical Report, University of Maryland (1987).

R. Ryne, "Lie Algebraic Treatment of Space Charge," Ph.D. Thesis, Department of Physics, University of Maryland (1987).

G. Turchetti, A. Bazzani, and A. Pisent, "Normal Forms for Hamiltonian Maps and Future Applications to Accelerators," Dipartimento di Fisica della Universita di Bologna, Bologna, Italy (1987).

K. Yokoya "Calculations of the Equilibrium Polarization of Stored Electron Beams Using Lie Algebra," Nucl. Instrum. Phys. Meth. Phys. Res. **A258**, p. 149 (1987).

M. Berz, "Part I, A Survey of Differential Algebra and Its Use for the Extraction of Maps to Arbitrary Order," Superconducting Super Collider Central Design Group Report, Lawrence Berkeley Laboratory (1988).

E. Forest et al., "Part II, Normal Form Methods for Complicated Periodic Systems: A Complete Solution Using Differential Algebra and Lie Operators," Superconducting Super Collider Central Design Group Report, Lawrence Berkeley Laboratory (1988).

A. Dragt and H. Moser, "Nonlinear Beam Optics with Real Fields in Compact Storage Rings," Nucl. Instrum. Method. Phys. Res. **B30**, p. 105 (1988).

A. Dragt et al., "Lie Algebraic Treatment of Linear and Nonlinear Beam Dynamics," to appear in Annual Review of Nuclear and Particle Science (1988).

A. Dragt, F. Neri, and G. Rangarajan, "Lie Algebraic Treatment of Moments and Moment Invariants," in preparation (1988).

L. Healy, "Concatenation of Lie Algebraic Maps," in preparation (1988).

F. Neri, "GENREC 5.0, a Program for Handling Rare Earth Cobalt Quads Through Fifth Order," in preparation (1988).

SELF-CONSISTENT TRANSFER MAPS FOR BEAMS WITH SPACE CHARGE USING LIE ALGEBRAIC METHODS

Robert D. Ryne
Lawrence Livermore National Laboratory
Livermore, CA 94550

ABSTRACT

We present a new approach to doing beam transport calculations in the presence of space charge. The approach is based on the self-consistent computation of transfer maps. In the linear approximation, the maps are represented by matrices. For the general case, the maps are represented and computed using Lie algebraic methods. This new approach is particularly useful for computing tunes and aberration coefficients in the presence of space charge.

INTRODUCTION

The purpose of this paper is to discuss the self-consistent computation of transfer maps in the presence of space charge [1].

The transfer map approach has been very useful in studying beam transport without space charge. Transfer map codes are routinely used in the fields of accelerator physics and magnetic optics to analyze and optimize beam transport systems.

The presence of space charge complicates beam transport analysis. In the past, attempts have been made to put space charge in transfer map codes by applying "space charge kicks" at various points along the beam transport system. This approach is a good one when the dynamics of the whole beam are of interest. However, it involves following a large number of particles, much like a particle simulation program.

Transfer map codes are often used to design and optimize beam transport systems. The strengths and positions of beamline elements are chosen and modified to produce desired tunes, minimize or cancel particular aberration coefficients, etc. Note that these are local quantities that depend on the beam dynamics near the origin. In contrast, particle simulation programs are usually used to study the global, large scale behavior of beams.

Is it possible to compute these local quantities (tunes and aberrations) in the presence of space charge without following particles? In other words, is it possible to compute transfer maps self-consistently without following particles? The answer is yes, as we will show below.

In the following sections, we will discuss the self-consistent computation of transfer maps. We believe it is worthwhile to present a simple, concrete example as soon as possible. This will demonstrate that our approach is a simple, obvious one. Therefore, after discussing some background material, and after outlining the steps for computing self-consistent transfer maps, we will perform a step-by-step analysis of a K-V distribution in a solenoid channel. This example will illustrate the main features of the self-consistent transfer map approach. Next we will discuss the treatment of more general distributions using Lie algebraic methods. In particular, we will discuss initially Gaussian distributions (in four-dimensional phase space). These distributions are treated by the computer program CHARLIE4F. Lastly, we will present a numerical example of a cold Gaussian beam expanding into a drift space.

PRELIMINARIES

First consider a system of noninteracting particles in a beam transport system. Suppose all the particles are governed by the same Hamiltonian, $H(x, p_x, y, p_y, z, p_z, t)$. The time evolution of the 6–vector $\zeta = (x, p_x, y, p_y, z, p_z)$ is governed by Hamilton's equations,

$$\frac{d\zeta}{dt} = -[H(\zeta, t), \zeta], \tag{1}$$

where [,] denotes the usual Poisson bracket. Now let t^i denote some initial time and let t^f denote some final time. We shall use the notation $\zeta^i = \zeta(t^i)$ and $\zeta^f = \zeta(t^f)$. Equation (1) may be regarded as defining a (generally nonlinear) mapping, $\mathcal{M}$, with that property that

$$\zeta^f = \mathcal{M}\zeta^i. \tag{2}$$

The quantity $\mathcal{M}$ is called the transfer map between t^i and t^f.

Now consider a beam of interacting particles. Let the beam evolve from t^i to t^f. During this time, the self-fields of the beam take on certain values. These fields (together with the external fields) determine the self-consistent Hamiltonian governing particle motion in the beam. As described in the previous paragraph, the Hamiltonian may be regarded as defining a transfer map, $\mathcal{M}$. In this case, it is a self-consistent transfer map.

Lastly, consider a system of particles that is characterized by a distribution function, $f(\zeta, t)$, with initial value

$$f(\zeta, t^i) = g(\zeta). \tag{3}$$

Further, suppose that f satisfies the equation

$$\frac{d}{dt} f(\zeta, t) = 0. \tag{4}$$

Then it is easy to show that the time-dependent distribution function is given by

$$f(\zeta, t) = g(\mathcal{M}^{-1}\zeta). \tag{5}$$

Since $\mathcal{M}$ determines the evolution of the particle beam, we will concentrate on calculating $\mathcal{M}$. The general problem of computing $\mathcal{M}$ for a Hamiltonian system using Lie algebraic methods has been treated by Dragt and Forest [2]. Using these results, computer programs have been developed to numerically compute $\mathcal{M}$ for a variety of (nonideal) beamline elements [3]. For example, the program GENREC 3.0 computes the third order transfer map for a rare earth cobalt quadrupole magnet including fringe fields; the program GENDIP 3.0 computes the third order map for a magnetic dipole with mid-plane symmetry and fringe fields. These same techniques can be used to compute self-consistent transfer maps. However, the analysis is complicated by the fact that the self-consistent Hamiltonian depends on $\mathcal{M}$ itself and on the initial distribution function.

In the next section, we outline the procedure for computing self-consistent transfer maps.

STEPS FOR COMPUTING A SELF-CONSISTENT TRANSFER MAP

1. Let the beam transport system be given.
2. Choose the initial distribution function, $g(x, p_x, y, p_y, z, p_z)$.
3. Choose a representation of the transfer map, $\mathcal{M}^{-1}$.
4. Analytically transform g by $\mathcal{M}^{-1}$ to obtain the evolved distribution function, $f(x, p_x, y, p_y, z, p_z, t)$.
5. Analytically integrate f over the momenta to obtain the charge density, $\rho(x, y, z)$, and integrate $f\vec{v}$ over the momenta to obtain the current density (if necessary).
6. Analytically solve the relevant field equations. For example, in the electrostatic approximation, solve Poisson's equation to obtain the scalar potential, $\psi(x, y, z)$, as a power series in x, y and z (with coefficients that depend on $\mathcal{M}$).
7. Obtain the self-consistent Hamiltonian, $H(x, p_x, y, p_y, z, p_z, t)$, as a power series in (x, p_x, y, p_y, z, p_z) (with coefficients that depend on $\mathcal{M}$).
8. Given the Hamiltonian, H, obtain the equations of motion for $\mathcal{M}$ and integrate numerically if necessary.

In the next section we will perform these steps explicitly for a particular problem.

EXAMPLE: A K-V DISTRIBUTION IN A SOLENOID CHANNEL

The steps 1 to 8 below correspond to the steps outlined in the previous section.

1. Consider an axially symmetric magnetic focussing channel (a solenoid channel) in the linear approximation. Let the external fields be associated with the following vector potential,

$$A_x = -\frac{B_0}{2}y, \qquad (6a)$$

$$A_y = +\frac{B_0}{2}x, \qquad (6b)$$

$$A_z = 0, \qquad (6c)$$

so that

$$\vec{B} = B_0(z)\hat{z}. \qquad (7)$$

2. Let the initial distribution function be a Kapchinskij-Vladimirskij (K-V) distribution. Specifically, let

$$g(x, p_x, y, p_y) = \frac{1}{\pi^2 \sigma^2 \lambda^2} \delta\left(\frac{r^2}{\sigma^2} + \frac{p^2}{\lambda^2} - 1\right), \qquad (8)$$

where

$$r^2 = x^2 + y^2, \qquad (9a)$$

$$p^2 = p_x^2 + p_y^2. \qquad (9b)$$

3. The scalar potential associated with a K-V distribution is proportional to r^2, so the self-forces are linear in x and y. Since we are considering a linear focussing channel, all the forces in the problem are linear. Thus, the transfer map $\mathcal{M}$ can be represented by a matrix M. Let the matrix M^{-1} be given by

$$M^{-1} = M_0^{-1} R, \tag{10}$$

where $R(z)$ is a rotation matrix and where

$$M_0^{-1} = \begin{pmatrix} a & b & 0 & 0 \\ c & d & 0 & 0 \\ 0 & 0 & a & b \\ 0 & 0 & c & d \end{pmatrix}. \tag{11}$$

The quantities a, b, c and d are functions of z, and they satisfy

$$ad - bc = 1. \tag{12}$$

4. It follows that the z-dependent distribution function is given by

$$f(x, p_x, y, p_y, z) = \frac{1}{\pi^2 \sigma^2 \lambda^2} \delta \left(\frac{a^2 r^2 + 2ab\vec{r}\cdot\vec{p} + b^2 p^2}{\sigma^2} \right.$$
$$\left. + \frac{c^2 r^2 + 2cd\vec{r}\cdot\vec{p} + d^2 p^2}{\lambda^2} - 1 \right). \tag{13}$$

5. Let Λ denote the charge per unit length. Integrating (13) over p_x and p_y, we obtain the charge density, $\rho(x, y)$,

$$\rho(x, y) = \begin{cases} \Lambda/(\pi R^2) & if \quad r < R \\ 0 & if \quad r > R, \end{cases} \tag{14}$$

where

$$R^2 = \sigma^2 d^2 + \lambda^2 b^2. \tag{15}$$

That is, R denotes the beam edge.

6. Using Gauss' law, the scalar potential ψ, valid inside the beam, is given by

$$\psi(x, y) = -\frac{\Lambda}{4\pi\varepsilon_o} \frac{r^2}{R^2}. \quad (r \leq R) \tag{16}$$

7. Using z as the independent variable, the Hamiltonian (in MKSA units) for a particle of charge q is given approximately by the following expression:

$$H = \frac{1}{2}(p_x^2 + p_y^2) - \frac{q}{p^o}(p_x A_x + p_y A_y)$$
$$+ \frac{1}{2}\left(\frac{q}{p^o}\right)^2 (A_x^2 + A_y^2) + \frac{q\psi}{p^o v_z^o (\gamma^o)^2}. \tag{17}$$

Here, $p^o = \gamma^o m v_z^o$ is the design momentum, that is assumed to be constant. The quantities p_x and p_y are the usual canonical momenta divided by p^o. The factor of $1/(\gamma^o)^2$ accounts for the self-magnetic field. Substitution of (6) and (16) into (17) yields

$$H = H_0 - \alpha J_z, \tag{18}$$

where

$$H_0 = \frac{1}{2}p^2 + \frac{1}{2}\alpha^2 r^2 - \frac{K}{2}\left(\frac{r^2}{R^2}\right), \qquad (19a)$$

$$J_z = (\vec{r} \times \vec{p}) \cdot \hat{z}, \qquad (19b)$$

and where

$$\alpha = \frac{qB_o}{2p^o}, \qquad (20)$$

$$K = \frac{q\Lambda}{2\pi\varepsilon_o p^o v_z^o (\gamma^o)^2}. \qquad (21)$$

Note that the Poisson bracket of H_0 with J_z is zero. This is the reason that the transfer matrix, M^{-1}, can be factored as in (10).

8. The equation of motion for M_0^{-1} is given by [2]

$$\left(M_0^{-1}\right)' = -M_0^{-1} J S, \qquad (22)$$

where a prime denotes d/dz. Here, the matrix J is given by

$$J = \begin{pmatrix} 0 & 1 & 0 & 0 \\ -1 & 0 & 0 & 0 \\ 0 & 0 & 0 & 1 \\ 0 & 0 & -1 & 0 \end{pmatrix}. \qquad (23)$$

The matrix S is obtained from the quadratic part of the H_0 (in this case H_0 itself):

$$S = \begin{pmatrix} \alpha^2 - \frac{K}{R^2} & 0 & 0 & 0 \\ 0 & 1 & 0 & 0 \\ 0 & 0 & \alpha^2 - \frac{K}{R^2} & 0 \\ 0 & 0 & 0 & 1 \end{pmatrix}. \qquad (24)$$

It follows that

$$a' = b\left(\alpha^2 - \frac{K}{R^2}\right), \qquad (25a)$$

$$b' = -a, \qquad (25b)$$

$$c' = d\left(\alpha^2 - \frac{K}{R^2}\right), \qquad (25c)$$

$$d' = -c. \qquad (25d)$$

These equations can be integrated numerically. The numerical solution of these equations, together with (13), completely solves the problem of determining the evolution of the initial K-V distribution, (8).

Last, we will show that (15) and (25) lead to the usual K-V envelope equation. Differentiating (15) with respect to z yields

$$RR' = \sigma^2 dd' + \lambda^2 bb', \qquad (26)$$

or, using (25b) and (25d),

$$R' = -\frac{1}{R}\left(\sigma^2 cd + \lambda^2 ab\right). \qquad (27)$$

Differentiating again, and using (25a) and (25c), it follows that

$$R'' = \frac{1}{R^3}\left[R^2\left(\sigma^2 c^2 + \lambda^2 a^2\right) - \left(\sigma^2 cd + \lambda^2 ab\right)^2\right] - R\left(\alpha^2 - \frac{K}{R^2}\right), \qquad (28)$$

or

$$R'' = \frac{\sigma^2 \lambda^2}{R^3}(ad - bc)^2 - R\left(\alpha^2 - \frac{K}{R^2}\right). \qquad (29)$$

By (12), we obtain

$$R'' + \alpha^2 R - \frac{K}{R} - \frac{\epsilon^2}{R^3} = 0, \qquad (30)$$

where $\epsilon = \sigma\lambda$. This is the usual K-V envelope equation for a cylindrically symmetric beam.

TREATMENT OF MORE GENERAL INITIAL DISTRIBUTIONS USING LIE ALGEBRAIC METHODS

The treatment of a K-V distribution in a linear transport system is simple because the transfer map can be represented by a matrix. However, in the general case the transfer map contains linear and nonlinear terms. This makes it algebraically difficult to perform some of the steps described above that are required to compute $\mathcal{M}$ self-consistently. For this reason, it is important to choose a representation of $\mathcal{M}$ that will make these analytic calculations as simple as possible. We believe that Lie algebraic methods are well suited for this purpose.

The most common method of representing $\mathcal{M}$ is as a Taylor series in the initial conditions:

$$\zeta_a^f = \sum_{b=1}^{6} M_{ab}\zeta_b^i + \sum_{1 \leq b \leq c}^{6} T_{abc}\zeta_b^i\zeta_c^i + \sum_{1 \leq b \leq c \leq d}^{6} U_{abcd}\zeta_b^i\zeta_c^i\zeta_d^i + \cdots. \qquad (31)$$

The Lie algebraic approach is to use a Lie series to represent $\mathcal{M}$:

$$\mathcal{M} = exp(:f_2:)\,exp(:f_3:)\,exp(:f_4:)\cdots. \qquad (32)$$

The quantities $f_n(\zeta)$ are homogeneous polynomials of degree n in ζ. They have a simple physical interpretation: f_2 governs the linear (paraxial) behavior, and the polynomials f_3, f_4, etc., govern the nonlinear behavior. The quantities $:f_n:$ are operators associated with f_n. By expanding the exponentials in (32), we obtain a power series that is completely equivalent to (31), as shown below:

$$\zeta_a^f = \sum_{b=1}^{6} M_{ab}\zeta_b^i + \sum_{1\le b\le c}^{6} T_{abc}\zeta_b^i\zeta_c^i + \sum_{1\le b\le c\le d}^{6} U_{abcd}\zeta_b^i\zeta_c^i\zeta_d^i + \cdots, \tag{33a}$$

$$\Updownarrow \qquad\qquad \Updownarrow \qquad\qquad \Updownarrow$$

$$\zeta_a^f = e^{:f_2:}\zeta_a^i \;+\; e^{:f_2:}(:f_3:)\zeta_a^i \;+\; e^{:f_2:}\left(\frac{1}{2}:f_3:^2 + :f_4:\right)\zeta_a^i + \cdots. \tag{33b}$$

An advantage of the Lie algebraic representation is that the coefficients of f_n are all independent, whereas the coefficients of T_{abc}, W_{abcd}, etc. are interrelated through the symplectic condition. Thus, by using the Lie algebraic representation, we handle the minimum number of quantities in our analytic calculations, and we numerically integrate the minimum number of quantities in our numerical calculations.

In the next section we will use a Lie algebraic representation to compute the transfer map through third order for an initially Gaussian distribution.

INITIALLY GAUSSIAN DISTRIBUTIONS

Consider a particle beam propagating on the axis of a perfectly aligned beam transport system consisting of drifts, quadrupoles, octupoles and combined function quadrupole/octupole magnets. (For simplicity, assume that there are no skew multipoles). Suppose the beam is initially represented by a Gaussian distribution in four-dimensional phase space. Let the initial distribution function, $g(\zeta)$, be given by

$$g(\zeta) = \frac{1}{(2\pi)^2 \sigma_x \lambda_x \sigma_y \lambda_y} exp\left[-\frac{1}{2}\left(\frac{x^2}{\sigma_x^2} + \frac{p_x^2}{\lambda_x^2} + \frac{y^2}{\sigma_y^2} + \frac{p_y^2}{\lambda_y^2}\right)\right]. \tag{34}$$

We will use a third order Lie algebraic representation of $\mathcal{M}$. Our calculation will be two dimensional (in x and y), and we will neglect chromatic effects. We will use z as the independent variable. It follows from the symmetry of the external and self fields that the Lie algebraic polynomial f_3 vanishes, and many of the coefficients of f_4 vanish. We will write

$$\mathcal{M}^{-1} = exp(:f_2:)\,exp(:f_4:), \tag{35}$$

where $exp(:f_2:)$ has a matrix representation M^{-1} given by

$$M^{-1} = \begin{pmatrix} a_x & b_x & 0 & 0 \\ c_x & d_x & 0 & 0 \\ 0 & 0 & a_y & b_y \\ 0 & 0 & c_y & d_y \end{pmatrix}, \tag{36}$$

and where

$$\begin{aligned}f_4 =& a_1 p_x^4 + a_2 p_y^4 + a_3 x^4 + a_4 y^4 + a_5 x^3 p_x + a_6 y^3 p_y + a_7 p_x^2 p_y^2 \\ &+ a_8 x^2 p_x^2 + a_9 y^2 p_x^2 + a_{10} x p_x^3 + a_{11} x^2 p_y^2 + a_{12} y^2 p_y^2 + a_{13} y p_y^3 \\ &+ a_{14} x^2 y^2 + a_{15} x p_x y^2 + a_{16} x p_x p_y^2 + a_{17} x^2 y p_y + a_{18} p_x^2 y p_y \\ &+ a_{19} x p_x y p_y. \end{aligned} \tag{37}$$

In this case, $\mathcal{M}^{-1}$ has 19+8=27 terms (though the matrix elements are not all independent). Given the above transfer map, it is possible to transform g by $\mathcal{M}^{-1}$ and obtain the scalar potential, ψ, of the resulting z-evolved distribution function as a power series in x and y. We will write

$$\psi(x,y) = C_1 x^2 + C_2 x^4 + C_3 y^2 + C_4 y^4 + C_5 x^2 y^2 + \cdots, \tag{38}$$

where the coefficients C_i are complicated functions of the 27 terms in $\mathcal{M}^{-1}$. Having obtained the scalar potential, we can write down the self-consistent Hamiltonian. Then, using standard Lie algebraic methods, we can obtain the differential equations of motion for $a_x, b_x, c_x, d_x, a_y, b_y, c_y, d_y$ and $a_i (i = 1, ..., 19)$.

The 19 first order differential equations for $\mathcal{M}^{-1}$ have been coded into a FORTRAN program called CHARLIE4F. (The "4F" refers to the 4–fold symmetry required of the initial distribution function and the beam transport system). CHARLIE4F numerically integrates these 19 equations. The resulting map can be used by a Lie algebraic beam transport code, such as MARYLIE 3.0 [4]. Since CHARLIE4F integrates only 19 equations, it is orders of magnitude faster than particle simulation codes that integrate the trajectories of thousands (or tens of thousands) of particles.

A version of CHARLIE4F has been modified to be used as a MARYLIE 3.0 subroutine. This makes it possible to fit and minimize combinations of matrix elements, monomial coefficients and user defined merit functions. In other words, this allows one to study and modify aberrations in beam transport systems with space charge (assuming the initial distribution is of the form (34)).

EXAMPLE: A COLD GAUSSIAN BEAM IN A DRIFT SPACE

Consider a long cylindrically symmetric beam launched in a drift space. Suppose the initial beam has a Gaussian profile in position space and suppose that all the particles have zero angular divergence. The initial distribution function $g(\zeta)$ is given by

$$g(x, p_x, y, p_y) = \frac{1}{2\pi a^2} \delta(p_x) \delta(p_y) exp\left(-\frac{x^2 + y^2}{2a^2}\right), \tag{39}$$

where a is the initial rms beam size in x and y.

We may regard the beam as a collection of concentric cylinders of charge expanding as they propagate along the drift space. Let $r(z)$ denote the radius of a cylinder (or particle) as a function of z, and let r_0 denote its initial radius:

$$r_0 = r(z = 0). \tag{40}$$

For a given drift length, there will be an interval I such that if $r_0 \in I$ then the trajectory $r(z)$ does not cross the trajectory of any other cylinder. In this case, it is possible to obtain the equation of motion for $r(z)$:

$$r''(z) = \frac{K}{r}\left[1 - exp\left(-\frac{r_0^2}{2a^2}\right)\right], \tag{41}$$

where K is the generalized perveance. Using this result, it is possible to obtain analytic formulae for the transfer map coefficients governing the motion of a beam particle. Since a beam particle has $(r')^i = 0$, we will write

$$r^f = \alpha_0 r^i + \alpha_1 \left(r^i\right)^3 + \cdots, \quad (42a)$$

$$\left(r'\right)^f = \beta_0 r^i + \beta_1 \left(r^i\right)^3 + \cdots. \quad (42b)$$

It turns out that the aberration coefficients are given by

$$Z = \frac{a}{\sqrt{K}} \int_1^{\alpha_0} \frac{ds}{\sqrt{\log s}}, \quad (43a)$$

$$\beta_0 = \alpha_0' = \frac{1}{a}\sqrt{K \log \alpha_0}, \quad (43b)$$

$$\alpha_1 = -\frac{Z}{8a^3}\sqrt{K \log \alpha_0}, \quad (43c)$$

$$\beta_1 = \frac{1}{2a}\sqrt{K \log \alpha_0}\left(\frac{\alpha_1/\alpha_0}{\log \alpha_0} - \frac{1}{4a^2}\right), \quad (43d)$$

where Z is the drift length.

The above results can be used to check CHARLIE4F. For the following numerical example, we considered a 200 ma beam of 50 MeV protons in a 2 m drift space, with a=.001 m. Numerical values of the exact aberration coefficients (using (43)) are shown below:

$$\alpha_0 = 1.33503776, \quad \beta_0 = 0.319380648$$
$$\alpha_1 = -79845.16, \quad \beta_1 = -72974.49$$

Correspondingly, we ran CHARLIE4F with $\sigma_x = \sigma_y = .001$. To simulate the cold beam, we set $\lambda_x = \lambda_y = 1 \times 10^{-6}$. The CHARLIE4F results are shown below:

$$\alpha_0 = 1.33503777, \quad \beta_0 = 0.319380645$$
$$\alpha_1 = -79845.21, \quad \beta_1 = -72974.51$$

The CHARLIE4F results agree with the analytic results (apart from errors associated with numerical integration).

SUMMARY

We have discussed a new approach to doing beam transport calculations with space charge. The approach is based on the self-consistent computation of transfer maps. As an illustration of the method, we have given a step-by-step analysis of a K-V distribution in a solenoid channel. We have also discussed the treatment of general distributions using Lie algebraic methods.

This new approach has some advantages over conventional methods:

1. The approach provides for *direct* computation of aberrations in the presence of space charge.
2. The approach is orders of magnitude faster than particle simulation. (This follows from the fact that we numerically integrate the equations of motion for the map, $\mathcal{M}$, rather than the equations of motion for thousands of particles.)

This approach also has some disadvantages:

1. The approach is only accurate near the beam core; the beam halo is poorly represented. The reason for this is that the approach uses an expansion of the self fields around the origin. For example, a third order code computes a map that is only valid for particles which stay sufficiently close to the origin that they are subject to linear, quadratic and cubic forces.
2. Much analytic work is required for each new initial distribution.

Above we mentioned that the self-consistent transfer map approach is much faster than particle simulation. However, this does not mean that self-consistent transfer maps can replace particle simulation. Instead, each approach is useful for studying certain beam behavior. Particle simulation programs are well suited to studying the global behavior of particle beams (though by increasing the number of particles, one could study local behavior). Transfer maps are well suited to studying the local behavior of beams (though by increasing the order of the calculation, one could study global behavior).

The self-consistent transfer map approach has been used to study K-V and Gaussian distributions. Work is currently in progress to analyze the Waterbag distribution and semi-Gaussian distributions.

REFERENCES

1. For a lengthier discussion of this subject, see R. Ryne, Lie Algebraic Treatment of Space Charge, Univ. of Maryland Ph.D. thesis (1987).
2. A. Dragt and E. Forest, J. Math. Phys. 24, 2734 (1983).
3. R. Ryne, Proceedings of the 1987 IEEE Particle Accelerator Conference, 1081 (1987)
4. A. Dragt et. al., MARYLIE 3.0, A Program for Charged Particle Beam Transport Based on Lie Algebraic Methods, Univ. of Maryland (1987).

275

DIFFERENTIAL ALGEBRAIC TREATMENT OF BEAM DYNAMICS TO VERY HIGH ORDERS INCLUDING APPLICATIONS TO SPACECHARGE

M. Berz
SSC Central Design Group, Universities Research Association
c/o Lawrence Berkeley Laboratory, Berkeley, CA 94720

ABSTRACT

The new method of Differential Algebra for the description of beam dynamics is presented. Differential Algebra allows a very efficient and elegant computation of transfer maps of accelerators, spectrometers and beam guidance systems. Contrary to existing techniques, the order of procedure is unlimited; in practice, maps of up to eleventh order have been computed.

Contrary to numerical differentiation techniques to obtain nonlinearities of the transfer map, the derivatives are accurate to machine precision. Space Charge effects of two and three dimensions can be included in the treatment. Finally, the method is also very helpful for purposes of analyzing the resulting maps. It can be used for basis changes, computation of the coefficients of the Lie algebraic representation, normal form problems and the computation of quantities of interest for accelerators such as tune shifts and chromaticities.

1. INTRODUCTION

The motion in an arrangement of electromagnetic fields can be described by a map relating final coordinates $\vec{z}_f$ of a particle to its initial coordinates $\vec{z}_i$ and some machine parameters $\vec{\delta}$ of interest as follows:

$$\vec{z}_f = \mathcal{M}(\vec{z}_i, \vec{\delta}) \qquad (1)$$

Having described the system by the map $\mathcal{M}$ in Equation (1), one can now expand the map in a power series around a reference trajectory. The higher the order to which the terms of this power series are taken, the more accurate the description.

However, only in special cases it is possible to obtain a closed solution for the expansion coefficients. In fact, in the important case that the fields depend on the independent variable, this is generally not the case.

© 1988 American Institute of Physics

The new method of Differential Algebra will allow us to compute the expansion coefficients for an arbitrary arrangement of (external and internal) fields in a very elegant and accurate way to arbitrary order. This entails detailed description of the map even for complicated systems. The accuracy of the expansion coefficients will not be limited as in the case of numerical differentiation techniques, and thanks to special software developed by the author, the order of the procedure is limited only by the power of the computer.

In the next section, we will provide some of the necessary mathematical background of the theory of Differential Algebra; for more detailed information the reader is referred to [1,2]. Section 3 will address the use of this method for the computation of nonlinearities under the assumption that the electromagnetic fields are known. Section 4 will describe the treatment of space charge effects in the Differential Algebraic context.

In Section 5, we will describe useful manipulations of the transfer map that include composition of maps, basis changes, inversions, generating functions, and normal forms as well as the computation of quantities relevant for accelerators like tune shifts and chromaticities.

2. DIFFERENTIAL ALGEBRAS

In the following subsection we will present parts of the mathematical theory of Differential Algebras to the extent that is required for the computation of the transfer map in its power series representation.

Differential Algebras are a subset of a generalization of the real numbers first introduced in the theory of Nonstandard Analysis [3,4]. In this generalization, infinitely small quantities, "infinitesimals", and infinitely large quantities are united with the real numbers in a consistent way. There is also some connection to the theories of formal power series [5] and automated differentiation [6]. The use of Differential Algebras for the purpose of computing a transfer map describing particle trajectories was first discussed in [7].

For the sake of clarity, we first address the simplest case of Differential Algebras, the structure $_1D_1$.

2.1 The Structure $_1D_1$

Consider the vector space R^2 of ordered pairs (q_0, q_1), $q_0, q_1 \in R$ in which an addition and a scalar multiplication are defined in the usual way:

$$(q_0, q_1) + (r_0, r_1) = (q_0 + r_0, q_1 + r_1) \qquad (2)$$

$$t \cdot (q_0, q_1) = (t \cdot q_0, t \cdot q_1) \qquad (3)$$

for $q_0, q_1, r_0, r_1, t \in R$. Besides the above addition and scalar multiplication a multiplication between vectors is introduced in the following way:

$$(q_0, q_1) \cdot (r_0, r_1) = (q_0 \cdot r_0, q_0 \cdot r_1 + q_1 \cdot r_0) \qquad (4)$$

for $q_0, q_1, r_0, r_1 \in R$. With this definition of a vector multiplication the set of ordered pairs becomes an algebra, denoted by $_1D_1$.

We introduce ordering relationships on $_1D_1$ in the following way. Let (a,b) and (c,d) be given; we say

$$\begin{aligned}(a,b) &< (c,d) \text{ if } a < c, \text{ or } (a = c \text{ and } b < d)\\(a,b) &> (c,d) \text{ if } a > c, \text{ or } (a = c \text{ and } b > d)\\(a,b) &= (c,d) \text{ if } a = c \text{ and } b = d\end{aligned} \qquad (5)$$

So in order to determine ordering, we look first only at the first components. If they already differ, they alone determine which of the numbers is larger. Only if they are the same do we compare the second components.

From this definition it is clear that for any (a,b) and (c,d), exactly one of the three properties always holds: $(a,b) = (c,d)$ or $(a,b) < (c,d)$ or $(a,b) > (c,d)$. Furthermore, it follows that if $(a,b) < (c,d)$, then for arbitrary (e,f) we have $(a,b)+(e,f) < (c,d)+(e,f)$, and for $(e,f) > 0$ we have $(a,b)\cdot(e,f) \leq (c,d)\cdot(e,f)$. Altogether, $_1D_1$ is a well-ordered set.

Looking at numbers $(a,0)$, we see that for addition, multiplication and ordering they behave like real numbers. So we can embed the real numbers into $_1D_1$ in the same way they could be embedded into the complex numbers.

Where in the complex numbers, $(0,1)$ was a root of -1, here it has another interesting property. Looking at the ordering relations, we infer that for $r > 0$,

$$(0,0) < (0,1) < (r,0) \qquad (6)$$

Hence $(0,1)$ lies "in between" 0 and every real number, i.e. $(0,1)$ is infinitely small.

Because of this we call $d = (0,1)$ the differential unit. The first component of the pair (q_0, q_1) is called the real part, and the second component is called the differential part.

It is easy to verify that $(1,0)$ is a neutral element of multiplication, because according to equation (4)

$$(1,0) \cdot (q_0, q_1) = (q_0, q_1) \cdot (1,0) = (q_0, q_1) \qquad (7)$$

It turns out that (q_0, q_1) has a multiplicative inverse if and only if q_0 is nonzero; so $_1D_1$ is not a field. In case $q_0 \neq 0$ the inverse is

$$(q_0, q_1)^{-1} = (\frac{1}{q_0}, -\frac{q_1}{q_0^2}) \qquad (8)$$

Using Equation (4), it is easy to check that in fact $(q_0, q_1)^{-1} \cdot (q_0, q_1) = (1,0)$.

The space $_1D_1$ is a subspace of the field R^* introduced in Nonstandard Analysis [3,4]. Besides the usual real numbers, R^* contains a variety of infinitely small and infinitely large quantities. The outstanding result of the theory of Nonstandard Analysis is that differentiation becomes an algebraic problem: a function f is differentiable if and only if for any arbitrarily small quantity δ, the real part of the quotient

$$\frac{f(x + \delta) - f(x)}{\delta} \qquad (9)$$

is independent of the choice of the specific δ. Thus, given any differentiable function f, we can compute its derivatives by just evaluating the formula for a special choice of δ. We choose $\delta = d = (0,1)$ and thus obtain

$$f'(x) = \mathcal{R}[\frac{f(x + d) - f(x)}{d}] \text{ or } f'(x) = \mathcal{D}[f(x + d)] \qquad (10)$$

where $\mathcal{R}$ denotes the real part and $\mathcal{D}$ denotes the differential part. Hence Differential Algebras are useful to compute derivatives directly, without requiring an analytic formula for the derivative and without the inaccuracies of numerical techniques.

The computation of derivatives shall be illustrated in an example using the following function:

$$f(x) = x^2 + \frac{1}{x} \tag{11}$$

Differentiating the function yields:

$$f'(x) = 2x - \frac{1}{x^2} \tag{12}$$

Suppose we are interested in the value and the derivative at x = 2. We obtain

$$f(2) = \frac{9}{2}, \ f'(2) = \frac{15}{4} \tag{13}$$

Now take the definition of the function f in Equation (11), replace all operations occuring in it by the corresponding ones in Differential Algebra, and evaluate it at $2 + d = (2, 1)$. One obtains:

$$\begin{aligned} f[(2,1)] &= (2,1)^2 + (2,1)^{-1} \\ &= (4,4) + (\frac{1}{2}, -\frac{1}{4}) \\ &= (\frac{9}{2}, \frac{15}{4}) \end{aligned} \tag{14}$$

As we can see, after the evaluation of the function the real part of the result is just the value of the function at $x = 2$, whereas the differential part is the derivative of the function at $x = 2$.

This is exactly what was to be expected from the theory of Nonstandard Analysis. However, for the sake of not relying on the quite advanced techniques of this relatively new field of mathematics, we also present an elementary but less elegant proof of this result.

By our choice of the starting vector $(2, 1)$, initially the vector contains the value $I(2)$ of the identity function $I : x \to x$ in the first component and the derivative of $I'(2) = 1$ in the second component.

Now assume that in an intermediate step two vectors of value and derivative $(g(2), g'(2))$ and $(h(2), h'(2))$ have to be added. According to (2) one obtains $(g(2) + h(2), g'(2) + h'(2))$. But according to the rule for the differentiation of sums, this is just the value and derivative of the sum function $(g + h)$ at $x = 2$.

The same holds for the multiplication: Suppose that two vectors of value and derivatives $(g(2), g'(2))$ and $(h(2), h'(2))$ have to be multiplied. Then according to (4) one obtains $(g(2) \cdot h(2), g(2) \cdot h'(2) + g'(2) \cdot h(2))$. But according to the product rule, this is just the value and derivative of the product function $(g \cdot h)$ at $x = 2$.

The evaluation of the function f at $(2,1)$ can now be viewed as successively combining two intermediate functions g and h, starting with the identity function and finally arriving at f. At each intermediate step the derivative of the intermediate function is automatically obtained as the differential part according to the above reasoning.

An interesting side aspect is that with the search for a multiplicative inverse in Equation (8) one has derived a rule to differentiate the function $f(x) = 1/x$ without explicitly using calculus rules.

After discussing the algebra ${}_1D_1$ and its virtues for the computation of derivatives, we now address the most general Differential Algebra, the structure ${}_nD_v$. It will eventually allow us to arithmetically compute partial derivatives of functions of v variables through order n.

2.2 The Structure ${}_nD_v$

In this section we describe a generalization of the Differential Algebra ${}_1D_1$ to higher orders and more than one variables. The resulting structure will then allow us to perform the advertised computation of transfer maps of trajectories in vacuum electronics devices.

Define $N(n, v)$ to be the number of monomials in v variables through order n. It can be shown that $N(n, v) = (n + v)!/n!v! = C(n + v, v)$ where $C(i, j)$ is the familiar binomial coefficient. Now assume that all these N monomials are arranged in a certain manner order by order. For each monomial M we call I_M the position of M according to the ordering. Conversely, with M_I we denote the Ith monomial of the ordering. Finally, for an I with $M_I = x_1^{i_1} \cdot \ldots \cdot x_v^{i_v}$ we define $F_I = i_1! \cdot \ldots \cdot i_v!$.

We now define an addition, a scalar multiplication and a vector multiplication on R^N in the following way:

$$(q_1, ..., q_N) + (r_1, ...r_N) = (q_1 + r_1, ..., q_N + r_N) \tag{15}$$

$$t \cdot (q_1, ..., q_N) = (t \cdot q_1, ..., t \cdot q_N) \tag{16}$$

$$(q_1, ..., q_N) \cdot (r_1, ...r_N) = (s_1, ..., s_N) \tag{17}$$

where the coefficients s_i are defined as follows:

$$s_i = F_i \cdot \sum_{\substack{0 \leq \nu, \mu \leq N \\ M_\nu \cdot M_\mu = M_i}} \frac{q_\nu \cdot r_\mu}{F_\nu \cdot F_\mu} \tag{18}$$

To help clarify these definitions, let us look at the case of two variables and second order. In this case, we have $n = 2$ and $v = 2$. There are $N = C(2+2, 2) = 6$ monomials in two variables, namely

$$1, x, y, xx, xy, yy \tag{19}$$

As an example, using the ordering in (19), we have $I_{xy} = 5$ and $M_3 = y$. Using the ordering in (19), we obtain for s_1 through s_6 in Equation (18):

$$\begin{aligned}
s_1 &= q_1 \cdot r_1 \\
s_2 &= q_1 \cdot r_2 + q_2 \cdot r_1 \\
s_3 &= q_1 \cdot r_3 + q_3 \cdot r_1 \\
s_4 &= 2 \cdot (q_1 \cdot r_4/2 + q_2 \cdot r_2 + q_4 \cdot r_1/2) \\
s_5 &= q_1 \cdot r_5 + q_2 \cdot r_3 + q_3 \cdot r_2 + q_5 \cdot r_1 \\
s_6 &= 2 \cdot (q_1 \cdot r_6/2 + q_3 \cdot r_3 + q_6 \cdot r_1/2)
\end{aligned} \tag{20}$$

Where in $_1D_1$, $d = (0, 1)$ was an infinitely small quantity, here we have a whole variety of infinitely small quantities that have the property that high enough powers of them vanish. We give special names to the ones in components I belonging to first order monomials, denoting them by dM_I. In the example of $_2D_2$, we have $dx = (0, 1, 0, 0, 0, 0)$ and $dy = (0, 0, 1, 0, 0, 0)$. It then follows from the theory of Nonstandard Analysis that instead of Equation (10) we obtain

$$f(x + dx, y + dy) = \\ (f, \frac{\partial f}{\partial x}, \frac{\partial f}{\partial y}, \frac{\partial^2 f}{\partial x^2}, \frac{\partial^2 f}{\partial x \partial y}, \frac{\partial^2 f}{\partial y^2})(x, y) \tag{21}$$

In the general case of v variables and order n, after evaluating f in the Differential Algebra one obtains:

$$\frac{\partial^{i_1+i_2+...+i_v} f}{\partial x_1^{i_1} \partial x_2^{i_2} ... \partial x_v^{i_v}} = s_\nu \qquad (22)$$

where $\nu = I_{(x_1^{i_1}....x_v^{i_v})}$ is the index of the monomial $(x_1^{i_1} \cdot ... \cdot x_v^{i_v})$, as defined in the beginning of the section.

As an example, consider the functions

$$\begin{aligned} x_f &= x_i + y_i^2 \\ y_f &= y_i - 2 \cdot x_i \end{aligned} \qquad (23)$$

Suppose, we are interested in value and derivatives at $x_i = 1, y_i = 2$. We have to substitute $x_i = (1, 1, 0, 0, 0, 0)$ and $y_i = (2, 0, 1, 0, 0, 0)$. Then we obtain

$$\begin{aligned} x_f &= (5, 1, 2, 0, 0, 1) \\ y_f &= (0, -2, 1, 0, 0, 0) \end{aligned} \qquad (24)$$

It is easy to verify that indeed the components contain the advertised derivatives.

We conclude this section with a short discussion of the generalization of frequently used functions to Differential Algebra. For details, the reader should consult [1]. As an example, here we will only address the case of the exponential $\exp(x)$.

First note that for any Differential Algebra vector of the form $(0, q_1, ..., q_N) \in {}_nD_v$, i.e. with a zero in the component belonging to the zeroth order monomial, we have the following property:

$$(0, q_1, ..., q_N)^i = (0, 0, ..., 0) \quad \text{for } i > n \qquad (25)$$

which follows directly from the definition of the multiplication in ${}_nD_v$ defined in Equation (17).

Assume now we have to compute the exponential of a Differential Algebra vector that has already been created by previous operations. Then we obtain for the exponential:

$$\begin{aligned}
&\exp[(q_0, q_1, q_2, ..., q_N)] \\
&= \exp(q_0) \cdot \exp[(0, q_1, q_2, ..., q_N)] \\
&= \exp(q_0) \cdot \sum_{i=0}^{\infty} \frac{(0, q_1, q_2, ..., q_N)^i}{i!} \\
&= \exp(q_0) \cdot \sum_{i=0}^{n} \frac{(0, q_1, q_2, ..., q_N)^i}{i!}
\end{aligned} \qquad (26)$$

In the last step use has been made of Equation (25) which entails that the sum has to be taken only through order n and thus allows the computation of the exponential in finitely many steps.

The fundamental operations of Differential Algebra as well as most of the functions supported in a FORTRAN computer environment have been implemented in a library. Special care has been taken to guarantee very efficient performance, particularly in the frequent case of sparse vectors in which many coefficients are zero.

For practical purposes it is of importance that in the FORTRAN environment Differential Algebraic operations can only be utilized by calls to subroutines. For this reason a precompiler [8] was developed that allows the use of a new data type "Differential Algebra" in regular FORTRAN formulas. The precompiler parses the entire program and transforms formulas containing Differential Algebraic quantities into subroutine calls.

3. THE COMPUTATION OF TRANSFER MAPS

3.1 Generation of Maps from Kicks

For pedagogical reasons, we begin this section with the simplest way to obtain transfer maps. In particular, in the case of many accelerators, the effect of individual beam line elements can often be described quite accurately by a kick in the center of the element. There is a large variety of codes using this technique to track particles. [9,10].

The whole system then is represented by a sequence of such kicks followed by drifts. The motion through a drift of length l can be described by

$$x_f = x_i + l \cdot a_i / \sqrt{\frac{p}{p_0} - a^2 - b^2}$$

$$y_f = y_i + l \cdot b_i / \sqrt{\frac{p}{p_0} - a^2 - b^2}$$
$$a_f = a_i$$
$$b_f = b_i \qquad (27)$$

On the other hand, the motion through a kick is given by

$$x_f = x_i$$
$$y_f = y_i$$
$$a_f = a_i + \frac{\partial}{\partial x} \Re \left(\sum_j M_j \cdot (x_i + i \cdot y_i)^j \right)$$
$$b_f = b_i + \frac{\partial}{\partial y} \Re \left(\sum_j M_j \cdot (x_i + i \cdot y_i)^j \right) \qquad (28)$$

here M_i are the normalized multipole strengths of the kick, and the quantity p/p_0 in Equation (27) is the momentum spread of the particle.

If a beam line is now described by a sequence of such drifts and kicks, it is quite straightforward how the transfer map expansion, i.e., the partial derivatives of final coordinates with respect to initial coordinates, can be obtained. One simply starts with $x, y, a, b = 0 + d_i$ and evaluates each drift (27) and kick (28) in Differential Algebra. In this process all multipole strengths are just constants, and the momentum spread is a Differential Algebraic quantity $1 + d_5$.

At the end, x, y, a, b contain the derivatives with respect to the final coordinates and the momentum spread p/p_0. Often it is of interest to describe the dependence of the map on some individual multipole strengths. This can be obtained by setting the multipole strengths of interest to $M_i + d_j$, $j = 6, 7, 8, ...$ Then the final x, y, a, b will also contain the derivatives with respect to these quantities.

Differential Algebraic techniques have been implemented into the kick codes TEAPOT [9,11] and THINTRAC [10]. This was facilitated by the use of the Differential Algebra precompiler [8]. Using this FORTRAN extension, almost all that is required to change a particle tracking code to a Differential Algebra map extraction code is to declare all quantities that depend on the particle coordinates as Differential Algebra, i.e., belonging to the new data type.

TEAPOT has been used for extensive checks of the Differential Algebra package and the "thick element" code COSY. For details about this comparison, see [12].

3.2 Generation of Maps Using Numerical Integration

In this section we will discuss how Differential Algebras can be used in combination with numerical integration techniques to obtain transfer maps. We will see that they can be obtained directly to arbitrary order for arbitrary field arrangements. Even though we do not have analytical formulas that relate the final coordinates to the initial coordinates, there is a way to computationally relate the final coordinates to the initial coordinates, by numerical integration of the equations of motion.

Suppose the system can be described by differential equations of the form

$$\frac{d}{dt}\vec{z} = \vec{F}(\vec{z}) \tag{29}$$

where $\vec{F}$ contains all the effects of the external and internal electromagnetic fields. For the equations of motion in normalized canonical curvilinear coordinates, see for instance [13].

Numerical integrators now use the right hand side of Equation (29) to obtain an approximate estimate for a new value of $\vec{z}$ at the time $t + \Delta t$. For instance, in the case of the well-known forth order Runge Kutta integrator, the procedure is as follows. One first computes quantities $k_1, ..., k_4$ recursively as

$$\begin{aligned}
\vec{k}_1 &= \vec{F}(\vec{z}, t) \cdot \Delta t \\
\vec{k}_2 &= \vec{F}(\vec{z} + \frac{1}{2}\vec{k}_1, t + \frac{1}{2}\Delta t) \cdot \Delta t \\
\vec{k}_3 &= \vec{F}(\vec{z} + \frac{1}{2}\vec{k}_2, t + \frac{1}{2}\Delta t) \cdot \Delta t \\
\vec{k}_4 &= \vec{F}(\vec{z} + \vec{k}_3, t + \Delta t) \cdot \Delta t
\end{aligned} \tag{30}$$

and from there obtains an approximate (and for reasonable Δt usually very good) estimate for the value of $\vec{z}$ at $t + \Delta t$ as

$$\vec{z}(t + \Delta t) = \vec{z}(t) + \frac{1}{6} \cdot (\vec{k}_1 + 2 \cdot \vec{k}_2 + 2 \cdot \vec{k}_3 + \vec{k}_4) + O(\Delta t^5) \tag{31}$$

Note that also in this case of a numerical integrator, the final coordinates are still computed from the initial coordinates using standard arithmetic and functions, even though the relations are more complex than in the trivial examples of the last section.

Now blindfoldedly performing all the operations in the steps of the Runge Kutta algorithm (31) in Differential Algebra instead of real arithmetic, one automatically obtains all desired derivatives of the transfer function, regardless of the form of the equations of motion.

Differential Algebraic techniques have been implemented in the program COSY [12,14]. They allow the computation of transfer maps of elements with a dependence on the independent variable for which an analytic solution cannot be obtained from the formula manipulator HAMILTON [12,13]. Using an eighth order Runge Kutta integrator, all operations required for a tracking of particles are performed in Differential Algebra [15]. This allows the computation of arbitrary fringing field effects as soon as the spacial distribution of the electromagnetic fields is known.

3.3 Hamiltonian Theory

In this subsection we will outline the usefulness of Differential Algebras for Hamiltonian systems. One of the most fundamental concepts of Hamiltonian theory is the Poisson bracket between two functions of phase space. This requires the differentiation with respect to phase space variables.

Suppose a Differential Algebra vector is given. Then it can be viewed as a descriptor of a function, containing its value and derivatives at a certain point. Hence the required differentiation is just a "bookkeeping operation" moving derivatives in the vector to different places. Thus all the required elements for a Poisson bracket are available, and a Poisson bracket for arbitrary order and arbitrary number of variables is contained in the Differential Algebraic package written by the author.

Lie operators, "Poisson brackets waiting to happen", can thus be computed. In fact, as soon as the generator $\vec{v}$ of the Lie operator $:\vec{v}:$ vanishes at the origin and has zero first derivatives, the process is closed in that no feeddown from higher orders occurs.

Using Lie operators, the transfer map or flow of a time independent Hamiltonian system can be computed as

$$\mathcal{M} = \exp(-t : H :) \qquad (32)$$

where H is the Hamiltonian of the system. Note that by the proper choice of the coordinates, it can always be achieved that $H(\vec{0}) = 0$ and also the first derivatives of H vanish. This entails that each summand in Equation (32) can be computed in a closed fashion. Taking enough terms of the power series (32), it is possible to compute the flow to any desired level of accuracy. As shown in [1], the series always converges and the maximum number of terms required for a certain accuracy can be determined rigorously. The treatment of time dependent Hamiltonians is also possible [1].

4. 2D and 3D SPACE CHARGE EFFECTS

In the previous section we have addressed the description of individual particles moving only in an external field. However, space charge fields produced by the particles themselves superimpose to the external fields and thus affect the particle motion in many cases. In order to include space charge effects, the following observations are of importance.

Firstly, the initial charge distribution has to be modelled in some way. Frequently this is done following a Monte Carlo procedure like in so-called particle in cell (PIC) codes [16,17,18]. Here particles describing the beam are selected statistically and their charges are chosen according to the density of the phase space distribution function. However, in this case the accuracy of the reproduction of the initially aimed for distribution is limited by statistical errors. Let N be the average number of particles that are inside one cell of the PIC code. Then the statistical error e of the Monte Carlo procedure can be estimated as

$$e = \sqrt{N} \qquad (33)$$

This shows that in order to obtain a smooth modelling of the distribution, very many simulation particles are required. So this suggests to arrange the simulation particles in a regular fashion, not statistically.

The second problem is the question what charge distribution to give to the individual regularly arranged particles. Even though there seems to be an abundance of possibilities, there is one choice superior to all others: the individual particles should have Gaussian charge and current distributions. As shown in detail in [19,20], the Gaussian is the best "micro" distribution to reproduce the original "macro" distribution.

The next aspect in the process of describing the space charge with Differential Algebra is to compute the fields required in the differential Equation (29). This is required in order to "push" the particles a certain Δt which is chosen small enough to approximate the fields as being constant.

In PIC codes the computation of these fields is usually done using a POISSON solver which computes the field on many grid points. As already pointed out in the last section, when a map with its linear and nonlinear effects is desired, all that is needed is the expansion of the field around the position of a reference particle. With Differential Algebra this expansion can easily be obtained.

We start by deriving closed formulas for the fields of Gaussian charge distributions in 2D and 3D. Assume that the charge distribution has the form

$$\rho(r) = \rho_0 \cdot \exp(-(\frac{r}{a})^2) \tag{34}$$

From Maxwell's Equations one obtains directly for the radially symmetric field:

$$\oint Eds = \frac{1}{\epsilon_0} \int_V \rho\, r dr d\phi \Rightarrow$$

$$E = \frac{\rho_0}{\epsilon_0} \cdot \frac{a^2}{2} \cdot \frac{1 - \exp(-(\frac{r}{a})^2)}{r} \tag{35}$$

And thus one obtains for the components E_x and E_y:

$$E_x = x \cdot \frac{\rho_0}{\epsilon_0} \cdot \frac{a^2}{2} \cdot \frac{1 - \exp(-(\frac{r}{a})^2)}{r^2} \tag{36}$$

$$E_y = y \cdot \frac{\rho_0}{\epsilon_0} \cdot \frac{a^2}{2} \cdot \frac{1 - \exp(-(\frac{r}{a})^2)}{r^2} \tag{37}$$

If the center of the Gaussian is located at (ξ, η) instead of $(0,0)$, then, of course, replace $\vec{r}$ by $\vec{r} - (\xi, \eta)$.

In the three dimensional case, the computation is similar. We obtain for the field of the distribution $\rho(r) = \rho_0 \cdot \exp(-(\frac{r}{a})^2)$:

$$\oint EdA = \frac{1}{\epsilon_0} \int_V \rho\, r^2 \sin(\theta)\, dr d\phi d\theta \Rightarrow$$

$$E = \frac{\rho_0}{\epsilon_0} \cdot \frac{1}{r^2} \cdot [-\frac{a^2}{2} \cdot \exp(-(\frac{r'}{a})^2) \cdot r'|_0^r + \int_0^r \frac{a^2}{2} \exp(-(\frac{r'}{a})^2) r'\, dr'] \tag{38}$$

Using the errorfunction $\text{erf}(x) = 2/\sqrt{\pi} \int_0^x \exp(-x'^2)\, dx'$, we thus obtain for the individual field components:

$$E_x = x \cdot \frac{\rho_0}{\epsilon_0} \cdot \frac{1}{r^3}[-\frac{a^2}{2} \cdot \exp(-(\frac{r}{a})^2) \cdot r + \frac{\sqrt{\pi}}{2} \cdot a \cdot \text{erf}(\frac{r}{a})]$$
$$E_y = y \cdot \frac{\rho_0}{\epsilon_0} \cdot \frac{1}{r^3}[-\frac{a^2}{2} \cdot \exp(-(\frac{r}{a})^2) \cdot r + \frac{\sqrt{\pi}}{2} \cdot a \cdot \text{erf}(\frac{r}{a})]$$
$$E_z = z \cdot \frac{\rho_0}{\epsilon_0} \cdot \frac{1}{r^3}[-\frac{a^2}{2} \cdot \exp(-(\frac{r}{a})^2) \cdot r + \frac{\sqrt{\pi}}{2} \cdot a \cdot \text{erf}(\frac{r}{a})] \qquad (39)$$

If desired, the magnetic fields of Gaussian currents can be obtained in a very similar way. Now that the fields of individual Gaussians are known, the total field in space is just the sum over all the Gaussians there are. Since all the functions contained in the analytical formulas for the fields are available in the FORTRAN Differential Algebra package written by the author (including the Errorfunction), all partial derivatives of the total field can be computed exactly. The only error is in the "discretization" of the initial distribution by representing it by Gaussians.

For the sake of computation speed, it is important to note that the time required by the above described process is proportional to the number of particles to the first power.

In the computation process some time can be saved because the field satisfies the Maxwell's equation

$$\vec{\nabla} \times \vec{E} = \vec{0}, \qquad (40)$$

which in the 2D case entails

$$\frac{d}{dy}E_x = \frac{d}{dx}E_y \qquad (41)$$

and in the 3D case entails

$$\frac{d}{dy}E_x = \frac{d}{dx}E_y, \ \frac{d}{dz}E_x = \frac{d}{dx}E_z, \ \frac{d}{dz}E_y = \frac{d}{dy}E_z, \qquad (42)$$

hence not all of the coefficients have to be determined and computer time can be saved.

To conclude this section, we discuss the treatment of possible wall interactions of space charge influenced beams. In the adiabatic (nonrelativistic) case such wall interactions produce image charges on metallic surfaces which have the property that the potential along these surfaces stays constant.

In order to compute these image charges, first determine the external and internal fields in the regular way. Then choose a set of n points P_i on the boundaries. For each of these discrete points P_i, compute the potential V_i induced at the point by the external and internal fields.

Now put a Gaussian charge distribution at each of the points P_i. Vary the charges q_i of these Gaussians such that the potential at each of the points P_i has the desired value. The computation of these boundary charges leads to a strongly diagonally dominant system of n times n equations which can be solved for the charges q_i located at P_i. After these charges are determined, the expansion of their electric field is computed exactly like the expansion of the field of the simulation particles, however requiring considerably less effort since the number of boundary charges is much less than the number of simulation charges. Note that this method is very general and does not require any specific form of the boundary conditions.

5. PROCESSING OF TRANSFER MAPS

5.1 Composition and Basis Changes

As we saw in Section 4, the map of a part of a system can be obtained using only Differential Algebra, without requiring time-intensive composition procedures as in thick element codes [21,22,23,14]. However, frequently a composer is still desirable, for instance in the case where the map for several turns through an accelerator is required. In this case it is more efficient to compose two one turn maps to obtain a two turn map than to start with the one turn map and apply all Differential Algebra operations for the second turn on it.

The Differential Algebra package contains a composer that is able to combine maps of arbitrary order and arbitrarily many variables. The most powerful feature of it is that it uses a strategy that saves execution time in the case that components of the map are zero. Since this is often the case, for instance due to certain symmetries, this usually entails considerable savings.

Frequently the map produced by a code is not in the coordinates in which a certain problem can be discussed most easily. For instance, while many codes work in the above discussed curvilinear canonical coordinates, others like TRANSPORT [21] and GIOS [22] use the slope instead of the normalized momentum. Again others use cartesian coordinates which are usually not very well suited for a discussion of properties of beam lines. The transformation of a transfer map to a different set of coordinates is quite straightforward using Differential Algebra. One simply evaluates the transformation formulas between the two sets of coordinates and their inverse formulas in Differential Algebra. Thus, one automatically obtains the coordinate switching maps to arbitrary orders. Composition with the transfer map yields the desired change of basis.

5.2 Inversion, Reversion

Frequently the inverse of a certain transfer map is required. This is for instance the case for the computation of the so-called reversed map described below. In order to compute the inverse, we first split the map into its linear and nonlinear parts:

$$A = A_1 + A_{\geq 2} \tag{43}$$

Furthermore, we write the inverse to be determined as a sum of maps of exact order ν:

$$A^{-1} = \sum_{\nu=1}^{\infty} M_\nu \tag{44}$$

Composing, we obtain

$$\begin{aligned}
(A_1 + A_{\geq 2}) \circ \sum_{\nu=1}^{\infty} M_\nu &= E \Rightarrow \\
A_1 \circ \sum_{\nu=1}^{\infty} M_\nu &= E - A_{\geq 2} \circ \sum_{\nu=1}^{\infty} M_\nu \Rightarrow \\
M_{\leq i} &= A_1^{-1} \circ (E - A_{\geq 2} \circ M_{\leq i-1})
\end{aligned} \tag{45}$$

Here "$\circ$" stands for the composition of maps. In the last step use has been made of the fact that to i-th order $A_{\geq 2} \circ M_{\leq i-1} = A_{\geq 2} \circ M_{\leq i}$. The necessary computation of A_1^{-1} is a linear matrix inversion and is performed by an off-the-shelf Gauss eliminator. Once this is done, equation (45) can be used in a recursive manner to compute the M_i order by order.

The reversed map of a system is the map obtained by going through the system in reverse direction. Such reversed maps are frequently of importance in the study of mirror symmetric systems and for theoretical reasons. The reversed motion can be described by first reversing time, i.e., switching all the signs of p_x

and p_y, then going through the inverse map, and finally re-reversing time. The time reversal operation can be performed easily using Differential Algebra, and the inversion of the transfer map is done as described above.

5.3 Generating Functions, Symplectic Tracking and Off-momentum fixed points

In many cases transfer maps are used for the tracking of individual particles. Using the map, tracking can be done significantly faster than by pushing the particles through the system. The accuracy of the final particle coordinates obviously depends on the accuracy of the map $\mathcal{M}$, which in general will increase with the order through which $\mathcal{M}$ is computed. However, for long term tracking of individual particles, it turns out, for not fully understood reasons of experience, that it is advantageous to modify the map such that it is automatically symplectic, i.e.,

$$M \cdot J \cdot M^t = J, \text{ or alternatively } M \cdot J = (M \cdot J)^t \tag{46}$$

where M is the Jacobian Matrix of partial derivatives of $\mathcal{M}$, and J has the form

$$J = \begin{pmatrix} 0 & 0 & 0 & -1 & 0 & 0 \\ 0 & 0 & 0 & 0 & -1 & 0 \\ 0 & 0 & 0 & 0 & 0 & -1 \\ 1 & 0 & 0 & 0 & 0 & 0 \\ 0 & 1 & 0 & 0 & 0 & 0 \\ 0 & 0 & 1 & 0 & 0 & 0 \end{pmatrix} \tag{47}$$

It can be shown [14] that any map obtained from a Hamiltonian system satisfies Equation (46). However, maps truncated after a certain order satisfy Equation (46) only to the accuracy to which the truncated map describes the system. This holds for all power series codes like COSY [14], GIOS [22], TRANSPORT [21] and MARYLIE [23].

Fortunately, there is a straightforward solution to this problem. Let $(\vec{q}, \vec{p})$ be the canonical positions and momenta, respectively. Instead of relating the final positions and momenta to the initial ones, one can produce "mixed" relationships (which are the gradient of the generating function multiplied with J):

$$\begin{aligned} (\vec{q}_f, \vec{p}_i) &= \mathcal{F}(\vec{q}_i, \vec{p}_f) \\ (\vec{q}_i, \vec{p}_f) &= \mathcal{F}(\vec{q}_f, \vec{p}_i) \end{aligned} \tag{48}$$

The process of obtaining these functions $\mathcal{F}$ (if posssible) from the transfer map $\mathcal{M}$ will be called "partial inversion", and will be discussed below. It is obvious that the higher the order of the transfer map, the more accurate the so obtained mixed relations obtained from the transfer map represents the true mixed relations.

However, regardless of how well the truncated generating function represents reality, it can be shown [24] that whenever the "mixed" relations derived from a generating function are used to implicitly search the proper final coordinates to the given initial coordinates, the result is always a symplectic map.

The procedure to perform symplectic tracking is as follows. First, one generates a transfer map describing the system as well as possible, i.e., to a high order. Second, one computes the "mixed" relations using the partial inversion described below. Third, one obtains an approximation of the final coordinates of the particles of interest using the transfer map. Finally, these approximative values are used as starting values of a Newton search to obtain a solution of the "mixed" equation.

To obtain the "mixed" relations from the transfer map, we proceed as follows. We denote with $\mathcal{M}_1$ the part of the transfer map describing the final positions, and with $\mathcal{M}_2$ the part describing the final momenta. Thus, we have $\mathcal{M} = (\mathcal{M}_1, \mathcal{M}_2)$. We do the same with the identity map: $\mathcal{E} = (\mathcal{E}_1, \mathcal{E}_2)$. In order to obtain the "mixed" relations $(\vec{q_f}, \vec{p_i}) = \mathcal{F}(\vec{q_i}, \vec{p_f})$, we start by setting $\mathcal{N} = (\mathcal{E}_1, \mathcal{M}_2)$. Then,

$$(\vec{q_i}, \vec{p_f}) = \mathcal{N}(\vec{q_i}, \vec{p_i}) \tag{49}$$

It turns out that the generating function exists if and only if $\mathcal{N}$ is invertible. In case $\mathcal{N}$ is invertible, we obtain

$$(\vec{q_i}, \vec{p_i}) = \mathcal{N}^{-1}(\vec{q_i}, \vec{p_f}) \tag{50}$$

composing the map $(\mathcal{M}_1, \mathcal{E}_2)$ and the map $\mathcal{N}^{-1}$, we finally obtain the desired "mixed" relations:

$$(\vec{q_f}, \vec{p_i}) = \left((\mathcal{M}_1, \mathcal{E}_2) \circ \mathcal{N}^{-1}\right)(\vec{q_i}, \vec{p_f}) = \mathcal{F}(\vec{q_i}, \vec{p_f}) \tag{51}$$

So the whole process of obtaining the gradient of the generating function can be performed to arbitrary order using only composition and inversion of transfer maps. The determination of the generating function is then only an integration.

The technique of generating functions is used virtually by every symplectic tracking technique. As it turns out, the ease of computing a generating function with Differential Algebra is one of the strong points of the power series representation of the map. In the Lie representation, the computation of the generating function can not be done in a straightforward pattern and gets increasingly cumbersome with high orders.

Now we will address quite a different problem which can be solved in a similar way. Circular particle accelerators are usually designed in such a way that particles with a higher momentum than that of a reference particle travel a longer way in order to obtain a similar flight time around the lattice, which is of importance in the accelerating RF cavities. As we will see, for every value of the momentum spread $p - p_0/p_0$ there will always be one phase space point that is invariant under the map. This point is called the off-momentum fixed point. As a generalization, we also address the question of fixed points not only depending on the momentum spread, but also on other parameters like quadrupole strengths. The momentum spread and all these other parameters are put into the vector of parameters $\vec{\delta}$.

Mathematically, the search for a fixed point can be described in the following way. Let $\mathcal{M}$ be the map describing the system (see Eq. (1)). For the following discussion we extend the map by components for the parameters $\vec{\delta}$; these components are just unity maps since the parameters are constant. We have

$$(\vec{x}_f, \vec{\delta}) = \mathcal{M}(\vec{x}_i, \vec{\delta}) \tag{52}$$

Then, the sought after fixed point $\vec{x}_F$ satisfies

$$(\vec{x}_F, \vec{\delta}) = \mathcal{M}(\vec{x}_F, \vec{\delta}) \tag{53}$$

or, letting $\mathcal{E}$ denote the unity map

$$(\vec{0}, \vec{0}) = (\mathcal{M} - \mathcal{E})(\vec{x}_F, \vec{\delta}) \tag{54}$$

Inspecting this equation, it turns out that it is a partial inversion problem like the ones discussed above in the process of determining the generating function. All that needs to be done is partially invert the equation for $\vec{x}_F$ using a "generating function" for the higher dimensional map $(\mathcal{M} - \mathcal{E})$ to obtain

$$(\vec{x}_F, \vec{0}) = \mathcal{F}(\vec{0}, \vec{\delta}) \tag{55}$$

5.4 Lie Coefficients, Normal Forms, Tuneshifts and Chromaticities

In this section we will describe how to switch between the map expansion representation of the transfer map and the Lie Algebraic representation used by Dragt, Forest and coworkers [25,26]. The fact that we are able to change between representations easily stresses the equivalence of the Physics described between both methods. We start our discussion with the simplest case, the computation of a transfer map from the Lie Algebraic representation. In this notation, the map is given by

$$\mathcal{M}(\vec{x}) = (L \circ \exp(:f_3:) \circ \exp(:f_4:) \circ ...)\vec{x} \qquad (56)$$

Here L is a linear matrix, and the f_i are polynomials of exact degree i. To compute the action of the Lie operators in equation (56), one first notes that since they all are of exact orders larger than 2, they all terminate after finitely many terms. So different from Section (3.3), all exponentials have to be taken only through a finite number of terms. In the last step, the previous result has to be multiplied with the Matrix L. This can be done by viewing L as a map which in this case has only linear terms, and using the map composer discussed in Section (5.1).

The reverse case of obtaining the Lie polynomials from the power series representation of the transfer map is a little more difficult. First we note that L is just the linear part of the map in power series expansion. Next we set

$$\mathcal{M}_1 = L^{-1} \circ \mathcal{M} \qquad (57)$$

Then $\mathcal{M}_1$ is symplectic as a composition of symplectic maps. Looking at Equation (56) through order 2, we find

$$\begin{aligned} \mathcal{M}_1 \vec{x} &= \vec{x} + [f_3, \vec{x}] \Rightarrow \\ (\mathcal{M}_1 - \mathcal{E}) &= \vec{\nabla} f_3 J \Rightarrow \\ \vec{\nabla} f_3 &= -(\mathcal{M}_1 - \mathcal{E}) J \end{aligned} \qquad (58)$$

Hence, we have to solve a potential problem to obtain f_3. First we must show that there is a potential for the "field" $(\mathcal{M}_1 - \mathcal{E})J$. As we know from simple potential theory, the potential exists if and only if the Jacobian of the field is symmetric. The Jacobian here is $M_1 J$, where M_1 is the Jacobian of $\mathcal{M}$. The identity map disappeared in the differentiation process. Hence, we have to fulfill

$$M_1 J = (M_1 J)^t \tag{59}$$

but this is simply one way to write the symplectic condition (46) for $\mathcal{M}_1$.

Now that we know that a potential exists, we can compute it by integrating the "field" on an arbitrary path. For simplicity we start at the origin and integrate along the diagonal to the point $\vec{x}$, and thus we obtain

$$f_3 = \int_0^{\vec{x}} -((\mathcal{M}_1 - \mathcal{E})) J(\vec{x}\,') d\vec{x}\,' \tag{60}$$

To compute f_4, we set $\mathcal{M}_2 = \exp(: -f_3 :) \circ L^{-1} \circ \mathcal{M}$.

Looking at Equation (56) through order 3, we obtain

$$\mathcal{M}_2 \vec{x} = \vec{x} + [f_4, \vec{x}] \tag{61}$$

Hence, we have the same situation as for the computation of f_3 in Equation (58). Proceeding in the same way as above for f_4 and then for f_5, f_6,..., we obtained a recursive procedure to compute all f_i.

While the author believes that the actual computation of transfer maps is most easily done in the map power series expansion using Differential Algebra and not using Lie Algebraic techniques, these techniques have a significant merit for the analysis of maps and normal form problems. Having the powerful tool of differential algebra available, it was possible [27,28] to implement the normal form algorithm of Deprit [29], which has been used before by Dragt, Forest and Michelotti [30], to arbitrary order for power series maps.

We do not go into details about the algorithm and its implementation but refer to [27,28]. The result of the procedure is a change of basis such that in the new positions and momenta $(q_x, q_y, q_z, p_x, p_y, p_z)$ the map has the very simple form

$$\mathcal{M} = exp(: f :), \text{ where } f = f(\frac{p_x^2 + q_x^2}{2}, \frac{p_y^2 + q_y^2}{2}, \frac{p_z^2 + q_z^2}{2}) \tag{62}$$

This change of basis is possible whenever the map is not on a linear resonance, i.e. for the three eigenvalues of the linear map ν_x, ν_y, ν_z there is no solution in integers k, l, m, n of $k \cdot \nu_x + l \cdot \nu_y + m \cdot \nu_z = n$. We note for the sake of completeness that in the case the linear motion is unstable, which is not of relevance in accelerators, a slight modification of the procedure is required ([27]).

In the transformed basis the exponent of the Lie map depends only on the "Actions" but not on the "Angles". Besides other significant advantages of this form, one consequence is that the map generated by such a Lie operator can be computed very easily. First we note that

$$[\frac{p_i^2+q_i^2}{2},\frac{p_j^2+q_j^2}{2}]=0 \tag{63}$$

Since $[a,b\cdot c]=[a,b]\cdot c+[a,c]\cdot b$, we infer

$$[f,\frac{p_i^2+q_i^2}{2}]=0 \text{ , and } [f,f_j]=0 \tag{64}$$

where $f_j=\partial f/\partial(p_j^2+q_j^2)/2$. Hence, we obtain

$$\begin{aligned}:f:q_j &= f_j p_j, & :f:^2 q_j &= -(f_j)^2 q_j, & :f:^3 q_j &= -(f_j)^3 p_j, \ldots \\ :f:p_j &= f_j q_j, & :f:^2 p_j &= -(f_j)^2 p_j, & :f:^3 p_j &= -(f_j)^3 q_j, \ldots \end{aligned} \tag{65}$$

Thus, after summing over all the terms of the exponential, we obtain for the final coordinates $(\vec{\bar{q}},\vec{\bar{p}})=\mathcal{M}(\vec{p},\vec{q})$

$$\begin{aligned}\bar{q}_j &= q_j\cdot\cos(f_j)+p_j\cdot\sin(f_j) \\ \bar{p}_j &= -q_j\cdot\sin(f_j)+p_j\cdot\cos(f_j)\end{aligned} \tag{66}$$

Hence, the motion in the new coordinates is described by circles whose frequencies are given by the values of the f_j. Thus, the quantities of interest for multipass systems, namely the dependencies of the tune on amplitude (tune shift), momentum spread (chromaticities) and perhaps some system parameters can all be read directly from the f_j.

Acknowledgements

For many fruitful discussions I am indebted to Drs. Etienne Forest, Edward Heighway, Walter Lysenko, Tom Mottershead and Professors Hermann Wollnik and Alex Dragt.

References

[1] M. Berz, "Differential Algebraic description of beam dynamics to very high orders," *Particle Accelerators*, in print (1988).

[2] M. Berz, "*Differential Algebraic Description of Beam Dynamics to Very High Orders,*" Technical Report SSC-152, SSC Central Design Group, Berkeley, Ca (1988).

[3] D. Laugwitz, "Ein Weg zur Nonstandard-Analysis," *Jahresberichte der Deutschen Mathematischen Vereinigung*, 75:66 (1973).

[4] A. Robinson, "Non-standard analysis," In *Proceedings Royal Academy Amsterdam, Series A*, page 432 (1961).

[5] I. Niven, "Formal power series," *American Mathematical Monthly*, 76-8:871 (1969).

[6] L.B. Rall, "The arithmetic of Differentiation," *Mathematics Magazine*, 59:275 (1986).

[7] M. Berz, "The method of power series tracking for the mathematical description of beam dynamics," *Nuclear Instruments and Methods*, A258:431 (1987).

[8] M. Berz, "*The Differential Algebra FORTRAN precompiler DAFOR,*" Technical Report AT-3:TN-87-32, Los Alamos National Laboratory (1987).

[9] L. Schachinger and R. Talman, "TEAPOT, a thin element program for optics and tracking," *Particle Accelerators*, 22:35 (1987).

[10] B. T. Leeman and E. Forest, "*Brief Description of the tracking codes THIN-TRACK and FASTRACK,*" Technical Report SSC-133, SSC Central Design Group, Berkeley, CA (1988).

[11] E. Forest, "Lie Algebraic maps and invariants produced by tracking codes," *Particle Accelerators*, 22:15 (1987).

[12] M. Berz, "Analytical solution of the equations of motion to high orders: the formula manipulator HAMILTON and the fifth order code COSY," *AIP Conference Proceedings*, these proceedings (1988).

[13] M. Berz and H. Wollnik, "The program HAMILTON for the analytic solution of the equations of motion in particle optical systems through fifth order," *Nuclear Instruments and Methods*, A258:364 (1987).

[14] M. Berz, H. C. Hofmann, and H. Wollnik, "COSY 5.0, the fifth order code for corpuscular optical systems," *Nuclear Instruments and Methods* A258:402 (1987).

[15] B. Hartmann, M. Berz, and H. Wollnik, "The computation of fringing fields using Differential Algebra," *Nuclear Instruments and Methods*, in preparation (1988).

[16] "PARMILA and PARMTEQ," widely used but undocumented codes.

[17] W. Lysenko, "*An RFQ Simulation Code*," Technical Report GSI-84-11, GSI Darmstadt (1984).

[18] H. Wollnik, J. Brezina, and M. Berz, "GIOS-BEAMTRACE, a computer code for the design of ion optical systems including linear or nonlinear space charge," *Nuclear Instruments and Methods*, A258:408 (1987).

[19] M. Berz and H. Wollnik, "Simulation of intense particle beams with regularly distributed gaussian subbeams," *Nuclear Instruments and Methods*, A267:25 (1988).

[20] M. Berz and H. Wollnik, "*Simulation of intense particle beams with regularly distributed Gaussian subbeams*," Technical Report AT-3:TN-87-29, Los Alamos National Laboratory (1987).

[21] K. L. Brown, *The Ion Optical Program TRANSPORT*," Technical Report 91, SLAC (1979).

[22] H. Wollnik, J. Brezina, and M. Berz, "GIOS-BEAMTRACE, a program for the design of high resolution mass spectrometers," In *Proceedings AMCO-7*, Darmstadt (1984).

[23] A. J. Dragt, L. M. Healy, F. Neri, and R. Ryne, "MARYLIE 3.0 - a program for nonlinear analysis of accelerators and beamlines," *IEEE Transactions on Nuclear Science*, Ns-3,5:2311 (1985).

[24] A. J. Dragt, "Lectures on nonlinear orbit dynamics," In *1981 Fermilab Summer School*, AIP Conference Proceedings Vol. 87 (1982).

[25] A.J.Dragt and J.M.Finn, *Journal of Mathematical Physics*, 17:2215, (1976).

[26] A.J.Dragt and E. Forest, "Computation of nonlinear behavior of Hamiltonian systems using Lie Algebraic methods," *Journal of Mathematical Physics*, 24(12):2734, (1983).

[27] E. Forest, M. Berz, and J. Irwin, "Normal form methods for complicated periodic systems: a complete solution using Differential Algebra and Lie operators," *Particle Accelerators*, in print, (1988).

[28] M. Berz, E. Forest, and J. Irwin, "*Exact Computation of Derivatives With Differential Algebra and Applications to Beam Dynamics,*" Technical Report SSC-166, SSC Central Design Group, Berkeley, Ca, (1988).

[29] A.Deprit, *Celestial Mechanics*, 1:12, (1969).

[30] L.Michelotti, "Deprit's algorithm, green's functions, and multipole perturbation theory," *Particle Accelerators*, 19:205, (1986).

ANALYTICAL SOLUTION OF SINGLE PARTICLE EQUATIONS OF MOTION TO HIGH ORDERS: THE FORMULA MANIPULATOR HAMILTON AND THE FIFTH ORDER CODE COSY 5.0

M. Berz
SSC Central Design Group, Universities Research Association
c/o Lawrence Berkeley Laboratory, Berkeley, Ca 94720
and II. Physikalisches Institut, 6300 Gießen, West Germany

and

H. Wollnik
II. Physikalisches Institut, 6300 Gießen, West Germany

ABSTRACT

The custom made formula manipulator HAMILTON is presented. HAMILTON was written for the purpose of obtaining analytical formulas for aberration coefficients of particle optical elements to very high orders. For reasons of speed, HAMILTON was written in FORTRAN.

It was used to derive formulas for the main fields of all commonly used particle optical elements through fifth order, which is far beyond the realm of off-the-shelf formula manipulators like REDUCE, MAXIMA or SMP.

The formalization of the problem, the internal data storage structure of HAMILTON and the required basic algebraic operations are discussed in detail. The results are generated directly as FORTRAN subroutines which were implemented into the fifth order code COSY 5.0. COSY 5.0 was verified in extensive checks against the results of codes upgraded for map extraction using Differential Algebra [1].

1. INTRODUCTION

When studying the motion through a section of a beamline element, often it is advantageous not only to look at the motion of individual phase space points through the system, but rather for some global structures like the map describing the system. Since many maps are predominantly linear with only small "aberrations" of this linearity, and because the map of all but the simplest systems can not be described analytically, one usually is interested in a power series expansion of the map up to some order.

As coordinates describing the system we use the following set of "scaled canonical" quantities:

$$\begin{aligned} r_1 &= x, & r_2 &= a = \frac{p_x}{p_0}, \\ r_3 &= y, & r_4 &= b = \frac{p_y}{p_0}, \\ r_5 &= l = v_0(t - t_0) \\ r_6 &= d = \frac{E - E_0}{E_0}, & r_7 &= g = \frac{m - m_0}{m_0} \end{aligned} \quad (1)$$

Here x and y are the horizontal and vertical distances to the optic axis, respectively. p_0, v_0, E_0 and m_0 denote momentum, velocity, energy and mass of the reference particle, whereas p, v, E and m stand for the same quantities of the particle under consideration.

The equations of motion in these new coordinates can be found for instance in [2]. The values of the coordinates in Equation (1) at the end of the beamline element are now expressed in terms of the values at the beginning in the advertised power series:

$$r_i^{(f)} = \sum_{j=1}^{7} \left(r_j \left((r_i r_j) + \sum_{k=j}^{7} r_k \left((r_i, r_j r_k) + \sum_{l=k}^{7} r_l (...) \right) \right) \right), \text{ i}=1,7 \quad (2)$$

There are two ways to accurately obtain the expansion coefficients of such power series. In the first approach, analytic solutions for the expansion coefficients are sought for certain well-defined sections of the system like individual drifts, sectors and multipoles. The map of the total system can then be obtained by composing the power series of all these elements to one total power series. This is the approach used by all existing power series codes [3,4,5,6]. Due to the rapid increase in complexity of the formulas for the coefficients of individual elements, this approach has until recently been limited to expansion coefficients of third order.

The second approach, which is far more general yet not as explicit, is based on the concept of Differential Algebra and discussed in detail in another paper of these proceedings [1]. Attempts also have been made to obtain the expansion coefficients by numerical differentiation techniques [7,4], but usually the accuracy of higher order derivatives is quite poor.

In this paper, we study in detail the first approach and focus on the analytical derivation of formulas for the nonlinearities of individual beam line elements to very high orders and the resulting power series code COSY 5.0 [8,6].

2. AN ALGORITHM FOR THE DETERMINATION OF ABERRATIONS

Inspecting the equations of motion in the coordinates discussed in the Introduction, which can be found in [6], it can be seen that they have the following form:

$$\begin{aligned}
x' &= a + f_x(x, a, y, b, g, d) \\
a' &= k_x^2 x + f_a(x, a, y, b, g, d) \\
y' &= b + f_y(x, a, y, b, g, d) \\
b' &= k_y^2 y + f_b(x, a, y, b, g, d) \\
l' &= f_l(x, a, y, b, g, d)
\end{aligned} \quad (3)$$

g and d affect the motion but are constant. $k_x(s)$ and $k_y(s)$ are the quadrupole strengths for the motion in the x and y planes that describe the linear motion. The functions f_x, f_a, f_y and f_b do not contain any linear terms in x, a, y and b any more. However, they may contain linear terms in g and d. We now expand the transfer map describing the motion from s_0 to s in a power series in the coordinates at s_0 and group the terms by orders:

$$\begin{aligned}
x(s) &= \sum_{i=1}^{\infty} x^{(i)}(s), \quad a(s) = \sum_{i=1}^{\infty} a^{(i)}(s) \\
y(s) &= \sum_{i=1}^{\infty} y^{(i)}(s), \quad b(s) = \sum_{i=1}^{\infty} b^{(i)}(s) \\
l(s) &= \sum_{i=1}^{\infty} l^{(i)}(s)
\end{aligned} \quad (4)$$

Here $x^{(i)}$, $a^{(i)}$, $y^{(i)}$, $b^{(i)}$ and $l^{(i)}$ are polynomials in the phase space coordinates at s_0 of exact degree i. In a similar way one expands the functions f from equation (3) and groups them by orders:

$$f_x = \sum_{i=1}^{\infty} f_x^{(i)}, \quad f_a = \sum_{i=1}^{\infty} f_a^{(i)},$$
$$f_y = \sum_{i=1}^{\infty} f_y^{(i)}, \quad f_b = \sum_{i=1}^{\infty} f_b^{(i)},$$
$$f_l = \sum_{i=1}^{\infty} f_l^{(i)} \tag{5}$$

As stated above, the linear terms $f_x^{(1)}$, $f_a^{(1)}$, $f_y^{(1)}$, and $f_b^{(1)}$ do not contain x, a, y and b, but may contain the chromatic quantities g and d. Inserting the expansions (4) into the differential equations (3) now yields a set of differential equations for the different orders:

$$\begin{aligned} x^{(i)\prime} &= a^{(i)} + F_x^{(i)} \\ a^{(i)\prime} &= k_x^2 x^{(i)} + F_a^{(i)} \\ y^{(i)\prime} &= b^{(i)} + F_y^{(i)} \\ b^{(i)\prime} &= k_y^2 y^{(i)} + F_b^{(i)} \\ l^{(i)\prime} &= F_l^{(i)} \end{aligned} \tag{6}$$

Note that the F's in this equation differ from the f's in equation (3). However, their linear properties are similar: $F_x^{(i)}$, $F_a^{(i)}$, $F_y^{(i)}$ and $F_b^{(i)}$ do not contain $x^{(i)}$, $a^{(i)}$, $y^{(i)}$ and $b^{(i)}$ any more since the f's in equation (3) do not have the corresponding linear terms. This implies that equation (3) now allows the iterative determination of the different orders $x^{(i)}$, $a^{(i)}, y^{(i)}$, $b^{(i)}$ in every order i since $F_x(i)$, $F_a^{(i)}$, $F_y^{(i)}$ and $F_b^{(i)}$ only contain already computed quantities. Once the terms $x^{(i)}$, $a^{(i)}, y^{(i)}$ and $b^{(i)}$ are computed, $l^{(i)}$ can be determined by a mere integration since $F_l^{(i)}$ does not contain l.

Since the transfer function from position s_0 to position s is the identical map, one obtains for the initial conditions

$$\begin{aligned} x^{(i)}(s=s_0) &= x_0 \cdot \delta_{i,1}, \quad a^{(i)}(s=s_0) = a_0 \cdot \delta_{i,1} \\ y^{(i)}(s=s_0) &= y_0 \cdot \delta_{i,1}, \quad b^{(i)}(s=s_0) = b_0 \cdot \delta_{i,1} \\ l^{(i)}(s=s_0) &= l_0 \cdot \delta_{i,1} \end{aligned} \tag{7}$$

Here δ stands for Kronecker's symbol. The resulting differential equations are now linear and inhomogeneous. Hence, by shifting our attention from particle trajectories with nonlinear, hard to solve equations to aberration coefficients we obtain linear, solvable differential equations. So on top of addressing more global concepts by inspecting aberrations instead of particles, the mathematics becomes significantly easier.

To solve the equations for the aberrations, one first determines the "homogeneous solution" of the two systems

$$\begin{aligned} x^{(i)\prime} &= a^{(i)} \\ a^{(i)\prime} &= k_x^2 x^{(i)} \\ y^{(i)\prime} &= b^{(i)} \\ b^{(i)\prime} &= k_y^2 y^{(i)} \end{aligned} \qquad (8)$$

We introduce the abbreviations c_x for the solution of the x-motion satisfying $x(0) = 1$, $a(0) = 0$ and s_x for the solution of the x-motion satisfying $x(0) = 0$, $a(0) = 1$. Similarly we introduce c_y and s_y for the y-motion.

Then we obtain for the general solution of the homogeneous system

$$\begin{aligned} x^{(i)} &= A_x c_x + B_x s_x, & a^{(i)} &= B_x c_x + k_x^2 A_x s_x \\ y^{(i)} &= A_y c_y + B_y s_y, & b^{(i)} &= B_y c_y + k_y^2 A_y s_y \end{aligned} \qquad (9)$$

The solution of the complete inhomogeneous differential equation will be obtained using the well-known method of the variations of the parameters. One makes a guess of the form in the last equations, except that the quantities A_x, B_x, A_y, and B_y now are viewed as s dependent. Differentiating these expressions with respect to s and comparing with the system of differential equations (6), one obtains the following equations:

$$\begin{aligned} A'_x c_x + B'_x s_x &= F_x^{(i)}, & B'_x c_x + k_x^2 A'_x s_x &= F_a^{(i)} \\ A'_y c_y + B'_y s_y &= F_y^{(i)}, & B'_y c_y + k_y^2 A'_y s_y &= F_b^{(i)} \end{aligned} \qquad (10)$$

This is just a set of linear equations in the unknowns A'_x, B'_x, A'_y and B'_y. Solving this system of equations for A' and B' and integrating finally yields

$$A_x = \int \left(c_x F_x^{(i)} - s_x F_a^{(i)} \right) ds, \quad B_x = \int \left(-k_x^2 s_x F_x^{(i)} + c_x F_a^{(i)} \right) ds$$
$$A_y = \int \left(c_y F_y^{(i)} - s_y F_b^{(i)} \right) ds, \quad B_y = \int \left(-k_y^2 s_y F_y^{(i)} + c_y F_b^{(i)} \right) ds \quad (11)$$

Inserting these expressions into (9) and choosing the integration constants such that the initial conditions are satisfied now yields the solution of the inhomogeneous system of differential equations.

Once the homogeneous, first order solution is known, the determination A_x, A_a, A_y, A_b is thus merely a quadrature.

So the only crucial part of the procedure is the first order solution. For cases in which k_x, k_y are constant, the first order solution can be obtained analytically:

$$c_x = \cosh(k_x s), \quad s_x = \frac{1}{k_x} \sinh(k_x s), \quad c_y = \cosh(k_y s), \quad s_y = \frac{1}{k_y} \sinh(k_y s) \quad (12)$$

In cases in which k_x and k_y are not constant, i.e., depend on s, usually no analytic solution can be obtained. In this case, which is of great importance for the case of fringe fields, one obtains integral formulas that relate the image aberrations to the first order solution. They usually have to be solved numerically or in some approximate manner. For the rest of the paper we restrict ourselves to the so-called main fields in which k_x and k_y are constant.

3. HAMILTON

In this section we will discuss the internal structure of the formula manipulator HAMILTON. We start with a discussion of the storage, followed by the necessary algebraic operations and the output.

3.1 The Storage of Data in the Program HAMILTON

According to the algorithm described in Section 2 which is used by HAMILTON, the general solution of any order consists of polynomials in the phase space coordinates. Each monomial in the phase space coordinates has the following standard form:

$$M = \frac{N}{D} \cdot x^{n_x} a^{n_a} y^{n_y} b^{n_b} g^{n_g} d^{n_d} \cdot s^{n_s} s_x^{n_{sx}} c_x^{n_{cx}} s_y^{n_{sy}} c_y^{n_{cy}}$$
$$\cdot k_x^{n_{kx}} k_y^{n_{ky}} \cdot k_1^{n_{k_1}} ... k_{32}^{n_{k_{32}}} \cdot d_1^{n_{d_1}} ... d_{24}^{n_{d_{24}}} \quad (13)$$

Here N and D are integers (numerator and denominator), x, a, y, b, g and d are the phase space coordinates, s is the independent variable, s_x, c_x, s_y, c_y are the functions defined in Section 2. k_x and k_y are the frequencies in x- and y- directions as defined in Section 2. k_1 through k_{32} can hold problem-related constants describing the reference particle, multipole strengths, the curvature of the optic axis, magnetic and electric rigidities, etc. d_1 through d_{24} are certain denominators that can occur when $s^i s_x^j c_x^k s_y^l c_y^n$ is integrated with respect to s. They have the form

$$d_\nu = \frac{1}{nk_x^2 - mk_y^2}, \quad 2 \leq n+m \leq 8 \tag{14}$$

Note that if we restrict ourselves to less than tenth order, all the exponents

$$n_x, ..., n_d, \; n_s, ..., n_{c_y}, \; n_{k_1}, ... n_{k_{32}}, \; n_{d_1}, ... n_{d_{24}} \tag{15}$$

are between 0 and 9, while the coefficients

$$n_{k_x}, n_{k_y} \tag{16}$$

can take larger positive and negative values.

The information describing one monomial is stored in 12 integer variables:

Number	Contents
1	Numerator
2	Denominator
3	$10^0 n_x + 10^1 n_a + 10^2 n_y + 10^3 n_b + 10^4 n_g + 10^5 n_d$
4	$10^0 n_{s_x} + 10^1 n_{c_x} + 10^2 n_{s_y} + 10^3 n_{c_y} + 10^4 n_s$
5	$1000100 + n_{k_x} + 10^4 n_{k_y}$
6	$10^0 n_{k_1} + ... + 10^7 n_{k_8}$
7	$10^0 n_{k_9} + ... + 10^7 n_{k_{16}}$
8	$10^0 n_{k_{17}} + ... + 10^7 n_{k_{24}}$
9	$10^0 n_{k_{25}} + ... + 10^7 n_{k_{32}}$
10	$10^0 n_{d_1} + ... + 10^7 n_{d_8}$
11	$10^0 n_{d_9} + ... + 10^7 n_{d_{16}}$
12	$10^0 n_{d_{17}} + ... + 10^7 n_{d_{24}}$

HAMILTON has space to store a maximum of about 50000 of these monomials. Each polynomial is represented by a string of monomials stored consecutively and in ascending order.

3.2 Additions, Multiplications and Integrations

Like all polynomials, two polynomials that have to be added are in ascending order. The addition is performed in a merging process with two pointers denoting the last elements processed of both monomials. To perform the addition, the next "candidate" monomials in the two polynomials are compared. If all the exponents, i.e., the characteristic integers 3 through 12 agree, a fraction addition of the numerators and denominators is performed and the new numerator and denominator are stored together with the other integers representing the exponents and both pointers are incremented by one. If the exponents do not agree, the monomial coming first is copied and the pointer of the respective polynomial is incremented by one. This is repeated until all monomials of both polynomials have been treated.

The multiplication is performed monomialwise with consecutive reordering. Note that the multiplication of two monomials is particularly simple since it just involves addition of the characteristic integers 3 through 12 (which contain the exponents of the occurring quantities) and a fraction multiplication of the numerators and the denominators.

Inspecting the algorithm described in Section 2, it turns out that any monomial whose phase space coordinates have an order larger than the one which is just being treated can be neglected. This considerably reduces the sizes of the polynomials after multiplication and avoids an exponential growth of the sizes.

The integration with respect to s uses a look up table containing all the integrals of the form

$$\int s^{i_s} \cdot s_x^{i_{sx}} \cdot c_x^{i_{cx}} \cdot s_y^{i_{sy}} \cdot c_y^{i_{cy}} \, ds \quad \text{for} \quad i_s + i_{sx} + i_{cx} + i_{sy} + i_{cy} \leq 8 \qquad (17)$$

This table is computer generated by the program INTEG. It first rewrites the sinhs and coshs in terms of exponentials, then integrates by parts all the terms of the form $\int s^n \cdot e^{c \cdot s} \, ds$, and finally re-expresses the result in terms of sinhs and coshs. A consequence is that the integrals are always expressable as a sum of terms as those in the integrand of equation (17), where in addition some of the denominators d_1 through d_{24} may occur.

Once a monomial has to be integrated with respect to s, the monomial is copied for each term occurring on the right hand side of the integral, the characteristic integers 4 and 10 through 12 are changed.

3.3 The Output of HAMILTON

As discussed earlier, the resulting analytic expressions for the aberrations can be very complex. For instance, many formulas for fifth order nonlinearities fill more then 20 lines. So in order to avoid copying mistakes, it is advisable to generate the formulas directly in computer usable form. In order to connect them to COSY, the output is performed in FORTRAN. Except for the use of more than 20 continuation lines in some instances, the output is compatible with FORTRAN 77.

Since the formulas for the nonlinearities are very complex, it is important to optimize the resulting code to as high a degree as possible. For this purpose, a special subroutine in HAMILTON looks at all formulas and tries to recognize operations that occur more than once. Such operations are then performed only once and the result is stored in a scratch variable which is reused for further occurrences of the term. With this technique, the required number of floating point multiplications can be reduced by more than a factor of 10.

COSY 5.0 contains a library with subroutines for the aberrations of all important beamline elements. The length of this library is about 30000 lines of code. Details about the use of COSY will be discussed in another paper of these proceedings [8].

4. VERIFICATION OF COSY AND DIFFERENTIAL ALGEBRA

The analytic solutions of the equations of motion generated by HAMILTON were verified in several ways. First, they were compared with the predictions of several third order codes. In general, very good agreement was found except for a few nonlinearities which gave different results with different codes. Since there is no other fifth order code besides COSY, higher order aberrations could not be verified in this way

However, as described in [1], it is quite easy to extract maps of arbitrary order from tracking codes using Differential Algebraic methods. One of the codes that was generalized to Differential Algebra map extraction is the SSC design code TEAPOT [9]. TEAPOT uses formulas that were developed completely independently of COSY. Since COSY (at least in the version used for the comparison) does not rely on Differential Algebraic techniques at all, there was no common part whatsoever in the two codes.

For a significant comparison, a lattice had to be taken that is relativistic but not fully relativistic, that contains many elements of interest and that has bending magnets with a relatively small radius of curvature. From several possible lattices, we picked the Los Alamos PSR as the reference for comparison.

Since COSY is a thick element code and TEAPOT a kick code, the thick elements had to be split into several hundred kicks in order to get good agreement. Of course this entailed a quite slow performance; TEAPOT was several hundred times slower than COSY. For most of the usual applications of TEAPOT, however, this does not present a problem since in those cases one kick per element is usually sufficient.

After extraction, the two maps were compared for all nonlinearities to fifth order. It turned out that all of the several hundred aberrations agreed to at least five significant digits. Sceptics can obtain a printout of the two transfer maps from the author.

This result is even more remarkable when considering that many third order codes that have been used for many years still are being corrected occasionally. The fifth order formulas are at least an order of magnitude more complex than the third order formulas, and hence it is quite clear that an efficient formula manipulator is the only conceivable way to obtain correct results.

5. OUTLOOK

Even with HAMILTON, it is not possible to determine analytic solutions much beyond fifth order. As mentioned in the last section, the length of the generated code is already a limiting factor. Also, running times of HAMILTON grow very rapidly with the order.

So in order to increase accuracy far beyond fifth order which is desirable in some cases, analytic solutions become too cumbersome. For this reason a version of COSY relying entirely on the Differential Algebraic computation of Hamiltonian flows [1,10,11] is under development. This code, under the tentative name COSY INFINITY, will work to arbitrary order and will be able to include dependencies on system parameters like multipole strengths in the transfer map. It will treat both main fields and fringing fields in the same framework. It will also be able to treat elements in their kick approximation.

In the case of very high order transfer maps, another phenomenon becomes much more important than in the case of low order maps. Frequently only a few terms of each new order yield significant contributions to the transfer map, while most others are negligible. This is in contrast to the rapid increase in the number of new aberrations in each new order. The internal structure of the Differential Algebra package is such that only nonzero terms or terms above a certain tolerance are stored and used. So both in storage and execution time a Differential Algebraic approach becomes more and more advantageous when calculating higher order terms as compared to analytic methods.

References

[1] M. Berz, "Differential Algebraic treatment of beam dynamics to very high orders including applications to spacecharge," *AIP Conference Proceedings*, these proceedings (1988).

[2] M. Berz and H. Wollnik, "The program HAMILTON for the analytic solution of the equations of motion in particle optical systems through fifth order," *Nuclear Instruments and Methods*, A258:364 (1987)

[3] K. L. Brown, "*The Ion Optical Program TRANSPORT*," Technical Report 91, SLAC (1979).

[4] H. Wollnik, J. Brezina, and M. Berz, "GIOS-BEAMTRACE, a computer code for the design of ion optical systems including linear or nonlinear space charge," *Nuclear Instruments and Methods*, A258:408 (1987).

[5] A. J. Dragt, L. M. Healy, F. Neri, and R. Ryne, "MARYLIE 3.0 - a program for nonlinear analysis of accelerators and beamlines," *IEEE Transactions on Nuclear Science*, Ns-3,5:2311 (1985)

[6] M. Berz, H. C. Hofmann, and H. Wollnik, "COSY 5.0, the fifth order code for corpuscular optical systems," *Nuclear Instruments and Methods*, A258:402 (1987).

[7] S. Kowalski and H. Enge, "*RAYTRACE* ," Technical Report, Cambridge, Massachussetts (1985).

[8] H. Wollnik, M. Berz and B. Hartmann, "Principles behind GIOS and COSY," *AIP Conference Proceedings*, these proceedings (1988).

[9] L. Schachinger and R. Talman, "TEAPOT, a thin element program for optics and tracking," *Particle Accelerators*, 22:35 (1987).

[10] M. Berz, "Differential Algebraic description of beam dynamics to very high orders," *Particle Accelerators*, in print (1988).

[11] M. Berz, "*Differential Algebraic Description of Beam Dynamics to Very High Orders,*" Technical Report SSC-152, SSC Central Design Group, Berkeley, CA (1988).

ALGORITHMS FOR THE SELF-CONSISTENT
SIMULATION OF HIGH POWER KLYSTRONS*

Kenneth Eppley
Stanford Linear Accelerator Center
Stanford, California, 94309

ABSTRACT

We discuss an improvement to the algorithm developed by Yu[1] for modelling rf cavities in klystrons using the port approximation. In this method, the cavity is simulated by imposing an rf voltage as a boundary condition across the outer wall. The voltage and phase are chosen to be consistent with the cavity impedance and with the rf current induced by the electron beam. In the original method, each cavity was calculated successively using either linear theory or an iterative method to achieve a self-consistent voltage. The new method relaxes the voltage and phase of several cavities simultaneously during the simulation. The time dependence of the voltages are calculated from a relaxation equation. The new algorithm reduces the total computation time by about a factor of five for a complete klystron.

INTRODUCTION

In the port approximation[1] to the modelling of rf cavities using an electromagnetic particle in cell code such as MASK[2], the cavities are simulated by imposing an rf voltage as a boundary condition across an opening or "port" in the outer wall of the drift tube (Figure 1). This method ignores the transient and looks only for the steady-state solution at a single operating frequency. In a real cavity, electromagnetic energy flows across the gap between the cavity and the electron beam in the drift tube. By writing the equations for energy flow across the gap, one splits the problem into two much simpler pieces:

$$V \cdot I_{ind} = \int E \cdot J dV \quad . \tag{1}$$

(Dot product involves the integration over rf cycle of the complex phase; for vector quantities it also subsumes the spatial dot product.)

Looking from the cavity side, in steady state the voltage across the gap and the current flowing in the walls uniquely determine the state of the cavity - thus they can be expressed in terms of the cold cavity impedance. From the drift tube side, the energy flow into or out of the beam is completely determined in steady state by the voltage and phase across the gap. The key trick is noticing

---

* Work supported by the Department of Energy, contract DE − AC03 − 76SF00515.

© 1988 American Institute of Physics

that the current flowing in the walls (the induced current) can be calculated simply in terms of the transform of the volume integral of $E \cdot J$. Of course, the current distribution is changed by the presence of the cavity voltage.

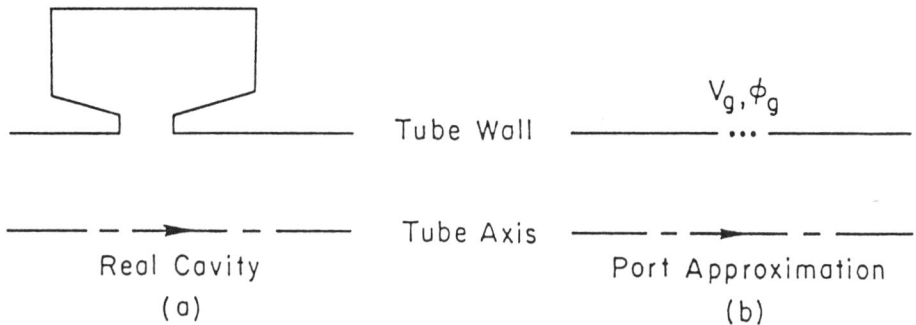

Figure 1. Simulation of a real cavity by a port boundary condition.

The voltage and phase must be chosen (by some means) to be consistent with the cavity impedance and with the rf current induced by the electron beam. Note that the induced current is not identical to the rf current flowing through the drift tube. The currents can be related through the transit angle factor, since the energy transfer to the beam is

$$P = V \cdot I_{rf} T \ . \tag{2}$$

Thus

$$I_{ind} = I_{rf} T \ . \tag{3}$$

The relation to cold cavity parameters comes through the relation

$$V = I_{ind} Z \ . \tag{4}$$

It is straightforward to relate Z to cavity Q, ω, and R/Q (taking care to be consistent if voltages are measured on axis or across the gap), using the relationship:

$$Z = e^{j\psi}/\alpha \ . \tag{5}$$

where

$$\alpha = I_{ind}/V = [Q_0^{-2} + 4(\Delta\omega/\omega)^2]^{1/2} \quad . \tag{6}$$

and

$$\psi = \phi_I - \phi_V = tan^{-1}[-2Q_0\Delta\omega/\omega] \quad . \tag{7}$$

The alternative to the port approximation would be to model the cavity boundaries in MASK. The advantage of using the port approximation is that it becomes possible, using current computational resources, to fully model the cavities of a klystron and their interaction with an electron beam. There are two reasons for preferring the port approximation:

1. To obtain the correct voltages. If the entire cavity were modelled, the frequencies of the gain cavities would need to be accurate to better than 0.1 percent, which would require prohibitively fine zoning.

2. To reduce computational time. Since the cavities typically have loaded Q's of order several hundred, this would necessitate excessive computation to reach steady state.

NEW PORT ALGORITHM

In Simon Yu's original method, each cavity was calculated successively using either linear theory or an iterative method to achieve a self-consistent voltage. We have devised a faster method which relaxes the voltage and phase of several cavities simultaneously. In steady state, the gap voltage should satisfy the condition

$$V_{ss} = I_{ss} \cdot Z \quad . \tag{8}$$

Here Z is the complex cavity impedance, and V_{ss} and I_{ss} are the Fourier components in steady state of the gap voltage and the induced current at the operating frequency. Now we assume a time dependence of the form:

$$V_t = V(t)e^{-j\omega t} \quad . \tag{9}$$

Here V_t is the instantaneous voltage across the port and V(t) is an envelope which varies slowly in an rf cycle. Asymptotically, we want $V(t)$ to converge to V_{ss}. We can achieve this by making $V(t)$ satisfy a relaxation equation, i.e.,

$$dV(t)/dt = -k \cdot [V(t) - I(t) \cdot Z] \quad . \tag{10}$$

Thus $V(t)$ will adjust itself until the impedence relation is satisfied self-consistently. We compute the induced current at the operating frequency by keeping a running table of the volume integral of $E \cdot J$. The equation converges faster if one takes into account the beam loading, assuming that the change in induced current is a linear function of the change in voltage, i.e.

$$\Delta I = \alpha_l \cdot \Delta V \quad . \tag{11}$$

Then

$$\Delta V = -k\Delta t(V - IZ) \div (1 - \alpha_l Z) \quad . \tag{12}$$

The constant α_l depends on dc current, frequency, and drift-tube size but is insensitive to gap width and beam profile.

APPLICATIONS

Applications of the original method have been described previously[1,3]. We now give an example of the application of the new algorithm to a similar problem. Figures 2 and 3 show a schematic of a SLAC XK-5 klystron, first with cavities drawn, then as it would be modelled using the port approximation in a MASK simulation.

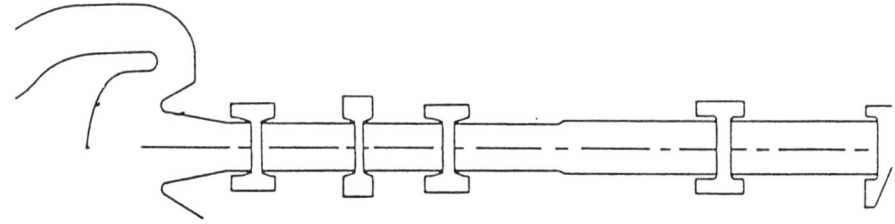

Figure 2. Schematic of the SLAC XK-5 klystron.

Here the electron gun is on the left, with the output cavity on the right hand side. Notice that there is an increase in tube radius between cavities three and four which was not modelled in the early simulation shown in Figure 3.

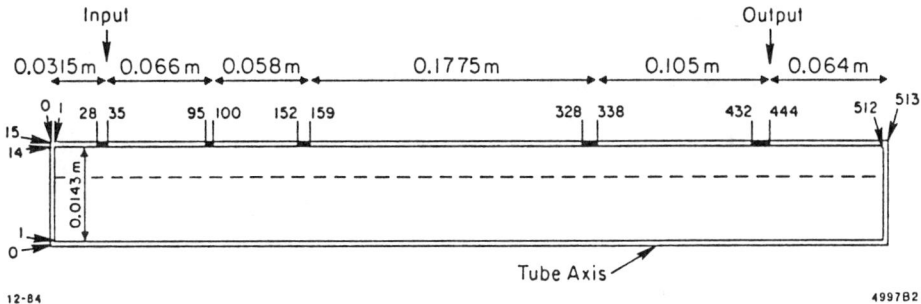

Figure 3. XK-5 klystron modelled in MASK using ports.

Figures 4 and 5 show position space and momentum space plots from a MASK simulation of a SLAC 5045 klystron. In Figure 4 the ports are located at the gaps in the blocks, with the output port just after the final block. The horizontal axis is Z, with the beam injected on the left, and the vertical axis is R, with the lower boundary being the symmetry axis R=0. The beam energy was 315 kV, and current was 393 amps. The magnetic field increased from zero at the cathode to a peak of 1200 gauss. The drift tube radius was 1.59 cm at the input and 1.75 cm at the output. The distance between input and output cavities was 56 cm.

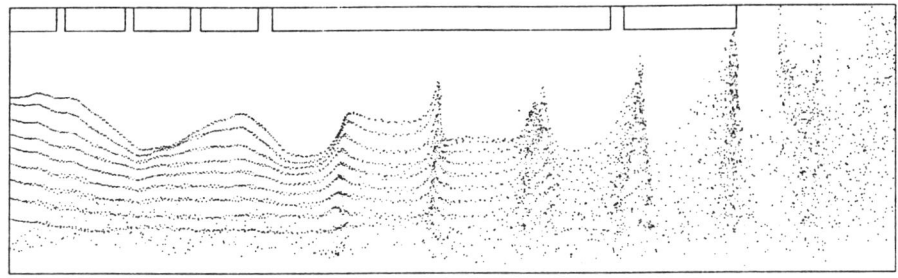

Figure 4. Position space distribution for the 5045.

In this tube the radius increased at the downstream nose of the output cavity. The varying radius was modelled with the conducting blocks shown in the figure. The input beam distribution was calculated using the EGUN code on the gun geometry and the magnetic field with POISSON from the electromagnet design.

The scalloping of the beam at the left is due to the magnetic field, while the longitudinal bunching downstream results from the rf modulation. Note that the bunching is nearly as much a tranverse effect as it is longitudinal.

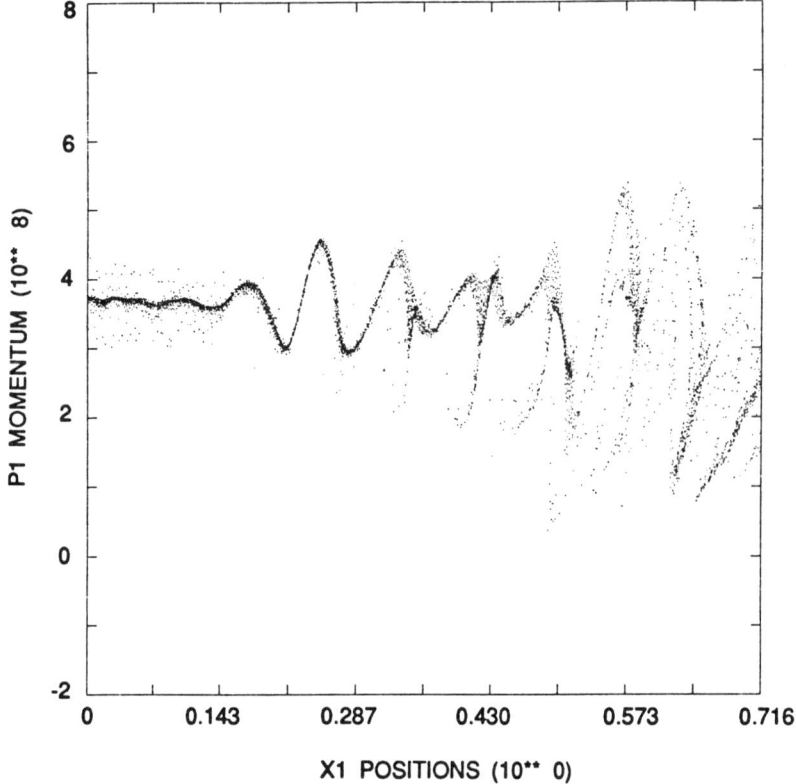

Figure 5. Longitudinal momentum (γv_z) for the 5045.

The initial modulation is sinusoidal. The trailing "fingers" are the result of space charge effects. The momentum profile "stands up" to produce maximum bunching at the output cavity. Because some current was left in the "antibunch," i.e., the portion of the rf cycle with the wrong phase to be decelerated by the output cavity, some electrons gain rather than lose energy there, limiting the efficiency to under 50 percent.

Figures 6 and 7 show the time history of the voltages and phases on each cavity using the algorithm described above as they relax to convergence. Cavities three through six were not turned on until cavity two had converged. (See discussion below.)

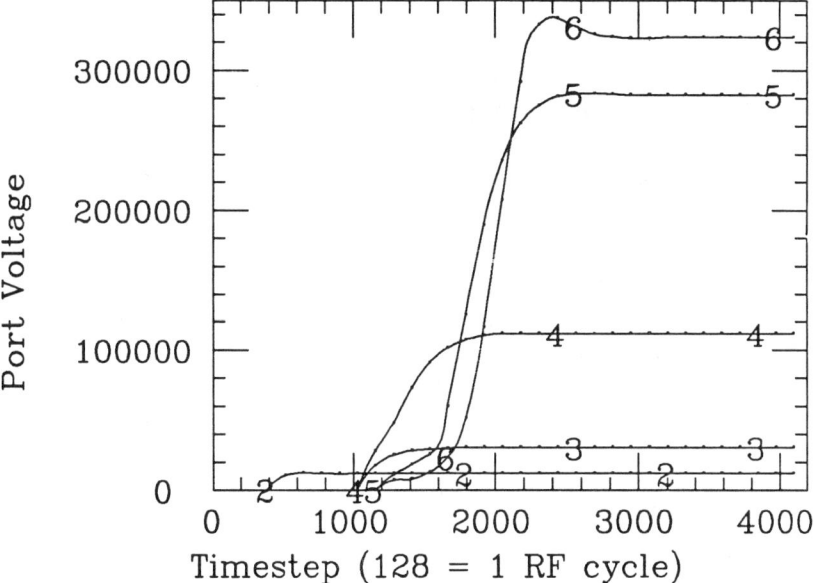

Figure 6. Port voltage vs. time in relaxation of the 5045.

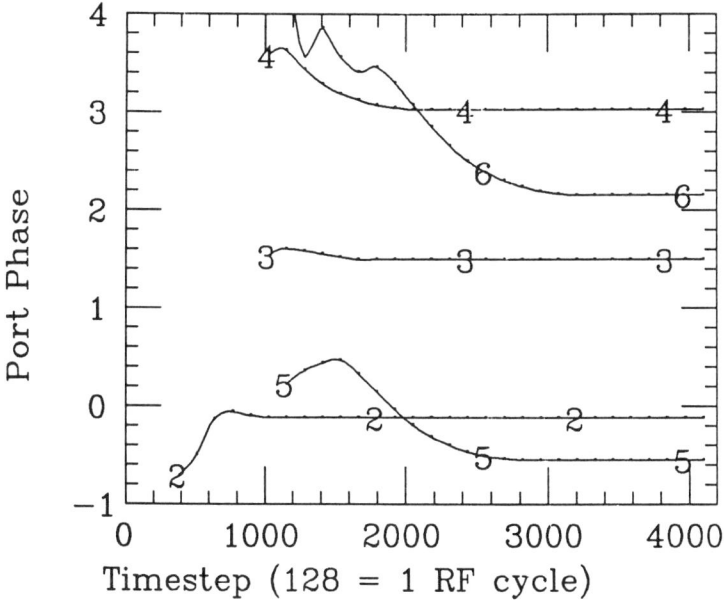

Figure 7. Port phase vs. time.

In this simulation cavity 2 was relaxed alone until it converged. Then cavities 3-6 were relaxed simultaneously. The reason this was required is that in this particular klystron, the detuning of the second cavity is very small. As a result, the voltage on the second cavity is very sensitive to noise, and the high intensity of the second harmonic component produced downstream can distort the calculation of the induced current for the second cavity. With a larger detuning on this cavity, it would be possible to relax all cavities simultaneously.

In the first run (relaxing all cavities) the solution converged after about 24 RF cycles. The solution for the low power cavities is linear. Thus once one has a solution for a particular input power, one can find values for other power levels by multiplying all gain cavity voltages by a constant, and relaxing only the high-power cavities (5 and 6). This is shown in Figures 8 and 9. Here the cavities converged after about 16 RF cycles.

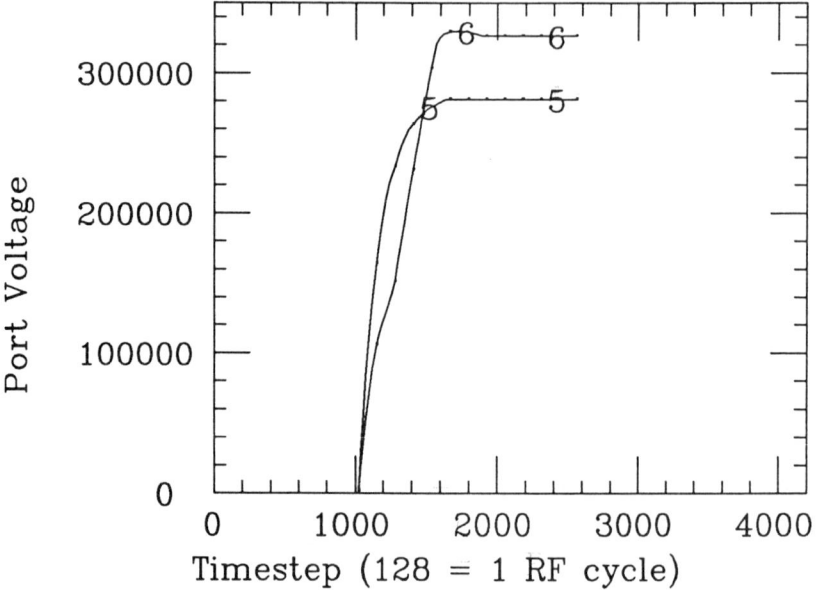

Figure 8. Port voltage vs. time for last two cavities, all others fixed.

This new algorithm reduces the total computation time (for a gain curve with three or four points) by a factor of about four to five for the complete klystron, compared to the original method. The voltages and phases agree to within a few percent of those obtained with the old algorithm. The new method actually is more accurate for the low power cavities, and will converge where the older method would sometimes fail. In a simulation of the SLAC 5045 S-band klystron, the old method predicted saturation at about 65 watts at a beam voltage of 315 kV, while the new algorithm gives saturation at about 120 watts, which is in better agreement to the data.

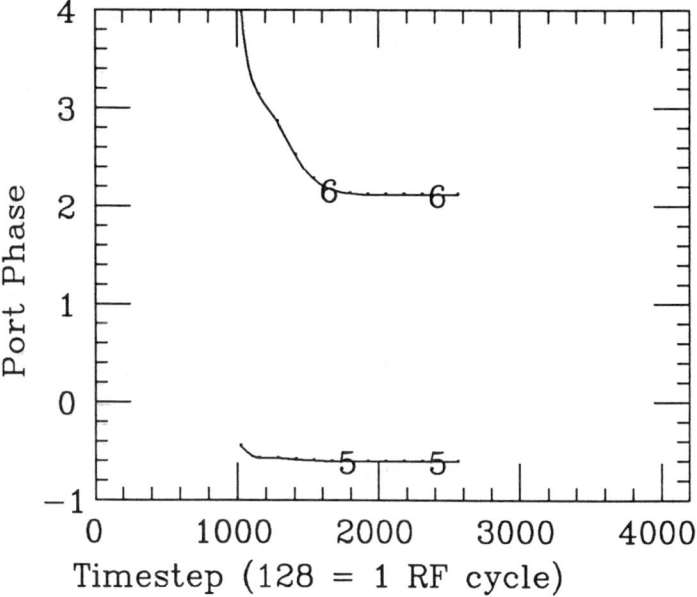

Figure 9. Port phase vs. time.

In some situations it is desirable to apply a fixed voltage and relax only the phase of the cavity. For example, for most non-externally loaded cavities the phase difference between voltage and induced current will be close to $-\pi/2$ (unless the detuning is very small), while for output cavities the phase difference is usually close to zero. If one is designing a klystron, it is often simpler to specify the voltages than the impedences. A modification of the algorithm described above relaxes the phase only, in a simple way:

$$d\phi_V/dt = k \cdot (\psi - \psi_0) \quad . \tag{13}$$

Here ϕ_V is the gap phase, ψ is the phase difference $\phi_I - \phi_V$, and ψ_0 is the desired phase difference.

CHAINING ALGORITHM

These new methods have been applied to the simulation of the "hybrid relativistic klystron" using magnetic induction technology, now being studied by SLAC and LLNL. The length of this device would have made the simulation prohibitively expensive with the older method. The length also required a method of chaining several simulations together, feeding the output from one segment into the input of a second. This required modification of the Neumann boundary conditions to prevent excessive transverse deflection at the junction between segments.

The chaining algorithm itself was straightforward. It assumes a beam injected from the left which does not reverse direction. All particles removed from the right boundary are put into a buffer, which is dumped at specified time intervals. The fields are calculated by assuming Neumann boundary conditions, so that the field quantities do not have to be dumped. A history of the number of particles removed on each cycle is stored in a separate array which is also dumped. This algorithm is useful either for a single pulse of finite length (in which case the entire pulse is written out) or for a steady-state problem. For a periodic problem an integral number of cycles can be dumped after steady state is reached, and then recycled periodically when they are read in. For a dc problem, the same procedure applies, with the "cycle" time being any interval long enough to produce a reasonable distribution.

The original Neumann boundary solver used in MASK was found to produce large transverse momentum disruptions in the reinjected beam whenever the space charge in the beam was significant. The algorithm solved for the correction potential whose gradient, when added to the fields, would produce a solution of Poisson's equation. However, the solution on the boundary is not the same as the one obtained by solving for the full potential. This is because the radial derivative of the radial fields on the left and right boundaries does not enter into the calculation of the Neumann boundary condition, and thus is not corrected by the solution. Another way of looking at this is that the solution is made to solve Poisson's equation in the interior, but not on the boundary itself.

To correct the problem, we solved Poisson's equation using the full charge distribution to get the total potential. Then we took the one dimensional (radial) gradient on the left and right boundaries to get the radial fields there. This technique greatly reduced the radial disruption across the boundaries.

REFERENCES

1. S. Yu, "Particle In Cell Simulation of High Power Klystrons," SLAC/AP-34, September 1984.

2. A. Palevsky and A. T. Drobot, "Application of E-M P.I.C. Codes to Microwave Devices" in Proceedings, Ninth Conference on Numerical Simulation of Plasmas, paper PA-2, Northwestern University, Evanston, Illinois, June 30-July 2, 1980.

3. K. Eppley, A. Drobot, W. Herrmannsfeldt, H. Hanerfeld, D. Nielsen, S. Brandon, R. Malendez, "Results of Simulations of High Power Klystrons" in Proceedings, 1985 Particle Accelerator Conference, Vancouver, British Columbia, Canada, May 13-16, 1985, p. 2903.

MOMENT INVARIANTS FOR PARTICLE BEAMS*

Walter P. Lysenko and Mark S. Overley

Los Alamos National Laboratory, AT-3, MS H808, Los Alamos, NM 87545

ABSTRACT

The rms emittance is a certain function of second moments in 2-D phase space. It is preserved for linear, uncoupled (1-D) motion. In this paper, we present new functions of moments that are invariants for coupled motion. These invariants were computed symbolically using a computer algebra system. Possible applications for these invariants are discussed. Also, approximate moment invariants for nonlinear motion are presented.

INTRODUCTION

Moments and Moment Equations

Let $g(\vec{x}, \vec{p}, t)$ be a function defined on six-dimensional phase space. We define the moment of g to be the integral of g over all of phase space, weighted by the single-particle phase-space distribution function. That is,

$$< g(\vec{x}, \vec{p}, t) > = \int_{-\infty}^{\infty} d^3p \int_{-\infty}^{\infty} d^3x \, f(\vec{x}, \vec{p}, t) \, g(\vec{x}, \vec{p}, t), \tag{1}$$

where f is the distribution function. For a discrete distribution consisting of N particles, with phase-space coordinates $\vec{x}_i, \vec{p}_i$, we have

$$< g(\vec{x}, \vec{p}, t) > = \frac{1}{N} \sum_{i=1}^{N} g(\vec{x}_i, \vec{p}_i, t). \tag{2}$$

This form is useful in computing moments in particle simulation codes. Note that these moments are functions of time only. A particle beam can be described by a set of moments of some basis functions such as the monomials in the variables x, p_x, y, p_y, z, p_z. The moment description has the advantage of being closely related to laboratory quantities. For example, the centroid of a beam in the x-direction is given by $< x >$ and the rms width in the x-direction is given by $\sqrt{< x^2 >}$.

---

*Work supported by Los Alamos National Laboratory Institutional Supporting Research, under the auspices of the U.S. Department of Energy.

For particle beams in most noncircular machines, the distribution function accurately satisfies the Vlasov equation

$$\frac{\partial f}{\partial t} + \frac{\partial f}{\partial \vec{x}} \cdot \frac{\vec{p}}{m} + \frac{\partial f}{\partial \vec{p}} \cdot \vec{F} = 0, \qquad (3)$$

where $\vec{F}$ is the sum of the external force and the space-charge force, whose potential ϕ satisfies the Poisson equation

$$\nabla^2 \phi = -\frac{1}{\epsilon_o} \int_{-\infty}^{\infty} d^3p\, f(\vec{x}, \vec{p}, t). \qquad (4)$$

If we are describing the beam with moments, we need a way to relate the integral in the above formula to the moments. (This is the major problem in any simulation scheme using moments as the dynamical variables. We will not be concerned with this here, however.) The rule for differentiating moments with respect to time is given by[1]

$$\frac{d}{dt} <g> = <\frac{d}{dt}g>, \qquad (5)$$

which says that the derivative of a moment is the moment of a derivative. Therefore, it is easy to get the equations of motion for the moments. For example, the equation for the $<xp_x>$ moment is

$$\begin{aligned}
\frac{d}{dt}<xp_x> &= <(\frac{d}{dt}x)p_x> + <x(\frac{d}{dt}p)> \\
&= \frac{1}{m}<p_x^2> + <xF_x> \\
&= \frac{1}{m}<p_x^2> + <x(a_0 + a_1 x + a_2 x^2 + \cdots)> \\
&= \frac{1}{m}<p_x^2> + a_0 <x> + a_1 <x^2> + a_2 <x^3> + \cdots.
\end{aligned} \qquad (6)$$

Because we expanded the force into a power series, the right hand side of the equation contains only moments. So the moment equations are a system of linear differential equations except that, in the case with space charge, the force coefficients will depend on the spatial moments.

Emittance and RMS Emittance

Emittance can be defined in various ways, but it is always some kind of area in a 2-D projection of phase space. If the x-degree of freedom is uncoupled from the y and z directions, then the total emittance in the x-direction is a conserved quantity because it is the same as the phase volume. It is an invariant that does not depend on the force.

The rms emittance in the x-direction is defined to be the following function of second moments:

$$\epsilon_x = \sqrt{<x^2><p_x^2> - <xp_x>^2}. \tag{7}$$

This definition assumes that the beam centroid is at the origin. Subtracting $<x>$ and $<p_x>$ from x and p_x, respectively, in the above definition will give the general formula. The rms emittance is conserved for linear, uncoupled motion. To see this, simply differentiate ϵ_x with respect to time, assuming the force is proportional to x, that is, assuming $dp_x/dt = a_1 x$.

For uncoupled motion, the rms emittance is somewhat like the total emittance in that it is also an invariant that does not depend on the force (value of constant a_1). But it is invariant only for linear forces. For coupled motion, there are no know invariants that are functions of emittances, even for linear forces. The problem we solve in this paper is to find other moment invariants, analogous to the rms emittance, that are valid for more general situations than linear, uncoupled motion.

FINDING NEW INVARIANTS

Setting Up the Problem

Consider the following homogeneous quadratic function of the second moments in one dimension (the x-direction).

$$\begin{aligned} I &= c_{11}<x^2><x^2> + c_{12}<x^2><xp_x> + c_{13}<x^2><p_x^2> \\ &+ c_{21}<xp_x><x^2> + c_{22}<xp_x><xp_x> \\ &+ c_{31}<p_x^2><x^2> \end{aligned} \tag{8}$$

For the following choice of the coefficients,

$$\begin{aligned} c_{13} &= 1, \\ c_{22} &= -1, \\ \text{other } c_{ij} &= 0, \end{aligned} \tag{9}$$

the function I is the square of the rms emittance in the x-direction and is therefore an invariant for linear, uncoupled motion.

Let us now find the moment invariant for linear motion in 2-D. To do this, let us start with an expression like Eq. (8) but include all second moments involving both the x and y directions. Then we will attempt to find a set of values for the coefficients that makes the expression an invariant.

First define the vector

$$\begin{aligned} X = \;&(<xx>, <xp_x>, <xy>, <xp_y>, \\ &<p_x x>, <p_x p_x>, <p_x y>, <p_x p_y>, \\ &<yx>, <yp_x>, <yy>, <yp_y>, \\ &<p_y x>, <p_y p_x>, <p_y y>, <p_y p_y>). \end{aligned} \tag{10}$$

This is a list of all the possible moments that our invariant can contain. Then the proposed invariant can be written as

$$I = X^T C X. \tag{11}$$

Remarkably enough, in writing Eq. (11), we have already sufficiently restricted the form of the invariant so that by applying the condition

$$\frac{d}{dt}I = 0, \tag{12}$$

we can solve for the coefficient matrix C and thereby determine the desired invariant. In taking the time derivative, we must assume that the forces are given by

$$\begin{aligned} F_x &= x a_{xx} + y a_{xy}, \\ F_y &= x a_{xy} + y a_{yy}. \end{aligned} \tag{13}$$

Notice the symmetry of the force coefficients, which is equivalent to requiring that the force be derivable from a potential. (Without this restriction, the invariants do not exist.)

Because of the size of this problem (we have to solve for 55 coefficients), we used the SMP (Symbolic Manipulation Program) computer algebra system to do the algebra symbolically.

Using the SMP Computer Algebra System

The computer algebra system SMP is a product of Inference Corporation (Los Angeles, CA). It is an interactive system that takes commands at the terminal and returns results on the screen. It can also read files containing commands. In this section, we show some general features of this system. In the next section, we will show how our actual problem was solved.

In SMP, most objects such as functions, matrices, and parts of expressions are all represented by things called projections. For example, if p is a polynomial, then the projection p[3] is the third term. To see how SMP works, consider an object f. Now define some properties of f. If we think of f as an array, then the command

```
f[1,4]:6
```

defines the $(1,4)$-th element to have value 6. Now add another property to f by issuing the following command:

```
f[x]:7
```

This causes f to map the quantity x into the value 7. If we now enter the command

```
f[$x]:$x^2
```

we define the function $f(x) = x^2$. (The $x is a dummy argument and is said to be generic.) Now when f is evaluated, an argument of 1,4 returns 6, an argument of x returns 7, and any other argument returns the square of the argument. So, for example, the value of f[x] is 7, but the value of f[y] is y^2.

More complicated definitions are possible. The following

```
f[$x _= ($x>7 | $x<-9)] : 75
```

says that if the argument is greater than 7 or less than -9, then the value of f is 75. The symbol _= should be thought of as "such that." Existing built-in functions can be redefined. For example, the following

```
D[h[$x],{$x,1,$y}]:g[$y]
```

adds to the differentiation function D the rule that the derivative of the function h is g. The list {$x,1,$y} means that we are taking the first derivative with respect to $x and evaluating at $y. After we make this definition, D[x^2,x] returns 2x so the old (built-in) definition is still there. But D[h[z],z] returns g[z], using the new rule.

THE NEW LINEAR INVARIANTS

To solve the problem of the invariant for the linear coupled motion, we had to define many SMP functions. We show some of these here. In the SMP code, the moment $<g>$ is denoted by mom[g]. The linearity property of moments is defined by the following SMP code.

```
mom[$x + $$y] :: mom[$x] + mom[$$y]
mom[$$y $x _= test[$x]] :: $x mom[$$y]
```

The $$y symbol is said to be doubly generic and stands for a complete expression, not just a single term. The function test is defined to return a true value if the argument is a constant. The following definitions cause <g> in the input commands to be translated into mom[g] and print out mom[g] as <g>.

```
"<" := "mom["
">" := "]"
_mom[Pr,$x] : Fmt[,"<",$x,">"]
```

This makes it possible to communicate with SMP using familiar notation.

We define moments to be the list of variables. Right now, we are doing the 2-D case, so we define this list as follows.

```
moments : {x,px,y,py}
```

Now define `der` to be the function that takes a time derivative. In addition to giving it the usual rules for differentiating possesed by the built-in D function, we add the rules for taking derivatives of coordinates.

```
der[x]  :: px
der[y]  :: py
der[px] :: fx
der[py] :: fy
```

Now we give the rule that the derivative of the moment is the moment of the derivative.

```
der[mom[$x]] :: mom[der[$x]]
```

We define a list of second moments, called `vector` by the following piece of SMP code.

```
vector2:Ldist[mom[Union[Flat[moments**moments]]]]
vector:Union[vector2]
```

This takes the outer product of `moments`, flattens this matrix into a list, and converts this list into a list of moments of these quantities.

Now letting `cmat` be some matrix, we define `eq`, the proposed invariant.

```
eq::Ex[vector.cmat.vector]
```

This is the same as Eq. (11), which says that our proposed invariant is a homogeneous quadratic combination of second moments. After defining the forces,

```
fx : x a[x,x] + y a[x,y]
fy : x a[x,y] + y a[y,y]
```

we differentiate our proposed invariant with the `der` function. Setting the coefficients of the moments and forces individually to zero leads to a linear system of equations for the coefficients. Using a built-in function in SMP to solve this system yields the coefficients in the invariant. Substituting these coefficients into the expression for `eq` gives us the desired invariant. The result, which we call I_2, is the following:

$$I_2 = \epsilon_x^2 + 2\epsilon_{xy}^2 + \epsilon_y^2, \qquad (14)$$

where

$$\epsilon_x^2 = <x^2><p_x^2> - <xp_x>^2,$$
$$\epsilon_y^2 = <y^2><p_y^2> - <yp_y>^2, \qquad (15)$$
$$\text{and} \quad \epsilon_{xy}^2 = <xy><p_xp_y> - <xp_y><yp_x>.$$

Equation (14) is our main result. It says that for two dimensions, the invariant is the sum of the square of the x-rms emittance, the square of the y-rms emittance, and a third term, which could be called a "cross-emittance." There is a similar result for three dimensions.

We also found that there are other invariants for linear motion. If we search for invariants that are homogeneous quadratic functions of *fourth* moments, we obtain (for 1-D)

$$I_4 = <x^4><p_x^4> - 4<x^3 p_x><x p_x^3> + 3<x^2 p_x^2>^2. \tag{16}$$

For sixth moments, we obtain

$$\begin{aligned} I_6 &= <x^6><p_x^6> - 6<x^5 p_x><x p_x^5> \\ &\quad + 15<x^4 p_x^2><x^2 p_x^4> - 10<x^3 p_x^3>^2. \end{aligned} \tag{17}$$

There are apparently an infinite number of such invariants. There are 1-D, 2-D, and 3-D versions of all of them.

There is a very simple way to generate all these invariants. They are all of the form

$$I_n = \frac{1}{2}<(X-P)^n>_{sym}, \tag{18}$$

where

$$\begin{aligned} X &= x+y+z, \\ P &= p_x+p_y+p_z, \end{aligned} \tag{19}$$

and where the "symmetrized" moment is defined by its linearity properties

$$\begin{aligned} <g+h>_{sym} &= <g>_{sym} + <h>_{sym} \\ <\alpha g>_{sym} &= \alpha <g>_{sym} \end{aligned} \tag{20}$$

and by its action on monomials

$$<x^i p_x^j \; y^k p_y^l \; z^m p_z^n>_{sym} = <x^i p_x^j \; y^k p_y^l \; z^m p_z^n><x^j p_x^i \; y^l p_y^k \; z^n p_z^m>. \tag{21}$$

This says that the symmetrized moment is just the usual moment, multiplied by a moment in which we interchange the powers of the coordinates and the conjugate momenta. The following, which computes the lowest-order invariant in 1-D, shows how this works.

$$\begin{aligned} 2I_2 &= <(x-p_x)^2>_s \\ &= <x^2 - 2xp_x + p_x^2>_s \\ &= <x^2>_s - 2<xp_x>_s + <p_x^2>_s \\ &= <x^2><p_x^2> - 2<xp_x><p_x x> + <p_x^2><x^2> \\ I_2 &= <x^2><p_x^2> - <xp_x>^2 \end{aligned} \tag{22}$$

NONLINEAR INVARIANTS

Consider some distribution in (x, p_x) phase space. Now consider a linear transformation to a new coordinate system $(\bar{x}, \bar{p}_x)$ in which the moment $<\bar{x}\bar{p}_x>$ vanishes. The rms emittance for this distribution is given by

$$\epsilon_x = \sqrt{<x^2><p_x^2> - <xp_x>^2} = \sqrt{<\bar{x}^2><\bar{p}_x^2>}. \tag{23}$$

In the new coordinates, the rms emittance is the product of the rms widths in the $\bar{x}$- and $\bar{p}_x$-directions. It is like an area in phase space. The cross term $-<xp_x>^2$ in the definition of the rms emittance serves to take out the effect of the "tilt" in the phase-space distribution.

If nonlinear forces are present, then the "twist" of the distribution can change. For example, if the initial distribution is a line in (x, p_x) space, the action of nonlinear forces will introduce an S-shape into the line. This causes the rms emittance to grow. If we seek a moment invariant that is preserved in the presence of nonlinear forces, then we need terms like $<xp_x^3>$ in the invariant that will take out the effect of the twist in the distribution. The higher invariants I_n do contain such terms; however, as we shall see, these are not nonlinear invariants.

But let us look for nonlinear invariants that are combinations of the I_n. Consider the following candidate for a nonlinear invariant.

$$J = I_4 + cI_2^2 \tag{24}$$

Because it is a function of linear invariants, J is constant for linear motion regardless of the value of c. Let us try to choose a value of c that will make J a nonlinear invariant. To do this, consider the distribution in (x, p_x) phase space that is constant in the region

$$\begin{aligned} -x_{max} &\leq x \leq x_{max} \\ -\Delta p_x &\leq p_x - (\alpha x + \beta x^3) \leq \Delta p_x \end{aligned} \tag{25}$$

and zero elsewhere. This distribution is shown in Fig. 1. This distribution has four parameters: x_{max}, Δp_x, α, and β. In order to get simplify the final result, let us eliminate the parameters Δp_x and β in favor of the rms emittance ϵ_x and another parameter δ, which measures the amount of twist in the distribution.

$$\Delta p_x = \frac{3\epsilon_x}{x_{max}}, \qquad \beta = \sqrt{\frac{525\,\delta\epsilon_x^4}{2\,x_{max}^4}} \tag{26}$$

Now compute the linear invariants I_2 and I_4 for the distribution parametrized by x_{max}, ϵ_x, α, and δ. This can be done using SMP to symbolically compute the required integrals over the distribution given by Eq. (25).

$$I_2 = \epsilon_x^2 + \delta^2 \epsilon_x^8 \tag{27}$$

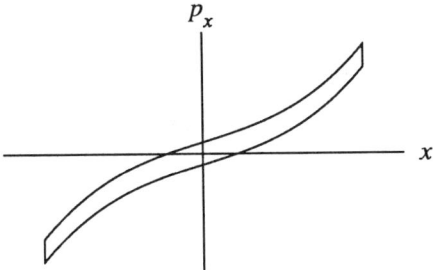

Fig. 1. This phase-space distribution was used to obtain a nonlinear invariant by requiring that the value of the invariant did not depend on the amount of "twist" in the distribution.

$$I_4 = \frac{156}{25}\epsilon_x^4 + 20\ \delta^2\epsilon_x^{10} + \mathcal{O}(\delta^4) \qquad (28)$$

Notice that these linear invariants depend only on the rms emittance ϵ_x and on the parameter δ, which measures the amount of twist in the distribution. These invariants do not depend on α, the amount of tilt in the distribution, nor on x_{max}, which measures the eccentricity of the distribution. Substituting the above I_2 and I_4 in the definition for J, we see that if $c = -10$, then J is independent of the twist parameter δ. So the desired invariant is

$$J = I_4 - 10I_2^2. \qquad (29)$$

The utility of this invariant was verified numerically for a situation in which a uniformly filled ellipse in phase space was transformed by a system containing linear and cubic forces. Figure 2 shows the initial and final phase-space distributions for our example. The strength of the nonlinearity is sufficient to put a noticeable twist in the distribution, which initially did not have any. Figure 3 shows some of the linear invariants and the new J invariant as functions of time, showing their evolution between the initial and final states shown in Fig. 2. In this graph, all invariants are normalized by their initial values. We see that the linear invariants are not constant and that the higher invariants are even more sensitive to the nonlinearity than is I_2. The new invariant J is more constant

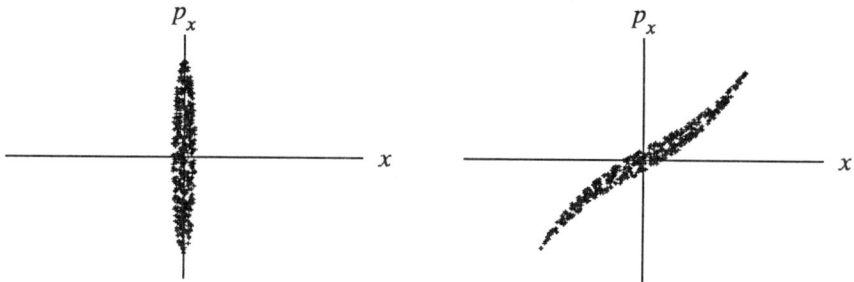

Fig. 2. The nonlinear invariant was checked by simulating the evolution of the initial distribution shown on the left in the presence of linear and cubic forces. The final distribution is shown on the right. The strength of the nonlinearity, for the given beam size and the elapsed time chosen, is large enough to introduce a noticeable twist in the distribution.

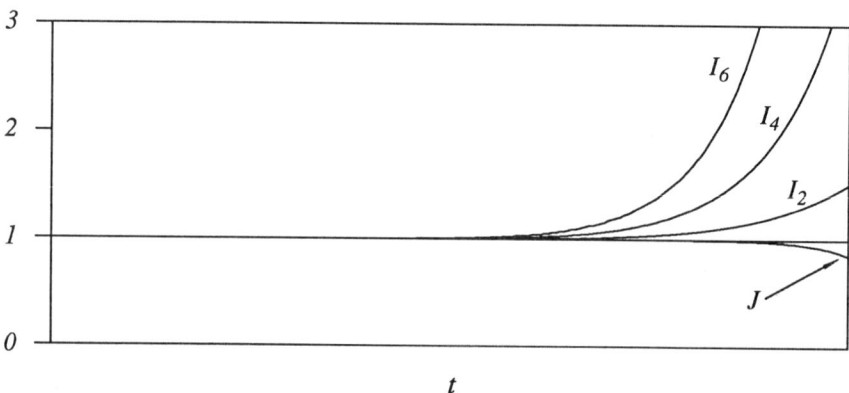

Fig. 3. The linear invariants I_n and the approximate nonlinear invariant J are shown as functions of time for the situation described in Fig. 2. All invariants are normalized by their initial values. Even though the quantity J is a linear combination of I_4 and the square of I_2, it is approximately constant.

than the rest, but it is not an exact invariant.

DISCUSSION

The moment description is useful because it is so closely related to laboratory quantities. The well-known rms emittances ϵ_x, ϵ_y, and ϵ_z are functions of second moments.

$$\begin{aligned}
\epsilon_x^2 &= <x^2><p_x^2> - <xp_x>^2 \\
\epsilon_y^2 &= <y^2><p_y^2> - <yp_y>^2 \\
\epsilon_z^2 &= <z^2><p_z^2> - <zp_z>^2
\end{aligned} \quad (30)$$

These quantities are invariant for linear, uncoupled (1-D) motion. If an accelerator or beamline has any coupling forces, even if they are linear, the rms emittance can change. Thus, the growth in an rms emittance is not necessarily an indication of nonlinearities in the machine.

We have shown in this paper a new moment invariant, I_2, that is conserved for linear forces even in 2- or 3-D. In 1-D it is the same as square of the rms emittance. In 2- or 3-D it is the sum of the squares of the rms emittances together with new cross terms. In three dimensions, we have

$$\begin{aligned}
I_2 =\ & <x^2><p_x^2> - <xp_x>^2 \\
+\ & <y^2><p_y^2> - <yp_y>^2 \\
+\ & <z^2><p_z^2> - <zp_z>^2 \\
+\ & 2<xy><p_xp_y> - 2<xp_y><yp_x> \\
+\ & 2<xz><p_xp_z> - 2<xp_z><zp_x> \\
+\ & 2<yz><p_yp_z> - 2<yp_z><zp_y>\ .
\end{aligned} \quad (31)$$

This invariant, or any of our other new invariants, is easily obtained using the formula given by Eq. (18). In this case, we just expanded

$$<(x + y + z - p_x - p_y - p_z)^2>_{sym}.$$

The utility of the new invariant is that it is conserved for any kind of linear motion, even with coupling present. Therefore, if a simulation or measurement showed that the quantity I_2 has increased, then we can conclude that the accelerator or beamline contains nonlinearities.

We have also found other moment invariants for linear motion that are homogeneous quadratic combinations of higher moments; these are the quantities I_2, I_4, I_6, etc. All these exist in 1-D, 2-D, and 3-D versions.

Consider, for a moment, uncoupled linear motion. We know that the beam current and the rms emittance are the only significant beam parameters in this situation*. Suppose we have some beam and some beamline. All we have to know is whether the beamline can transport the emittance of our beam. Neither the details of the machine acceptance nor of our beam are important. We know that we can always make a linear matching section to match our beam into the machine. This is true because any beam of a given emittance can be transformed by a linear transformation into any other beam of the same emittance.

The above is only true if our knowledge is limited to second moments. That is, we consider two beams identical if they have the same collection of moments, up to second order. If we describe our beam to higher order, then we need new invariants. This is how the higher invariants I_n presented in this paper can be used. For example, if we describe a beam by moments up to fourth order, then the invariant beam parameters are I_2 and I_4. (I_2 is the square of the rms emittance, of course.)

In the more general situation of linear coupled motion, the same ideas apply, but we need additional invariants. In a second-order description in 2-D, we need two invariants, but we have only one, the 2-D version of I_2. We need further work to determine these invariants. Before we discuss extending these results to nonlinear motion, let us consider other possible uses for the linear invariants.

In numerical simulations, it is useful to use slow variables. For example, for nonlinear oscillatory single-particle motion, it is better to use amplitude-phase variables rather than x, p_x because we are solving directly for the effect of the nonlinearity. If the nonlinearity is zero, then the amplitude and phase are constant. In the same way, we could use our moment invariants as the dynamical variables in a simulation code. To do this, we have to determine all the independent invariants up to a given order. In linear single-particle motion there is a simple relation between the amplitude-phase variables and the action-angle variables. If the new moment invariants are like the action variables, perhaps we can find a Hamiltonian formulation of the moment equations. Therefore, we believe the linear moment invariants may have significance both in numerical simulation and in theoretical analysis of beam motion.

---

*The beam current depends on the charge in the beam, which is the zeroth phase-space moment and could be considered to be the invariant I_0 in our scheme. We prefer to not consider here moments below second order and to consider space charge as a source of forces in addition to those of the beamline elements.

Because the linear moment invariants respond only to nonlinearities, they are useful in studying nonlinear motion as well as linear motion. However, determining nonlinear invariants would be even more useful. For example, we would like an invariant that is constant if no fifth- or higher-order forces are present. Such invariants would help us to analyze aberrations. We found the approximate nonlinear invariant J by assuming a certain model for a beam. The difficulty with working with nonlinear invariants concerns the handling of correlations. For example, if we differentiate the expression for J, we get many terms that do not appear to cancel. However, in a particular beam, such as in the example used in the numerical simulation shown in this paper, high-order correlations are absent. Furthermore, some of this lack of correlation is preserved throughout the motion. (Exactly what happens depends on the initial condition and the forces.) This means that some of the higher-order moments are actually functions of lower-order moments. The behavior of these correlations must be understood before we can derive more useful nonlinear moment invariants in the way we did for the linear case. Fortunately, the evolution of the correlations can be studied by the easy-to-use moment equations. This is because the correlations are just combinations of moments; they consist of differences of moments and the expression describing the moments in the absence of correlations.

REFERENCE

1. P. J. Channell, "The Moment Approach to Charged Particle Beam Dynamics," IEEE Trans. Nucl. Sci. **30** (4), (August 1983), p. 2607.

THE CORRECTION OF ABERRATIONS IN BEAMS FILLING ELLIPTICAL PHASE–SPACE AREAS

H. Wollnik

2. Physikalisches Institut, Universität Giessen, 6300 Giessen, W–Germany

ABSTRACT

For the optimization of an optical system it is advantageous to amend the system by a virtual object lens so that the calculation always starts from an upright phase-space distribution. Furthermore, in case of a beam filling an elliptical phase–space volume, the most extreme rays of a beam, filling a parallelogram–like phase–space volume, do not exist, so that the corresponding sum of aberrations is smaller. For an optimization thus corresponding attenuation factors should be taken into account.

1. INTRODUCTION

The optical properties of a particle analyzer or transport system are characterized by the differences between the trajectory of an arbitrary particle and the "optic axis" along which a reference particle moves with kinetic energy K_0, rest mass m_{00} and charge q_0. The arbitrary particle here is assumed to be characterized by its kinetic energy K, rest mass m_0 and charge q with

$$\frac{K}{q} = \frac{K_0}{q_0}(1+\delta_K) \qquad \frac{m_0}{q} = \frac{m_{00}}{q_0}(1+\delta_m) .$$

The differences between these two trajectories are described by their distances (x, y) and inclinations ($a = p_x/p_0$, $b = p_y/p_0$) as well as the flight-time differences $T - T_0$ between the arbitrary particle and the reference particle moving along the optic axis with

$$T = T_0(1+\delta_t) .$$

Here p_0 is the momentum of the reference particle and p_x, p_y are the components of the momentum of the arbitrary particle in x- and y-directions, i.e., directions perpendicular to the optic axis. Altogether, the trajectory of an arbitrary particle is characterized by phase–space coordinates[1]

$$x, a, y, b, \delta_t, \delta_K$$

which form a set of canonically conjugate coordinates where the independent variable is the z-coordinate measured along the optic axis. If particles of different masses must be taken into account, a seventh coordinate (δ_m) must be added.

2. THE DESCRIPTION OF IMAGE ABERRATIONS

Knowing r_{i0} at $z = z_0$ for $r_i \in \{x, a, y, b, \delta_t, \delta_K\}$, the solution of the differential equations of motion describes $r_{in}(z_n)$ as:

$$r_{in}(z_n) = \sum_{k=1}^{6} \frac{r_{k0}}{1} \left\{ (r_i|r_k) + \sum_{\ell=1}^{6} \frac{r_{\ell 0}}{2} \left\{ (r_i|r_k r_\ell) + \sum_{m=1}^{6} \frac{r_{m0}}{3} \{(r_i|r_k r_\ell r_m) + \ldots \} \right\} \right\} \quad (1)$$

with $r_k, r_\ell, r_m \ldots \in \{x, a, y, b, \delta_t, \delta_K\}$. Denoting the powers of $x_0, a_0, y_0, b_0, \delta_{t0}, \delta_{K0}$ in Eq. (1) by $n_x, n_a, n_y, n_b, n_{\delta t}, n_{\delta K}$, one can rewrite r_{in} as a sum of terms

$$A_j(n_x, n_a, n_y, n_b, n_t, n_K) = \frac{1}{j!}(r_i|x^{n_x}a^{n_a}y^{n_y}b^{n_b}\delta_t^{n_t}\delta_K^{n_K})x_0^{n_x}a_0^{n_a}y_0^{n_y}b_0^{n_b}\delta_{t0}^{n_t}\delta_{K0}^{n_K}(2a)$$

with $j = n_x + n_a + n_y + n_b + n_t + n_K$. For each value of j there are thus several combinations of $x^{n_x}, a^{n_a}, y^{n_y}, b^{n_b}, \delta_t^{n_t}, \delta_K^{n_K}$ and corresponding aberration coefficients $(r_i|x^{n_x}a^{n_a}y^{n_y}b^{n_b}\delta_t^{n_t}\delta_K^{n_K})$ as shown in Table I where the coefficients $f_{n_x n_a n_y n_b n_t n_K}$ are scaling factors which for the moment are unity.

Table I. Aberrations of Second– and Third–Order

second order	third order
$f_{20}(r_i\|r_k r_k)$	$f_{300}(r_i\|r_k r_k r_k)$
$f_{11}(r_i\|r_k r_\ell)$	$f_{210}(r_i\|r_k r_k r_\ell)$
	$f_{111}(r_i\|r_k r_\ell r_m)$

Instead of any $x_0, a_0, y_0, b_0, \delta_{t0}, \delta_{K0}$ in Eq. (2a), one can also choose the largest possible value of any of these quantities: $x_{00}, a_{00}, y_{00}, b_{00}, \delta_{t00}, \delta_{K00}$ finding

$$A_{jM}(n_x, n_a, n_y, n_b, n_t, n_K) = \frac{1}{j!}(r_i|x^{n_x}a^{n_a}y^{n_y}b^{n_b}\delta_t^{n_t}\delta_K^{n_K})x_{00}^{n_x}a_{00}^{n_a}y_{00}^{n_y}b_{00}^{n_b}\delta_{t00}^{n_t}\delta_{K00}^{n_K} \quad (2b)$$

where each A_j is a certain percentage of A_{jM} with

$$f_{n_x n_a n_y n_b n_t n_K} = \frac{A_j(n_x, n_a, n_y, n_b, n_t, n_K)}{A_{jM}(n_x, n_a, n_y, n_b, n_t, n_K)}$$

$$= \left(\frac{x_0}{x_{00}}\right)^{n_x}\left(\frac{a_0}{a_{00}}\right)^{n_a}\left(\frac{y_0}{y_{00}}\right)^{n_y}\left(\frac{b_0}{b_{00}}\right)^{n_b}\left(\frac{\delta_{t0}}{\delta_{t00}}\right)^{n_t}\left(\frac{\delta_{K0}}{\delta_{K00}}\right)^{n_K}$$

$$= (\overline{x})^{n_x}(\overline{a})^{n_a}(\overline{y})^{n_y}(\overline{b})^{n_b}(\overline{\delta_t})^{n_t}(\overline{\delta_K})^{n_K} \leq 1. \quad (3)$$

If —as shown in Fig. 1— the phase–space ellipse at the object is not upright or the phase–space parallelogram is not degenerated to an upright rectangle, the ranges of x_0 or y_0 do not vary between $\pm x_{00}$ and $\pm y_{00}$ but between different limits for different choices of a_0 or b_0. In this case the image aberrations cause an

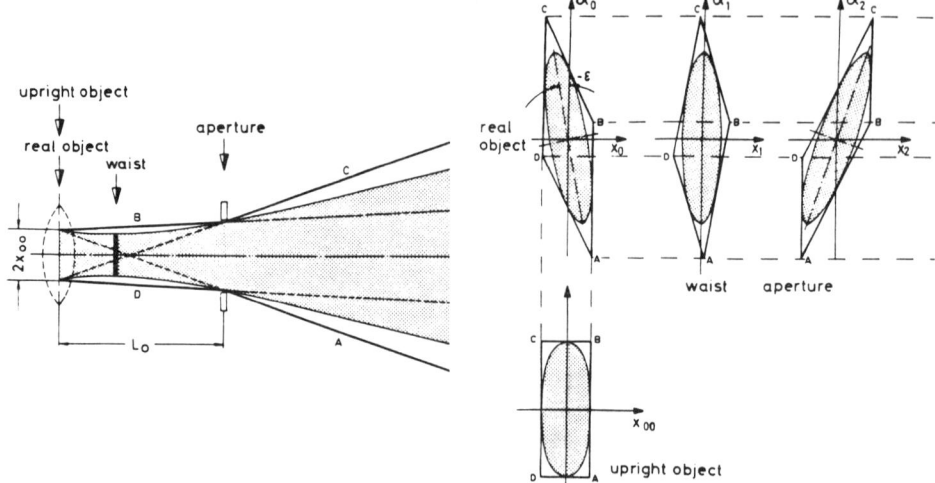

Figure 1. Particle trajectories starting from some object and limited by some aperture are shown in real space and in phase space. Note that the optical system this beam is sent into is implemented by a "virtual thin object lens" placed at the position of the object. This lens transforms the assumed upright phase–space area into that phase–space area the particle beam occupies at the object in reality.

asymmetric widening of the beam. To illustrate this situation, one may look at the triplet of second–order aberration coefficients $(x|xx), (x|xa), (x|aa)$ which — for this discussion— we assume to be all positive. For an upright parallelogram–like phase–space area at the object of Fig. 1 the four corners of the parallelogram are characterized by $A(x_{00}, a_A)$, $B(x_{00}, a_B)$, $C(-x_{00}, -a_A)$, $D(-x_{00}, -a_B)$ with

$$a_A = -a_{00} - x_{00}/L_x \qquad (4a)$$
$$a_B = +a_{00} - x_{00}/L_x . \qquad (4b)$$

where L_x is the distance between the object and the entrance pupil[1,2]. Trajectories which start at $+x_{00}$ or at $-x_{00}$ under all possible angles ranging from a_A to a_B or from $-a_A$ to $-a_B$, respectively, finally fill a region in the image deviating from the optic axis by

$$x_{max_A} = (x|xx)x_{00}^2 + (x|xa)x_{00}a_A + (x|aa)a_A^2 \qquad (5a)$$
$$x_{max_B} = (x|xx)x_{00}^2 + (x|xa)x_{00}a_B + (x|aa)a_B^2 . \qquad (5b)$$

These relations can be rewritten using Eqs. (4) as

$$x_{max_A} = (x|xx)_n x_{00}^2 - (x|xa)_n x_{00}a_{00} + (x|aa)_n a_{00}^2 \qquad (6a)$$
$$x_{max_B} = (x|xx)_n x_{00}^2 + (x|xa)_n x_{00}a_{00} + (x|aa)_n a_{00}^2 \qquad (6b)$$

with

$$(x|xx)_n = (x|xx) - (x|xa)/L_x + (x|aa)/L_x^2$$
$$(x|xa)_n = (x|xa) - 2(x|aa)/L_x$$
$$(x|aa)_n = (x|aa).$$

The advantage of Eqs. (6) over Eqs. (5) is, that the second and third terms in Eqs. (6) are equal in magnitude while they are unequal in Eqs. (5) because of $a_A \neq a_B$. Thus, the contribution of these aberrations can be expressed by one number instead of two. Though only the geometric x–aberrations are discussed here, the definition of "normalized aberration coefficients" applies to aberrations in y–direction as well.

In GIOS[3] as well as in COSY[4], the "normalized aberration coefficients" of Eqs. (6) are used. These "normalized aberration coefficients" are calculated most easily as the matrix coefficients for the optical system under consideration amended by virtual thin object lenses[1,2] of focal lengths L_x and L_y, where L_x and L_y are the distances between the object and the corresponding entrance pupils. Applying the coordinates x_{00}, a_A and x_{00}, a_B to the coefficients of Eqs. (5), as calculated, for instance, by TRANSPORT[5] or MARYLIE[6], thus yields the same result as applying the points $x_{00}, \pm a_{00}$ to the matrix coefficients of Eqs. (6), as calculated, for instance, by GIOS[3] or by COSY[4]. Note that for $L_x = L_y = \infty$ the matrix coefficients of TRANSPORT, MARYLIE, COSY and GIOS in principle are identical.

3. THE MAGNITUDE OF IMAGE ABERRATIONS FOR ELLIPTICAL PHASE–SPACE VOLUMES

In the case of an upright rectangular phase–space distribution, we have

$$|\overline{r}_q| \leq 1 \qquad for \qquad \overline{r}_q \in \{\overline{x}, \overline{a}, \overline{y}, \overline{b}, \overline{\delta_t}, \overline{\delta_K}\}. \tag{7}$$

In the case of an upright elliptical phase–space distribution in which a beam is described that has passed through many symmetrically arranged rectangular apertures in an optical system (see Ref. 1), we find always

$$\overline{x}^2 + \overline{a}^2 \leq 1 \qquad and \qquad \overline{y}^2 + \overline{b}^2 \leq 1, \tag{8}$$

and frequently, $\overline{\delta}_t^2 + \overline{\delta}_K^2 \leq 1$. When the above mentioned apertures have elliptical and circular shapes, we find

$$\overline{x}^2 + \overline{a}^2 + \overline{y}^2 + \overline{b}^2 \leq 1, \tag{9}$$

and frequently, $\bar{\delta}_t^2 + \bar{\delta}_K^2 \leq 1$ and when time–dependent accelerating devices are distributed in the mentioned optical system, we find

$$\bar{x}^2 + \bar{a}^2 + \bar{y}^2 + \bar{b}^2 + \bar{\delta}_t^2 + \bar{\delta}_K^2 \leq 1. \quad (10)$$

In the cases of Eqs. (8,9 or 10), many extreme particles do not exist, for instance, the one characterized by $x_{00}, a_{00}, y_{00}, b_{00}, \delta_{t00}, \delta_{K00}$. Thus, the resultant image aberrations are smaller than the ones expected for a parallelogram–like phase–space distribution.

For an elliptical phase–space distribution the most extreme particle is found by determining the $f_{n_x n_a n_y n_b n_t n_K}$ in Eq. (3) such that the corresponding surface osculates the circles of Eq. (8), the four–dimensional sphere of Eq. (9) or the six–dimensional sphere of Eq. (10). The osculation point is found by postulating r_k as well as $\partial \bar{r}_k / \partial \bar{r}_\ell, \partial \bar{r}_k / \partial \bar{r}_m, \partial \bar{r}_k / \partial \bar{r}_n, \partial \bar{r}_k / \partial \bar{r}_o, \partial \bar{r}_k / \partial \bar{r}_p$ to be equal when determined from Eq. (3) or one of the Eqs. (8,9 or 10) where $\bar{r}_k, \bar{r}_\ell, \bar{r}_m, \bar{r}_n, \bar{r}_o, \bar{r}_p \in \{\bar{x}, \bar{a}, \bar{y}, \bar{b}, \bar{\delta}_t, \bar{\delta}_K\}$.

3.1 THE MAGNITUDE OF IMAGE ABERRATIONS FOR THREE TWO–DIMENSIONAL PHASE–SPACE AREAS

Determining the osculating surface $f_{n_x n_a n_y n_b n_t n_K} = const.$, one finds from Eq. (8)

$$\bar{x}^2 = \frac{n_x}{n_x + n_a} \quad \bar{y}^2 = \frac{n_y}{n_y + n_b} \quad \bar{\delta}_t^2 = \frac{n_t}{n_t + n_K}$$
$$\bar{a}^2 = \frac{n_a}{n_x + n_a} \quad \bar{b}^2 = \frac{n_b}{n_y + n_b} \quad \bar{\delta}_K^2 = \frac{n_K}{n_t + n_K} \quad (11)$$
$$f_{n_x n_a n_y n_b n_t n_K} = \left(\frac{\sqrt{n_x n_a}}{n_x + n_a}\right)\left(\frac{\sqrt{n_y n_b}}{n_y + n_b}\right)\left(\frac{\sqrt{n_t n_K}}{n_t + n_K}\right).$$

when at least one coordinate is uncorrelated to others, one must distinguish 3 cases:

1. for $n_x = 0$ (or $n_a = 0$), the first bracket must be replaced by 1,

2. for $n_y = 0$ (or $n_b = 0$), the second bracket must be replaced by 1,

3. for $n_t = 0$ (or $n_K = 0$), the third bracket must be replaced by 1. Note here that for time-independent fields, i.e.: $n_t = 0$, the third bracket is always 1.

For $j = n_x + n_a + n_y + n_b + n_t + n_K \leq 5$ the $f_{n_x n_a n_y n_b n_t n_K}$ are listed in column 2 of Table II.

3.2 THE MAGNITUDE OF IMAGE ABERRATIONS FOR ONE FOUR–DIMENSIONAL AND ONE TWO–DIMENSIONAL PHASE–SPACE VOLUME

Determining the osculating surface $f_{n_x n_a n_y n_b n_t n_K} = const.$, one finds from Eqs. (9)

$$\overline{x}^2 = \frac{n_x}{n_x + n_a + n_y + n_b} \quad \overline{y}^2 = \frac{n_y}{n_x + n_a + n_y + n_b} \quad \overline{\delta}_t^2 = \frac{n_t}{n_t + n_K}$$

$$\overline{a}^2 = \frac{n_a}{n_x + n_a + n_y + n_b} \quad \overline{b}^2 = \frac{n_b}{n_x + n_a + n_y + n_b} \quad \overline{\delta}_K^2 = \frac{n_K}{n_t + n_K} \quad (12)$$

$$f_{n_x n_a n_y n_b n_t n_K} = \left(\frac{\sqrt{n_x n_a n_y n_b}}{n_x + n_a + n_y + n_b} \right) \left(\frac{\sqrt{n_t n_K}}{n_t + n_K} \right).$$

When at least one coordinate is uncorrelated to others, one must distinguish four cases with $n_k, n_\ell, n_m, n_n \in \{n_x, n_a, n_y, n_b\}$:

1. for $n_t = 0$ (or $n_K = 0$), the last bracket must be replaced by 1,

2. for $n_k \neq 0$, $n_\ell \neq 0$, $n_m \neq 0$, $n_n = 0$, one finds the first bracket as:

$$\left(\frac{\sqrt{n_x n_a n_y n_b}}{(n_x + n_a + n_y + n_b)^2} \right) = \frac{\sqrt{n_k n_\ell n_m}}{(n_k + n_\ell + n_m)^{3/2}}$$

3. for $n_k \neq 0$, $n_\ell \neq 0$, $n_m = n_n = 0$, one finds the first bracket as:

$$\left(\frac{\sqrt{n_x n_a n_y n_b}}{(n_x + n_a + n_y + n_b)^2} \right) = \frac{\sqrt{n_k n_\ell}}{n_k + n_\ell}$$

4. for $n_k \neq 0$, $n_\ell = n_m = n_n = 0$, the first bracket must be replaced by 1.

For $j = n_x + n_a + n_y + n_b + n_t + n_K \leq 5$, the, $f_{n_x n_a n_y n_b n_t n_K}$ are listed in column 3 of Table II.

3.3 THE MAGNITUDE OF IMAGE ABERRATIONS FOR A SIX–DIMENSIONAL PHASE–SPACE VOLUME

Determining the osculating surface $f_{n_x n_a n_y n_b n_t n_K} = const.$, one finds from Eq. (10) for $n_k, n_\ell, n_m, n_n, n_o, n_p \in \{n_x, n_a, n_y, n_b, n_t, n_K\}$ and $\overline{r}_k, \overline{r}_\ell, \overline{r}_m, \overline{r}_n, \overline{r}_o, \overline{r}_p \in \{\overline{x}, \overline{a}, \overline{y}, \overline{b}, \overline{\delta}_t, \overline{\delta}_k\}$

$$\overline{r}_k^2 = \frac{n_k}{n_k + n_\ell + n_m + n_n + n_o + n_p}$$

$$f_{n_x n_a n_y n_b n_t n_K} = \frac{\sqrt{n_x n_a n_y n_b n_t n_K}}{(n_x + n_a + n_y + n_b + n_t + n_K)^3}. \quad (13)$$

When at least one coordinate is uncorrelated to others, one must distinguish five cases with $n_k, n_\ell, n_m, n_n, n_o, n_p \in \{n_x, n_a, n_y, n_b, n_t, n_K\}$

1. for $n_k \neq 0$, $n_\ell \neq 0$, $n_m \neq 0$, $n_n \neq 0$, $n_o \neq 0$, $n_p = 0$, one finds

$$f_{n_x n_a n_y n_b n_t n_K} = \frac{\sqrt{n_k n_\ell n_m n_n n_o}}{(n_k + n_\ell + n_m + n_n n_o)^{5/2}}$$

2. for $n_k \neq 0$, $n_\ell \neq 0$, $n_m \neq 0$, $n_n \neq 0$, $n_o = n_p = 0$, one finds

$$f_{n_x n_a n_y n_b n_t n_K} = \frac{\sqrt{n_k n_\ell n_m n_n}}{(n_k + n_\ell + n_m + n_n)^2}$$

3. for $n_k \neq 0$, $n_\ell \neq 0$, $n_m \neq 0$, $n_n = n_o = n_p = 0$, one finds

$$f_{n_x n_a n_y n_b n_t n_K} = \frac{\sqrt{n_k n_\ell n_m}}{(n_k + n_\ell + n_m)^{3/2}}$$

4. for $n_k \neq 0$, $n_\ell \neq 0$, $n_m = n_n = n_o = n_p = 0$, one finds

$$f_{n_x n_a n_y n_b n_t n_K} = \frac{\sqrt{n_k n_\ell}}{(n_k + n_\ell)}$$

5. for $n_k \neq 0, n_\ell = n_m = n_n = n_o = n_p = 0$, one finds $f_{n_x n_a n_y n_b n_t n_K} = 1$

For $\jmath = n_x + n_a + n_y + n_b + n_t + n_K \leq 5$ the $f_{n_x n_a n_y n_b n_t n_K}$ are listed in column 4 of Table II.

4. THE GOAL FUNCTION FOR THE OPTIMIZATION OF AN OPTICAL SYSTEM

When optimizing an optical system by varying its geometry or its multipole components, one thus —in case of a beam filling an elliptical phase–space volume— should optimize the function

$$A_{elliptical} = \Sigma (r_i | x^{n_x} a^{n_a} y^{n_y} b^{n_b} \delta_t^{n_t} \delta_K^{n_K})^2 f_{n_x n_a n_y n_b n_t n_K}^2 x_{00}^{2n_x} a_{00}^{2n_a} y_{00}^{2n_y} b_{00}^{2n_b} \delta_{t0}^{2n_t} \delta_{K0}^{2n_K} \quad (14a)$$

with $f_{n_x n_a n_y n_b n_t n_K}$ being determined from Eqs. (11,12 or 13). The result is better than the one obtained by optimizing the function

$$A_{parallelogram} = \Sigma (r_i | x^{n_x} a^{n_a} y^{n_y} b^{n_b} \delta_t^{n_t} \delta_K^{n_K})^2 x_{00}^{2n_x} a_{00}^{2n_a} y_{00}^{2n_y} b_{00}^{2n_b} \delta_{t0}^{2n_t} \delta_{K0}^{2n_K} \quad (14b)$$

which would have been appropriate for the case of a beam filling a parallelogram-like phase–space volume. However, both Eqs. (14a) and (14b) yield better results than the one obtained by optimizing the function

$$A_{coefficients} = \Sigma (r_i | x^{n_x} a^{n_a} y^{n_y} b^{n_b} \delta_t^{n_t} \delta_K^{n_K})^2 \quad (14c)$$

Optimizing a system by using Eq. (14c) is equivalent to minimizing the sum of Lie polynomials[7] though Eq. (14c) contains more elements since several of the aberration coefficients of Eq. (14c) depend on others because of the symplectic condition[1,8]. In cases in which one can make all coefficients $(r_i|x^{n_x}a^{n_a}y^{n_y}b^{n_b}\delta_t^{n_t}\delta_K^{n_K})$ zero, there naturally is no difference between the finally obtained system whether Eqs. (14a, 14b or 14c) have been used for an optimization procedure. However, under normal conditions in which one can only minimize the sum of the aberrations, an optimization using Eq. (14a) is definitely superior, if the beam fills an elliptical phase-space volume.

Table II. Numerical Values of the Coefficients $f_{n_k n_\ell n_m n_n n_o n_p}$ to Fifth-Order

	3*2-dim.	4+2-dim.	6-dim.
f_{20}	1.00000	1.00000	1.00000
f_{11}	0.50000	0.50000	0.50000
f_{300}	1.00000	1.00000	1.00000
f_{210}	0.47140	0.47140	0.47140
f_{111}	0.50000	0.19245	0.19245
f_{4000}	1.00000	1.00000	1.00000
f_{3100}	0.43301	0.43301	0.43301
f_{2200}	0.50000	0.50000	0.50000
f_{2110}	0.47140	0.17678	0.17678
f_{1111}	0.25000	0.06250	0.06250
f_{50000}	1.00000	1.00000	1.00000
f_{41000}	0.40000	0.40000	0.40000
f_{32000}	0.48990	0.48990	0.48990
f_{31100}	0.43301	0.15492	0.15492
f_{22100}	0.50000	0.17888	0.17888
f_{21110}	0.23570	0.05657	0.05657
f_{11111}	0.25000	0.04000	0.01789

5. REFERENCES

1. H. Wollnik, "Optics of Charged Particles", Acad. Press, Orlando, USA (1987).
2. H. Wollnik, Nucl. Instrum. and Meth. **137** p. 169 (1976).
3. H. Wollnik et al., GSI–Rep. THD–26 p. 679 (1984);
   Nucl. Instrum. and Meth. **A258** p. 408 (1987).
4. M. Berz et al., Nucl. Instrum. and Meth. **A258** p. 402 (1987).
5. K. L. Brown et al., CERN–Rep. 80–04 (1980).
6. A.Dragt and L. Haley, Proc. IEEE Part. Acc. Conf. p. 1062 (1987)
7. A. Dragt, et al., Ann. Rev. Nucl. and Particle Science, in print.
8. H. Wollnik and M. Berz, Nucl. Instrum. and Meth. **238** p. 127 (1985).

SUMMARY FOR THE WORKING GROUP ON MAGNET CODES

Sergio Pissanetzky, Texas Accelerator Center
2319 Timberlock, The Woodlands, Texas 77380

The working group consisted of Vernon Bailey, Bill Herrmannsfeldt, Andrew Jason, Mary Menzel, Hans Nestle, Sergio Pissanetzky, Walt Trzeciak, Alan Wadlinger, and Andrew Wilson.

The group discussed the status of development of magnet codes for accelerator design which included:

POISSON: A two-dimensional code very popular in the U.S. and other countries.

FAMA: A two-dimensional code developed at University of Giessen. This code is the result of incorporating an algorithm for successive mesh refinement into POISSON, and executes an order of magnitude faster than POISSON.

MAGNUS-3D: A commercial three-dimensional code. Popular in the U.S. and well known in Europe. It can be compared to the British code TOSCA, but MAGNUS-3D is more modern and specialized for accelerator work.

Presently a standard version of the POISSON Group of Codes resides at the Los Alamos Accelerator Code Center. A user and a reference manual have been written by Los Alamos National Laboratory with DOE-NP and DOE-HEP providing the financial support. These documents and tapes of the codes are distributed to all requestors. The code center at Los Alamos National Laboratory strives, within its limited available resources, to correct all known errors, to provide assistance to users and to keep users informed of errors and changes. However, DOE feels that these codes are obsolete and thus has not provided LANL with the funds needed to make major changes and improvements. Even though these codes are old, they are still the most widely used 2-D magnet codes, especially in the accelerator community. We therefore recommend that DOE or DOD provide funding to implement all significant improvements made by others into the standard version, as well as investigate the possibility of writing a new POISSON code as outlined below.

Since no other 2-D codes seem to have gained popularity in accelerator work, the need for a new 2-D magnet code is recognized. Such a code should have its input, as well as its output, based on a CAD system and could be obtained by further development of the code MAGNUS-2D (previously known as CUARM), existing at Texas Accelerator Center, or by offering appropriate incentives to commercial software developers, or both. It is suggested and recommended that DOE or DOD should provide funding for this purpose.

A three-dimensional code is also necessary. It should also be based on a CAD system for input and output. It is highly desirable that the 3-D code be a natural extension of the 2-D code, using a similar technique for input and similar

© 1988 American Institute of Physics

editing procedures. This would greatly simplify the training of 3-D code users. This is a very desirable feature because of the complexity of 3-D codes. It is recommended that this be done, and funding be provided from DOE or DOD.

Future issues should include: code validation, benchmarks, code modularization, animation, time dependent codes, a National Center for accelerator magnet code maintenance. It is recommended that other workshops on magnet codes be held in the near future. These workshops should be scheduled in conjunction with major magnet and/or accelerator conferences.

SUMMARY FOR THE WORKING GROUP ON LINAC CODES*

Thomas P. Wangler
Los Alamos National Laboratory

and

John Staples
Lawrence Berkeley Laboratory

The linac codes group discussed two main topics: (1) the PARMTEQ codes and their environment and (2) limitations of the linac simulation codes.

THE PARMTEQ CODES AND THEIR ENVIRONMENT

The RFQ design program PARMTEQ[1] is used at many laboratories, both for the detailed cell generation and for multiparticle simulation to test the design. However, RFQ beam-dynamics design procedure requires the determination of a large number of parameters to produce the required focusing, adiabatic bunching, and acceleration of the beam. Frequently, both high beam intensity and low emittance are desired at minimum values for peak surface field, vane length, and rf power. RFQ designers have found that preprocessor design codes are almost a necessity for these cases. Although the best design technique is not yet known, codes such as CURLI and RFQUIK, written originally by T. Wangler at Los Alamos, allow the designer to determine the space-charge limits and to achieve an acceptable compromise among the design parameters. The purpose of CURLI is to evaluate the transverse and longitudinal current limits as a function of the RFQ parameters, thereby allowing the designer to choose an RFQ design with adequate focusing for the anticipated space-charge forces. The program RFQUIK is then used by the designer to choose a complete set of input data for PARMTEQ. The input data contain the adiabatic bunching prescriptions and other rules of thumb for choosing parameters. Tradeoffs between beam performance and RFQ length and power can be chosen by the designer. Many preprocessor codes have been written at other laboratories, some with rather different objectives, such as the design of short, heavy-ion linacs where space charge can be ignored.

---

*This work was supported by Los Alamos National Laboratory Program Development funds, under the auspices of the United States Department of Energy.

The PARMTEQ code itself, universally used, serves as a useful basis and standard of comparison among various groups designing RFQs. Recent work has allowed different pole-tip designs to be treated with more accuracy, using a 4- or even an 8-term potential function. Some versions have included a more accurate representation of the radial matcher and a variation of the voltage along the vane.

Output processor codes, which use the output results from PARMTEQ as their input, are useful after an RFQ beam-dynamics design has been chosen. At Los Alamos, such a program was written by K. Crandall to perform a variety of functions as chosen by the designer. The most important function is the choice of actual vane geometry and the adjustment of RFQ parameters to produce the design values of the lowest-order multipoles. An additional output program is required to produce the vane construction information for the numerically controlled milling machine.

A program of value for RFQ cavity design and tuning has been written by R. Hutcheons at Chalk River.[2] This program is based on a coupled circuit model and contains enough information for studies of the effects of higher-order modes.

The Working Group discussed the value of maintaining a standard version of the PARMTEQ code to allow easy comparison of any RFQ design at all laboratories. The concensus was that the existence of a standard code is valuable, but should not hinder new code developments.

LIMITATIONS OF THE LINAC SIMULATION CODES

The number of particles required for adequate calculation of space-charge effects is an important consideration for numerical simulation studies. The consensus of the group was that for detailed transfer maps with space charge, many particles are required. It was reported at this meeting[3] that 90 000 particles for programs like CHARLIE may even be inadequate for accurate transfer maps that allow determination of space-charge effects. However, for describing the average properties of the beam, including the second moments of the distribution and the rms emittance, many fewer particles, often no more than about 10^3, are adequate for obtaining reproducible and accurate results. PIC-type codes can calculate second moments to good (1%) accuracy, but when good resolution of sharp beam edges is necessary, PIC codes do not perform adequately.

The physics contained in linac codes like PARMILA,[4] PARMELA,[5] and PARMTEQ[1] was felt to be adequate for description of the small amplitude particles in the beam, but probably was inadequate for the larger amplitude particles that contribute to beam halo. Accurate description of the halo will require putting all sources of nonlinear fields into the codes, including both space-charge and image forces. The effect of the image forces is not simple to analyze because it depends on the detailed geometry and on the relative alignment of the beam and accelerator. No bunched-beam image-force analyses have been applied to these codes. An encouraging result about the image-charge effect comes from the studies of Celata,[6] which indicate that even for off-axis beams, image-charge effects may only become important at very low values of the space-charge tune depression. Neutralization may persist from the LEBT into the RFQ. At present, we assume no neutralization, and it is not clear how to handle this possible effect.

A few comparisons of these linac codes with experimental measurements have been made, and more should be available within the next few years. R. Jameson described the comparison of PARMILA with measurements on the CERN linac,[7] where it was concluded that the experimental emittance results could be explained from the PARMILA runs, with reasonable assumptions about beam misalignment and mismatch. PARMTEQ calculations have agreed well with measured longitudinal distributions,[1] even with excitation errors. Comparison with transverse measurements have also been made, and the agreement has been good.

REFERENCES

1. Thomas P. Wangler and Kenneth R. Crandall, "PARMTEQ--a Beam- Dynamics Code for the RFQ Linear Accelerator," these proceedings.

2. R. M. Hutcheons, "A Modeling Study of the Four-Rod RFQ," Proc. 1984 Linac Conf., GSI Laboratory report GSI-84-11, 94 (1984), and D. Dohan, F. Fong, and R. Hutcheons, "Modeling the TRIUMF RF Cavity Using the Code RFQ3D," IEEE Trans. Nucl. Sci. 32 (5), 2933 (1985).

3. R. Ryne, "Lie Algebraic Treatment of Space Charge," these proceedings.

4. G. Boicourt, "PARMILA," these proceedings.

5. L. Young, "PARMELA," these proceedings.

6. C. M. Celata, I. Haber, L. J. Laslett, L. Smith, and M. G. Tiefenbach, "The Effect of Induced Charge at Boundaries on Transverse

Dynamics of a Space-Charge-Dominated Beam," IEEE Trans. Nucl. Sci. **32**(5), 2480 (1985).

7. R. A. Jameson, "Emittance Growth in the New CERN Linac—Transverse Plane Comparison Between Experimental Results and Computer Simulation," Los Alamos National Laboratory, AT-Division memorandum AT-DO-377 (U), January 15, 1979.

Summary of the Working Group on Wakefield Codes

K.C.D. Chan and R. Cooper, Chairman

The computational tools for calculating wakefields are well developed and are currently complimented with experimental measurement.

There are various computer codes used in calculating wakefield effects nowadays. Among them, the most notable and complete one is the MAFIA group of codes. This code group is being supported by a collaboration between the Deutsches Elektronen-Synchrotron (DESY) and the Los Alamos National Laboratory (LANL). It consists of codes that are used for time-domain and frequency-domain calculations. Calculations with 3D and 2D geometries are possible. An introductory talk on 3D calculations using MAFIA codes was given by R. K. Cooper of LANL showing the present capabilities of the codes. Up-to-date information was reviewed during discussions with T. Weiland of DESY.

At the Argonne National Laboratory (ANL), an Advanced Accelerator Test Facility (AATF) has been commissioned that will allow the measurement of wakefield effects of beamline components up to a length of 1 m. The transverse and longitudinal electromagnetic fields induced by a primary electron bunch are probed using a second collinear electron bunch that trails the primary bunch at a variable time delay. This facility will be extremely useful for benchmarking results from computer codes. A talk introducing the facility was given by P. Schoessow of ANL. During the talk, the measured transverse and longitudinal wake effects of several iris-loaded RF-structure were presented, showing good agreements with calculations made with TBCI, a code from the MAFIA code group.

SUMMARY FOR THE WORKING GROUP ON TRANSPORT CODES

C.R. Eminhizer, Physical Dynamics, Inc., 7855 Fay Ave., Suite 200
La Jolla, California 92038

The thirty-two participants, including many authors and primary users of beam transport codes represented six national laboratories, two universities, and six corporations. They were Martin Berz, Barbara Blind, Harold Butler, Mark Campbell, Dave Carey, Judy Colman, Walter Davies, Bill DeHope, Charles Eminhizer, John Fields, Etienne Forest, Bob Hardekopf, Jim Hurd, Christoph Iselin, Ronald Kashuba, Frank Krienen, Walter Lysenko, Elliott McCrory, Philip Meads, Jr., Tom Mottershead, James Niederer, Filippo Nero, Anna Porto, Gordon D. Pusch, Odeal W. Richardson, Robert Ryne, Ray Schmitt, Martin Schulze, Richard Silbar, Ronald Sverdlove, Alan Todd and Hermann Wollnik.

During the transport codes session, the current status of the codes used to design beam transport systems was reviewed, and in a few cases, the advantages and disadvantages of the codes were discussed, and plans for modifications and enhancements were described. Most of the codes discussed in the session were described in detail during talks and poster sessions and, in many cases, are described in articles in these proceedings, which should be consulted for more detail.

The discussion began with Dave Carey who has made many improvements to the code TRANSPORT reported on its status. Earlier he presented a talk on TRANSPORT with MAD input. Recent changes included the addition of fringe fields with the exception of the effect of steering magnets, and the ability to calculate the effect of many different errors. CDC, IBM, and VAX versions of the code with documentation, which is being revised, are available from Dave Carey.

Judy Colman asked Dave if TRANSPORT could be used to optimize multiple beams. He replied that he would have to check. She announced an IBM PC version of TRANSPORT.

Professor Wollnik stated that he has cases where TRANSPORT had failed to find a solution. Dave Carey indicated that he had not seen any examples and would be interested in seeing them.

Robert Ryne reported on the status of MARYLIE 3.0. Talks on MARYLIE and CHARLIE were presented by Alex Dragt and Robert Ryne. In Alex's absence Robert reported that recent additions to the code included analysis and normal form routines, an optimizer developed by Thomas Mottershead and fitting. VAX and CRAY versions of MARYLIE exist and under certain conditions may be available. Robert reported that CHARLIE, can calculate 2-D spare charge effects, which has been compared with PIC code. Calculational methods implemented in CHARLIE are described in a separate article. Robert indicated that a 3-D version is planned, but could give no firm estimate when it would be available.

Professor Wollnik, who presented a talk on GIOS and GOSY, summarized their status. GIOS, version 2, includes linear space charge and two fitting routines, (conjugate gradient method and the symplectic method). VAX, CDC 164, IBM, and a reduced PC version exist. Documentation is available. Version 3 includes inhomogeneous space charge in 2-D and boundary effects. COSY can calculate through fifth order and contains the Mottershead optimizer. It does not have standard graphics, but it has been modified at Los Alamos National Laboratory (LANL) to use the plotting routines of MARYLIE. The code can use one of two methods, a formula manipulator or differential algebra to calculate to fifth order. The differential algebra method, described in a talk by Martin Berz, has been checked using the code TEAPOT. The fringe field treatment is being improved and space charge effects are being added to COSY. During his talk, Martin described briefly the method he is using to calculate space charge effects. It was noted that GIOS differs from CHARLIE in that GIOS uses a ray tracing method and advances the particles, where CHARLIE advances maps.

Ronald Kashuba and Ray Schmitt presented a poster on the status of PATH (a PARMILA, TURTLE hybrid) developed by John Farrell. They are using it to design a beam expander. Presently, PATH contains 2-1/2D nonlinear space charge with plans to go to 3D, a third order quadrupole model with Halbach fringe fields and the Mottershead optimizer. The fringe fields need to be generalized. The code is documented and has interactive input. It can be obtained from the Los Alamos Accelerator Code Group.

Christopher Iselin reported on the status of MAD. This code is used in the design of high energy circular machines where fringe fields are assumed to be less important, being calculated to second order. It contains the MARYLIE Lie Algebraic manipulators and the optics and tracking routines use accelerator data organized as an object oriented relational data base. The MAD program is portable among IBM, CRAY, and VAX computers and Unix workstations. The documentation for the program is prepared in the IBM SMGL typesetting format. The MAD input language is described in the 1984 Snow Mass Conference Proceedings and was included by Dave Cary in his talk on TRANSPORT. The code has twenty thousand lines of comments in seventy-five thousand lines of code. James Niederer indicated that although it is used extensively in Europe it has not been used much in the United States.

James Niederer explained that the objective in developing MAD was to provide a common, easy to use language for describing accelerators and program commands defining it and for the program decoder/interpreter in the form of a dictionary which is itself data to the program. One decoder thus serves all present and future commands. Data is stored in the form of objects or data modules in the data base. These modules are modified during the computations accumulating the computational results which are then used to create intermediate tables. These can then be used to generate graphics and other presentation media. He indicated some hope that this same approach will be extended directly to very similar roles in accelerator control systems.

Professor Wollnik indicated that the code MIRKO, which is a kick code similar to TRACE, is used in Europe. It is interactive with interactive graphics, but has a poor treatment of fringe fields. There is a VAX version at Los Alamos. Documentation is available.

Thomas Mottershead referred to RAYTRACE, the code developed by Stan Kowalski and described by him in a separate talk. It numerically integrates the differential equations describing the motion of particles through real fields and can be used to trace rays through an optical system. It is used after using TRANSPORT to calculate higher order aberrations, but does not do fitting, optimization or treat space charge effects. It is written in FORTRAN and runs on a VAX or a PC. Also, he indicated that the code MOTER was used by LANL to design the telescope tested at Argonne National Laboratory. It is a ray tracing code through fifth order which has evolved from RAYTRACE. It has an automatic optimizer and is used after TRANSPORT, TURTLE, and RAYTRACE. Documentation is available.

Phil Meads reported briefly on the code LATTICE, which was described in a talk by John Staples, in which both he and John have made many improvements. The code has been converted from FORTRAN to PASCAL to run on a VAX GKS workstation. It is highly interactive and has very good graphics output to an Apple Laserwriter. It is a first order envelope code, without x-y coupling or space charge. The manual is available over Bitnet in TROFF.

During the general discussion, it was mentioned that TRANSOPTR, a transport code developed by Ed Heighway and Mark de Jong, can do parameter optimization for 2-D and 3-D space charge problems and can be run to second order without space charge.

Thomas Mottershead was asked to expand on some ideas for a method to integrate the transport codes into one master code. It was originally presented briefly at the end of Ed Heighway's talk on computer tools. The method would require that modularization of the existing codes so that different input and output formats and devices could be used. By making the wide variety of codes available to the system designer in one integrated package, the design process would be simplified. There was general agreement that such a system would be a valuable design tool, but there was uncertainty as to how it would be developed and who would do the work. It is an ambitious effort which would be expensive and require the cooperation of both computer scientists and accelerator code developers and users. The enhancements, such as better graphics and I/O and conversion to PCs indicate that there is considerable interest in using the most recent advancements in computer hardware and software technology, however, the type of system advocated by Tom Mottershead goes beyond these efforts. A workshop which would bring together code developers and computer scientists could be the next step in realizing this goal.

LIST OF PARTICIPANTS

Richard Abbott
U.S. Army SDC
Attn: CSSD-H-D
P.O. Box 1500
Huntsville, Alabama 35807
(205) 895-3220

Oscar Anderson
Lawrence Berkeley Laboratory
1 Cyclotron Road
Berkeley, California 94720
(415) 486-5011

Karl Bane
Bin 26
Stanford Linear Accelerator Center
P.O. Box 4349
Stanford, California 94305
(415) 854-3300   X-3342

Reinhardt Becker
Institut Fur Angewandte Physik
Robert Mayer
Street 2-4 D-6000
Frankfurt Am Main, DBR-West Germany
69-798-3488

Martin Berz
Mail Stop 90/4040
Lawrence Berkeley Laboratory
1 Cyclotron Road
Berkeley, California 94720
(415) 486-6781

Barbara Blind
AT-3 MS H808
Los Alamos National Laboratory
Los Alamos, New Mexico 87545
(505) 667-9130

Jack Boers
Varian-Extrion
Blackburn Industrial Park
Gloucester, Massachusetts 01930
(617)-281-2000

Grenfell Boicourt
MS H829
Accelerator Technology Division
Los Alamos National Laboratory
Los Alamos, New Mexico 87545
(505) 667-1965

Mildred J. Browman
AT-3 Division MS H824
Los Alamos National Laboratory
Los Alamos, New Mexico 87545
(505) 667-3108

Harold Butler
Medium Energy Physics Division
Los Alamos National Laboratory
Los Alamos, New Mexico 87545
(505) 667-8876

Mark Campbell
Mission Research Corporation
1720 Randolph SE
Albuquerque, New Mexico 87106
(505) 768-7600

Dave Carey
Research Division
Fermi National Accelerator Lab
P.O. Box 500
Batavia, Illinois 60510
(312) 840-3639

Shlomo Caspi
MS 46-161
Lawrence Berkeley Laboratory
1 Cyclotron Road
Berkeley, California 94720
(415) 486-7244

Christine Celata
47-112
Lawrence Berkeley Laboratory
1 Cyclotron Road
Berkeley, California 94720
(415) 486-7220

Dominic Chan
H-829
Los Alamos National Laboratory
Los Alamos, New Mexico 87545
(505) 665-0376

Weiren Chou
APS Division
Building 360
Argonne National Laboratory
9700 South Cass Avenue
Argonne, Illinois 60439
(312) 972-6286

Judith Colman
Brookhaven National Laboratory
Upton, New York 11973
(516) 282-2590

Richard Cooper
Los Alamos National Laboratory
Los Alamos, New Mexico 87545
(505) 667-2839

George Craig
Lawrence Livermore National Laboratory
University of California
Livermore, Califoria 94550
(415) 422-4683

Kenneth R. Crandall
AccSys Technology Inc.
P.O. Box 5247
Pleasanton, California 94566
(415) 462-6949

Michael Curtin
Booz-Allen & Hamilton Inc.
1725 Jefferson Davis Highway, Suite 1100
Arlington, Virginia 22202
(703) 769-7936

Walter G. Davies
Chalk River Nuclear Lab
Chalk River
Ontario K0J1J0, Canada
(613) 584-3311 or 4103

John DeFord
L-626
Lawrence Livermore National Laboratory
P.O. Box 808
Livermore, California 94550
(415) 423-5118

William J. DeHope
MS 2/362
GA Technologies Inc.
P.O. Box 85608
San Diego, California 92138
(619) 455-2312

N. J. Diserens
Rutherford Appleton Laboratory
Chilton Didcot OX1 10QY
England
(235) 21900

Don Dobrott
Science Applications International
10210 Campus Point Drive
San Diego, California 92121
(619) 546-6452

Alex Dragt
Department of Physics and Astronomy
University of Maryland
College Park, Maryland 20742
(301) 454-7324

Shimon Eckhouse
Maxwell Laboratories
8888 Balboa Avenue
San Diego, California 92123
(619) 576-7782

Charles Eminhizer
Physical Dynamics Inc.
7855 Fay Ave., Suite 200
La Jolla, California 92037
(619) 456-5481

Kenneth Eppley
Bin 26
Stanford Linear Accelerator Center
P.O. Box 4349
Stanford, California 94305
(415) 854-3300   X-3381

Zoltan D. Farkas
Bin 26
Stanford Linear Accelerator Center
P.O. Box 4349
Stanford, California 94305
(415) 854-3300   X-3387

John R. Fields
CN5300
David Sarnoff Research Center
Princeton, New Jersey 08543-5300
(609) 734-2530

Etienne Forest
SSC/CDG MS 90-4040
Lawrence Berkeley Laboratory
1 Cyclotron Road
Berkeley, California 94720
(415) 486-6580
or 486-7215

George Gillespie
G.H. Gillespie Associates
P.O. Box 2961
Del Mar, California 92014
(619) 259-6552

Robert L. Gluckstern
Department of Physics
University of Maryland
College Park, Maryland 20742
(301) 454-7476

Brendan Godfrey
Mission Research Corporation
1720 Randolph Road SE
Alburquerque, New Mexico 87106
(505) 768-7704

Terry Godlove
Naval Research Laboratory
9713 Manteo Court
Fort Washington, Maryland 20744
(202) 767-0610

Michael Green
Synchrotron Radiation Center
University of Wisconsin
3731 Schneider Drive
Stoughton, Wisconsin 53589
(608) 873-6651

Frank W. Guy
AT-1 MS H817
Los Alamos National Laboratory
Los Alamos, New Mexico 87545
(505) 667-9137

Irving Haber
Code 4791
Naval Research Laboratory
4555 Overlook Avenue SW
Washington, DC 20375-5000
(202) 767-3198

Kyoung Hahn
FM-15
Department of Physics
University of Washington
Seattle, Washington 98195
(206) 543-6133

Harold Hanerfeld
Bin 26
Stanford Linear Accelerator Center
P.O. Box 4349
Stanford, California 94305
(415) 854-3300 x-2068

Robert A. Hardekopf
MS H808
Los Alamos National Laboratory
Los Alamos, New Mexico 87545
(505) 667-2846

Edward A. Heighway
AT-3 MS H808
Los Alamos National Laboratory
Los Alamos, New Mexico 87545
(505) 667-1543

William Herrmannsfeldt
Bin 26
Stanford Linear Accelerator Center
Stanford University
P.O. Box 4349
Stanford, California 94305
(415) 854-3300 ext. 3342

Adolf R. Hochstim
President
La Jolla Institute
7855 Fay Ave., Suite 320
La Jolla, California 92037
(619) 456-5476

Andrew J. Holmes
Applied Physics & Technology Division
UKAEA/Culham Laboratory
Abingeton Oxfordshire OX14-3DB, England
(235) 21840

Eric Horowitz
Energy Research Building
University of Maryland
College Park, Maryland 20742
(301) 454-7100

Richard F. Hubbard
Code 4790
Naval Research Laboratory
Washington, DC 20375-5000
(202) 767-2927

Stanley Humphries Jr.
Department of Chemical & Nuclear Engineering
University of New Mexico
Albuquerque, New Mexico 87131
(505) 277-5422

James W. Hurd
MP-5 MS H838
Los Alamos National Laboratory
Los Alamos, New Mexico 87545
(505) 667-6413

Christoph Iselin
LEP Division
CERN
Geneva 23, CH-1211 Switzerland
(22) 83 36 57

Robert H. Jackson
Code 6842
Naval Research Laboratory
Washington, DC 20375
(202) 767-6656

Mark Jakobson
Department of Physics & Astronomy
University of Montana
Missoula, Montana 59812
(406) 243-6221

Robert Jameson
AT Division MS H811
Los Alamos National Laboratory
Los Alamos, New Mexico 87545
(505) 667-5634

Andrew Jason
AT-3 Division MS H808
Los Alamos National Laboratory
Los Alamos, New Mexico 87545
(505) 667-2842

Les Johnson
General Research Corporation
635 Discovery Drive
Huntsville, Alabama 35806
(205) 721-1941

Michael Jones
MS E-531
Los Alamos National Laboratory
Los Alamos, New Mexico 87545
(505) 667-7760

Glenn Joyce
Code 4790
Naval Research Laboratory
Washington, DC 20375-5000
(202) 767-6785

Thomas Jurgens
MS 307
Fermi National Laboratory
Batavia, Illinois 60510
(312) 840-2383

Ronald Kashuba
E463/92/2/251
McDonnell Douglas Astronautics Co.
P.O. Box 516
Saint Louis, Missouri 63166
(314) 925-7837

Charles Kim
MS 47-112
Lawrence Berkeley Laboratory
One Cyclotron Road
Berkeley, California 94720
(415) 486-7218

Harold Kirk
Department of Physics
Bldg. 510C
Brookhaven National Laboratory
Upton, New York 11973
(516) 282-3780

R. K. Koul
Building 510
Brookhaven National Laboratory
Upton, New York 11973
(516) 282-7830

Stanley Kowalski
Bldg. 26-505
Laboratory of Nuclear Science
Massachusetts Institute of Technology
Cambridge, Massachusetts 02139
(617) 253-4288

Jonathan Krall
Code 4790
Naval Research Laboratory
Washington DC 20375
(202) 767-4252

Frank Krienen
Brookhaven National Laboratory
Upton, New York 11973
(516) 282-2327

Bruce Langdon
L-472
Livermore National Laboratory
P.O. Box 808
Livermore, California 94550
(415) 422-5444

Theodore L. Lavine
Bin 26
Stanford Linear Accelerator Center
P.O. Box 4349
Stanford, California 94305
(415) 854-3300

Martin Lee
Bin 26
Stanford Linear Accelerator Center
P.O. Box 4349
Stanford, California 94305
(415) 854-3300 x-2851

K. M. Ling
CTR-6 MS-642
Los Alamos National Laboratory
Los Alamos, New Mexico 87545
(505) 667-7062

Michael L. Lively
AFSTC/NPT
Kirtland Air Force Base
Albuquerque, New Mexico 87117-6008
(505) 846-5802

Gary K. Loda
Beta Development Corporation
6780-R Sierra Court
Dublin, California 94568
(415) 828-0555

Walter Lysenko
AT-3 Division MS H808
Los Alamos National Laboratory
Los Alamos, New Mexico 87545
(505) 667-7431

James MacLachlan
Research Division
Fermi National Accelerator Lab
P.O. Box 500
Batavia, Illinois 60510
(312) 840-4484

Alan Mankofsky
Science Applications International
1710 Goodridge Drive
McLean, Virginia 22102
(703) 734-5596

Steven S. McCready
AFWL/AWYW
Air Force Weapons Laboratory
Kirtland AFB, New Mexico 87117
(505) 844-1871

Elliott McCrory
MS 307
Fermilab
P.O. Box 500
Batavia, Illinois 60510
(312) 840-4414

Kirk McDonald
Jadwin Hall
Princeton University
P.O. Box 708
Princeton, New Jersey 08544
(609) 452-6608

R. B. McKenzie-Wilson
Building 902C
Brookhaven National Laboratory
Upton, New York 11973
(516) 282-4710

Gerry McMichael
Atomic Energy of Canada Ltd.
Chalk River Nuclear Laboratories
Chalk River
Ontario KOJ 1JO, Canada
(613) 584-3311

Philip F. Meads Jr.
Argonne National Laboratory
7053 Shirley Drive
Oakland, California 94611
(415) 531-8172

M.T. Menzel
AT-6 Division INS H829
Los Alamos National Laboratory
Los Alamos, New Mexico 87545
(505) 667-2841

Jean Merson
AT-6 Division MS H829
Los Alamos National Laboratory
Los Alamos, New Mexico 87545
(505) 665-0954

Zoran Mikic
Science Applications International
10260 Campus Point Drive
San Diego, California 92121
(619) 546-6934

Alfred Mondelli
Science Applications International
1710 Goodridge Drive
McLean, Virginia 22102
(703) 734-4066

Warren Mori
1-130 Knudsen Hall
University of California
Los Angeles, California 90024
(213) 825-7514

C. Thomas Mottershead
AT-3 MS 808
Los Alamos National Laboratory
Los Alamos, New Mexico 87544
(505) 667-9730

Filippo Neri
Department of Physics and Astronomy
University of Maryland
College Park, Maryland 20742
(301) 454-8973

Hans Nestle
2 Physikal Institute
University of Giessen
6300 Giessen, West Germany

George Neuschaefer
Group AT-1 MS H817
Los Alamos National Laboratory
Los Alamos, New Mexico 87545
(505) 667-1958

James Niederer
Brookhaven National Laboratory
Building 515
Upton, New York 11973
(516) 282-4178

M.P.G. Nightingale
Applied Physics & Technology Division
UKAEA/Culham Laboratory
Abingeton Oxfordshire OX14-3DB
England
(235) 21840

Les Oleksiuk
Fermilab
P.O. Box 500
Batavia, Illinois 60510
(312) 840-4308

John Orthel
G.H. Gillespie Associates
P.O. Box 2961
Del Mar, Calfironia 92014
(619) 259-6552

Carl C. Paulson
Grumman Space Systems Division
107 Morgan Lane
Plainsboro, New Jersey 08536
(609) 275-2691

Serge Pissanetsky
Texas Accelerator Center
2319 Timberlock Place
The Woodlands, Texas 77380
(713) 363-7925

Gordon D. Pusch
AT-3 MS-H808
Los Alamos National Laboratory
Los Alamos, New Mexico 87545
(505) 667-3204

Odeal W. Richardson
U.S. Army SDC
Attn: CSSD-H-D
P.O. Box 1500
Huntsville, Alabama 35807
(205) 895-4735

Thomas G. Roberts
Physical Dynamics Inc.
P.O. Box 7207
Huntsville, Alabama 35807
(205) 880-1707

Helen Rudd
Code 4790
Naval Research Laboratory
4555 Overlook Avenue SW
Washington, DC 20375-5000
(202) 767-3296

Robert Ryne
L626
Lawrence Livermore National Laboratory
P.O. Box 808
Livermore, California 94550
(415) 423-4387

Raymond J. Schmitt
MDAC-STL E438/92/2W/255
McDonnell Douglas Astronautics Company
P.O. Box 516
St. Louis, Missouri 63166
(314) 925-7481

Paul Schoessow
362-HEP
Argonne National Laboratory
9700 S. Cass Avenue
Argonne, Illinois 60439
(312) 972-6280

Stanley Schriber
AT-DO MS H811
Los Alamos National Laboratory
Los Alamos, New Mexico 87545
(505) 667-5634

Petra Schuett
Department of Physics and Astronomy
University of Maryland
College Park, Maryland 20742
(301) 454-7243

Martin E. Schulze
Science Applications International Corporation
Research Park
227 and 230 Wall Street
Princeton, New Jersey 08540
(617) 245-6600 or (609) 619-9030

John Seeman
Stanford Linear Accelerator Center
P.O. Box 4349
Stanford, California 94305
(415) 854-3300 x-3566

Frank Selph
MS 71-259
Lawrence Berkeley Laboratory
Berkeley, California 94720
(415) 486-6902

Richard R. Silbar
T5 MS B283
Los Alamos National Laboratory
Los Alamos, New Mexico 87545
(505) 667-5253

James D. Simpson
Argonne National Laboratory
9700 Cass Avenue
Argonne, Illinois 60439
(312) 972-6587

Kenneth Smith
Group AT-2 MS H818
Los Alamos National Laboratory
Los Alamos, New Mexico 87545
(505) 667-3509

John Staples
Lawrence Berkeley Laboratory
One Cyclotron Road, Bldg. 64
Berkeley, California 94720
(415) 486-5831

Paul Steen
Maxwell Laboratories
8888 Balboa Avenue
San Diego, California 92123
(619) 576-7782

Bradford C. Stockwell
Mail Stop A-109
Varian Associates
611 Hansen Way
Palo Alto, California 94304
(415) 424-5622

Ronald Sverdlove
CN5300
David Sarnoff Research Center
Princeton, New Jersey 08543-5300
(609) 734-2517

Lester Thode
MS A11O
Los Alamos National Laboratory
Los Alamos, New Mexico 87545
(505) 667-0402

Charles B. Thorington
Electron Dynamics Division
Hughes Aircraft Company
P.O. Box 2999
Torrance, California 90509-2999
(213) 517-6906

Alan Todd
Grumman Space Systems
107 Morgan Lane
Plainsboro, New Jersey 08536
(609) 275-2675

Richard True
Litton Electron Devices
960 Industrial Road
San Carlos, California 94070
(415) 591-8411

Walter Trzeciak
Synchrotron Radiation Center
University of Wisconsin
3731 Schneider Drive
Stoughton, Wisconsin 53589
(608) 873-6651

Anahid Veremian
MS 1E-87
Boeing Aircraft Corporation
P.O. Box 3999
Seattle, Washington 98124
(206) 773-9399

E. Alan Wadlinger
AT Division MS H808
Los Alamos National Laboratory
Los Alamos, New Mexico 87545
(505) 667-2797

Thomas P. Wangler
AT-1 MS H817
Los Alamos National Laboratory
Los Alamos, New Mexico 87545
(505) 667-3200

Gary Warren
Mission Research Corporation
Capitol II Office Building, Suite 201
5503 Cherokee Avenue
Alexander, Virginia 22312
(703) 750-3556

Thomas Weiland
DESY-MPY
Notkestr 95
D2000 Hamburg 52
DBR, West Germany
(40) 8998-3196

John Whealton
MS-1
Oak Ridge National Laboratory
Martin Marietta Energy Systems
P.O. Box Y Bldg. 9201-2
Oak Ridge, Tennessee 37831
(615) 574-1130

Kenneth Whitham
L-627
Lawrence Livermore National Laboratory
P.O. Box 808
Livermore, California 94550
(415) 422-3613

Andrew Wilson
Maxwell Laboratories
8888 Balboa Avenue
San Diego, California 92123
(619) 576-7782

Norman D. Winarsky
CN5300
David Sarnoff Research Center
Princeton, New Jersey 08543-5300
(609) 734-2872

Herman Wollnik
2 Physikalisches Institute
University of Giessen
Heinrich-Buff-Ring 14-16
6300 Giessen
West Germany
641-702-2770

Lloyd Young
AT-1 MS H817
Los Alamos National Laboratory
Los Alamos, New Mexico 87545
(505) 667-1951

David Yu
Duly Consultants
1912 Mac Arthur Street
Rancho Palos Verdes, California 90732
(213) 548-7123

AIP Conference Proceedings

		L.C. Number	ISBN
No. 1	Feedback and Dynamic Control of Plasmas – 1970	70-141596	0-88318-100-2
No. 2	Particles and Fields – 1971 (Rochester)	71-184662	0-88318-101-0
No. 3	Thermal Expansion – 1971 (Corning)	72-76970	0-88318-102-9
No. 4	Superconductivity in d- and f-Band Metals (Rochester, 1971)	74-18879	0-88318-103-7
No. 5	Magnetism and Magnetic Materials – 1971 (2 parts) (Chicago)	59-2468	0-88318-104-5
No. 6	Particle Physics (Irvine, 1971)	72-81239	0-88318-105-3
No. 7	Exploring the History of Nuclear Physics – 1972	72-81883	0-88318-106-1
No. 8	Experimental Meson Spectroscopy –1972	72-88226	0-88318-107-X
No. 9	Cyclotrons – 1972 (Vancouver)	72-92798	0-88318-108-8
No. 10	Magnetism and Magnetic Materials – 1972	72-623469	0-88318-109-6
No. 11	Transport Phenomena – 1973 (Brown University Conference)	73-80682	0-88318-110-X
No. 12	Experiments on High Energy Particle Collisions – 1973 (Vanderbilt Conference)	73-81705	0-88318-111-8
No. 13	$\pi\text{-}\pi$ Scattering – 1973 (Tallahassee Conference)	73-81704	0-88318-112-6
No. 14	Particles and Fields – 1973 (APS/DPF Berkeley)	73-91923	0-88318-113-4
No. 15	High Energy Collisions – 1973 (Stony Brook)	73-92324	0-88318-114-2
No. 16	Causality and Physical Theories (Wayne State University, 1973)	73-93420	0-88318-115-0
No. 17	Thermal Expansion – 1973 (Lake of the Ozarks)	73-94415	0-88318-116-9
No. 18	Magnetism and Magnetic Materials – 1973 (2 parts) (Boston)	59-2468	0-88318-117-7
No. 19	Physics and the Energy Problem – 1974 (APS Chicago)	73-94416	0-88318-118-5
No. 20	Tetrahedrally Bonded Amorphous Semiconductors (Yorktown Heights, 1974)	74-80145	0-88318-119-3
No. 21	Experimental Meson Spectroscopy – 1974 (Boston)	74-82628	0-88318-120-7
No. 22	Neutrinos – 1974 (Philadelphia)	74-82413	0-88318-121-5
No. 23	Particles and Fields – 1974 (APS/DPF Williamsburg)	74-27575	0-88318-122-3
No. 24	Magnetism and Magnetic Materials – 1974 (20th Annual Conference, San Francisco)	75-2647	0-88318-123-1
No. 25	Efficient Use of Energy (The APS Studies on the Technical Aspects of the More Efficient Use of Energy)	75-18227	0-88318-124-X

No. 26	High-Energy Physics and Nuclear Structure – 1975 (Santa Fe and Los Alamos)	75-26411	0-88318-125-8
No. 27	Topics in Statistical Mechanics and Biophysics: A Memorial to Julius L. Jackson (Wayne State University, 1975)	75-36309	0-88318-126-6
No. 28	Physics and Our World: A Symposium in Honor of Victor F. Weisskopf (M.I.T., 1974)	76-7207	0-88318-127-4
No. 29	Magnetism and Magnetic Materials – 1975 (21st Annual Conference, Philadelphia)	76-10931	0-88318-128-2
No. 30	Particle Searches and Discoveries – 1976 (Vanderbilt Conference)	76-19949	0-88318-129-0
No. 31	Structure and Excitations of Amorphous Solids (Williamsburg, VA, 1976)	76-22279	0-88318-130-4
No. 32	Materials Technology – 1976 (APS New York Meeting)	76-27967	0-88318-131-2
No. 33	Meson-Nuclear Physics – 1976 (Carnegie-Mellon Conference)	76-26811	0-88318-132-0
No. 34	Magnetism and Magnetic Materials – 1976 (Joint MMM-Intermag Conference, Pittsburgh)	76-47106	0-88318-133-9
No. 35	High Energy Physics with Polarized Beams and Targets (Argonne, 1976)	76-50181	0-88318-134-7
No. 36	Momentum Wave Functions – 1976 (Indiana University)	77-82145	0-88318-135-5
No. 37	Weak Interaction Physics – 1977 (Indiana University)	77-83344	0-88318-136-3
No. 38	Workshop on New Directions in Mossbauer Spectroscopy (Argonne, 1977)	77-90635	0-88318-137-1
No. 39	Physics Careers, Employment and Education (Penn State, 1977)	77-94053	0-88318-138-X
No. 40	Electrical Transport and Optical Properties of Inhomogeneous Media (Ohio State University, 1977)	78-54319	0-88318-139-8
No. 41	Nucleon-Nucleon Interactions – 1977 (Vancouver)	78-54249	0-88318-140-1
No. 42	Higher Energy Polarized Proton Beams (Ann Arbor, 1977)	78-55682	0-88318-141-X
No. 43	Particles and Fields – 1977 (APS/DPF, Argonne)	78-55683	0-88318-142-8
No. 44	Future Trends in Superconductive Electronics (Charlottesville, 1978)	77-9240	0-88318-143-6
No. 45	New Results in High Energy Physics – 1978 (Vanderbilt Conference)	78-67196	0-88318-144-4
No. 46	Topics in Nonlinear Dynamics (La Jolla Institute)	78-57870	0-88318-145-2
No. 47	Clustering Aspects of Nuclear Structure and Nuclear Reactions (Winnipeg, 1978)	78-64942	0-88318-146-0
No. 48	Current Trends in the Theory of Fields (Tallahassee, 1978)	78-72948	0-88318-147-9

No. 49	Cosmic Rays and Particle Physics – 1978 (Bartol Conference)	79-50489	0-88318-148-7
No. 50	Laser-Solid Interactions and Laser Processing – 1978 (Boston)	79-51564	0-88318-149-5
No. 51	High Energy Physics with Polarized Beams and Polarized Targets (Argonne, 1978)	79-64565	0-88318-150-9
No. 52	Long-Distance Neutrino Detection – 1978 (C.L. Cowan Memorial Symposium)	79-52078	0-88318-151-7
No. 53	Modulated Structures – 1979 (Kailua Kona, Hawaii)	79-53846	0-88318-152-5
No. 54	Meson-Nuclear Physics – 1979 (Houston)	79-53978	0-88318-153-3
No. 55	Quantum Chromodynamics (La Jolla, 1978)	79-54969	0-88318-154-1
No. 56	Particle Acceleration Mechanisms in Astrophysics (La Jolla, 1979)	79-55844	0-88318-155-X
No. 57	Nonlinear Dynamics and the Beam-Beam Interaction (Brookhaven, 1979)	79-57341	0-88318-156-8
No. 58	Inhomogeneous Superconductors – 1979 (Berkeley Springs, W.V.)	79-57620	0-88318-157-6
No. 59	Particles and Fields – 1979 (APS/DPF Montreal)	80-66631	0-88318-158-4
No. 60	History of the ZGS (Argonne, 1979)	80-67694	0-88318-159-2
No. 61	Aspects of the Kinetics and Dynamics of Surface Reactions (La Jolla Institute, 1979)	80-68004	0-88318-160-6
No. 62	High Energy e^+e^- Interactions (Vanderbilt, 1980)	80-53377	0-88318-161-4
No. 63	Supernovae Spectra (La Jolla, 1980)	80-70019	0-88318-162-2
No. 64	Laboratory EXAFS Facilities – 1980 (Univ. of Washington)	80-70579	0-88318-163-0
No. 65	Optics in Four Dimensions – 1980 (ICO, Ensenada)	80-70771	0-88318-164-9
No. 66	Physics in the Automotive Industry – 1980 (APS/AAPT Topical Conference)	80-70987	0-88318-165-7
No. 67	Experimental Meson Spectroscopy – 1980 (Sixth International Conference, Brookhaven)	80-71123	0-88318-166-5
No. 68	High Energy Physics – 1980 (XX International Conference, Madison)	81-65032	0-88318-167-3
No. 69	Polarization Phenomena in Nuclear Physics – 1980 (Fifth International Symposium, Santa Fe)	81-65107	0-88318-168-1
No. 70	Chemistry and Physics of Coal Utilization – 1980 (APS, Morgantown)	81-65106	0-88318-169-X
No. 71	Group Theory and its Applications in Physics – 1980 (Latin American School of Physics, Mexico City)	81-66132	0-88318-170-3
No. 72	Weak Interactions as a Probe of Unification (Virginia Polytechnic Institute – 1980)	81-67184	0-88318-171-1
No. 73	Tetrahedrally Bonded Amorphous Semiconductors (Carefree, Arizona, 1981)	81-67419	0-88318-172-X

No. 74	Perturbative Quantum Chromodynamics (Tallahassee, 1981)	81-70372	0-88318-173-8
No. 75	Low Energy X-Ray Diagnostics – 1981 (Monterey)	81-69841	0-88318-174-6
No. 76	Nonlinear Properties of Internal Waves (La Jolla Institute, 1981)	81-71062	0-88318-175-4
No. 77	Gamma Ray Transients and Related Astrophysical Phenomena (La Jolla Institute, 1981)	81-71543	0-88318-176-2
No. 78	Shock Waves in Condensed Matter – 1981 (Menlo Park)	82-70014	0-88318-177-0
No. 79	Pion Production and Absorption in Nuclei – 1981 (Indiana University Cyclotron Facility)	82-70678	0-88318-178-9
No. 80	Polarized Proton Ion Sources (Ann Arbor, 1981)	82-71025	0-88318-179-7
No. 81	Particles and Fields –1981: Testing the Standard Model (APS/DPF, Santa Cruz)	82-71156	0-88318-180-0
No. 82	Interpretation of Climate and Photochemical Models, Ozone and Temperature Measurements (La Jolla Institute, 1981)	82-71345	0-88318-181-9
No. 83	The Galactic Center (Cal. Inst. of Tech., 1982)	82-71635	0-88318-182-7
No. 84	Physics in the Steel Industry (APS/AISI, Lehigh University, 1981)	82-72033	0-88318-183-5
No. 85	Proton-Antiproton Collider Physics –1981 (Madison, Wisconsin)	82-72141	0-88318-184-3
No. 86	Momentum Wave Functions – 1982 (Adelaide, Australia)	82-72375	0-88318-185-1
No. 87	Physics of High Energy Particle Accelerators (Fermilab Summer School, 1981)	82-72421	0-88318-186-X
No. 88	Mathematical Methods in Hydrodynamics and Integrability in Dynamical Systems (La Jolla Institute, 1981)	82-72462	0-88318-187-8
No. 89	Neutron Scattering – 1981 (Argonne National Laboratory)	82-73094	0-88318-188-6
No. 90	Laser Techniques for Extreme Ultraviolt Spectroscopy (Boulder, 1982)	82-73205	0-88318-189-4
No. 91	Laser Acceleration of Particles (Los Alamos, 1982)	82-73361	0-88318-190-8
No. 92	The State of Particle Accelerators and High Energy Physics (Fermilab, 1981)	82-73861	0-88318-191-6
No. 93	Novel Results in Particle Physics (Vanderbilt, 1982)	82-73954	0-88318-192-4
No. 94	X-Ray and Atomic Inner-Shell Physics – 1982 (International Conference, U. of Oregon)	82-74075	0-88318-193-2
No. 95	High Energy Spin Physics – 1982 (Brookhaven National Laboratory)	83-70154	0-88318-194-0
No. 96	Science Underground (Los Alamos, 1982)	83-70377	0-88318-195-9

No. 97	The Interaction Between Medium Energy Nucleons in Nuclei – 1982 (Indiana University)	83-70649	0-88318-196-7
No. 98	Particles and Fields – 1982 (APS/DPF University of Maryland)	83-70807	0-88318-197-5
No. 99	Neutrino Mass and Gauge Structure of Weak Interactions (Telemark, 1982)	83-71072	0-88318-198-3
No. 100	Excimer Lasers – 1983 (OSA, Lake Tahoe, Nevada)	83-71437	0-88318-199-1
No. 101	Positron-Electron Pairs in Astrophysics (Goddard Space Flight Center, 1983)	83-71926	0-88318-200-9
No. 102	Intense Medium Energy Sources of Strangeness (UC-Sant Cruz, 1983)	83-72261	0-88318-201-7
No. 103	Quantum Fluids and Solids – 1983 (Sanibel Island, Florida)	83-72440	0-88318-202-5
No. 104	Physics, Technology and the Nuclear Arms Race (APS Baltimore –1983)	83-72533	0-88318-203-3
No. 105	Physics of High Energy Particle Accelerators (SLAC Summer School, 1982)	83-72986	0-88318-304-8
No. 106	Predictability of Fluid Motions (La Jolla Institute, 1983)	83-73641	0-88318-305-6
No. 107	Physics and Chemistry of Porous Media (Schlumberger-Doll Research, 1983)	83-73640	0-88318-306-4
No. 108	The Time Projection Chamber (TRIUMF, Vancouver, 1983)	83-83445	0-88318-307-2
No. 109	Random Walks and Their Applications in the Physical and Biological Sciences (NBS/La Jolla Institute, 1982)	84-70208	0-88318-308-0
No. 110	Hadron Substructure in Nuclear Physics (Indiana University, 1983)	84-70165	0-88318-309-9
No. 111	Production and Neutralization of Negative Ions and Beams (3rd Int'l Symposium, Brookhaven, 1983)	84-70379	0-88318-310-2
No. 112	Particles and Fields – 1983 (APS/DPF, Blacksburg, VA)	84-70378	0-88318-311-0
No. 113	Experimental Meson Spectroscopy – 1983 (Seventh International Conference, Brookhaven)	84-70910	0-88318-312-9
No. 114	Low Energy Tests of Conservation Laws in Particle Physics (Blacksburg, VA, 1983)	84-71157	0-88318-313-7
No. 115	High Energy Transients in Astrophysics (Santa Cruz, CA, 1983)	84-71205	0-88318-314-5
No. 116	Problems in Unification and Supergravity (La Jolla Institute, 1983)	84-71246	0-88318-315-3
No. 117	Polarized Proton Ion Sources (TRIUMF, Vancouver, 1983)	84-71235	0-88318-316-1

No. 118	Free Electron Generation of Extreme Ultraviolet Coherent Radiation (Brookhaven/OSA, 1983)	84-71539	0-88318-317-X
No. 119	Laser Techniques in the Extreme Ultraviolet (OSA, Boulder, Colorado, 1984)	84-72128	0-88318-318-8
No. 120	Optical Effects in Amorphous Semiconductors (Snowbird, Utah, 1984)	84-72419	0-88318-319-6
No. 121	High Energy e^+e^- Interactions (Vanderbilt, 1984)	84-72632	0-88318-320-X
No. 122	The Physics of VLSI (Xerox, Palo Alto, 1984)	84-72729	0-88318-321-8
No. 123	Intersections Between Particle and Nuclear Physics (Steamboat Springs, 1984)	84-72790	0-88318-322-6
No. 124	Neutron-Nucleus Collisions – A Probe of Nuclear Structure (Burr Oak State Park - 1984)	84-73216	0-88318-323-4
No. 125	Capture Gamma-Ray Spectroscopy and Related Topics – 1984 (Internat. Symposium, Knoxville)	84-73303	0-88318-324-2
No. 126	Solar Neutrinos and Neutrino Astronomy (Homestake, 1984)	84-63143	0-88318-325-0
No. 127	Physics of High Energy Particle Accelerators (BNL/SUNY Summer School, 1983)	85-70057	0-88318-326-9
No. 128	Nuclear Physics with Stored, Cooled Beams (McCormick's Creek State Park, Indiana, 1984)	85-71167	0-88318-327-7
No. 129	Radiofrequency Plasma Heating (Sixth Topical Conference, Callaway Gardens, GA, 1985)	85-48027	0-88318-328-5
No. 130	Laser Acceleration of Particles (Malibu, California, 1985)	85-48028	0-88318-329-3
No. 131	Workshop on Polarized ^{3}He Beams and Targets (Princeton, New Jersey, 1984)	85-48026	0-88318-330-7
No. 132	Hadron Spectroscopy–1985 (International Conference, Univ. of Maryland)	85-72537	0-88318-331-5
No. 133	Hadronic Probes and Nuclear Interactions (Arizona State University, 1985)	85-72638	0-88318-332-3
No. 134	The State of High Energy Physics (BNL/SUNY Summer School, 1983)	85-73170	0-88318-333-1
No. 135	Energy Sources: Conservation and Renewables (APS, Washington, DC, 1985)	85-73019	0-88318-334-X
No. 136	Atomic Theory Workshop on Relativistic and QED Effects in Heavy Atoms	85-73790	0-88318-335-8
No. 137	Polymer-Flow Interaction (La Jolla Institute, 1985)	85-73915	0-88318-336-6
No. 138	Frontiers in Electronic Materials and Processing (Houston, TX, 1985)	86-70108	0-88318-337-4
No. 139	High-Current, High-Brightness, and High-Duty Factor Ion Injectors (La Jolla Institute, 1985)	86-70245	0-88318-338-2

No. 140	Boron-Rich Solids (Albuquerque, NM, 1985)	86-70246	0-88318-339-0
No. 141	Gamma-Ray Bursts (Stanford, CA, 1984)	86-70761	0-88318-340-4
No. 142	Nuclear Structure at High Spin, Excitation, and Momentum Transfer (Indiana University, 1985)	86-70837	0-88318-341-2
No. 143	Mexican School of Particles and Fields (Oaxtepec, México, 1984)	86-81187	0-88318-342-0
No. 144	Magnetospheric Phenomena in Astrophysics (Los Alamos, 1984)	86-71149	0-88318-343-9
No. 145	Polarized Beams at SSC & Polarized Antiprotons (Ann Arbor, MI & Bodega Bay, CA, 1985)	86-71343	0-88318-344-7
No. 146	Advances in Laser Science–I (Dallas, TX, 1985)	86-71536	0-88318-345-5
No. 147	Short Wavelength Coherent Radiation: Generation and Applications (Monterey, CA, 1986)	86-71674	0-88318-346-3
No. 148	Space Colonization: Technology and The Liberal Arts (Geneva, NY, 1985)	86-71675	0-88318-347-1
No. 149	Physics and Chemistry of Protective Coatings (Universal City, CA, 1985)	86-72019	0-88318-348-X
No. 150	Intersections Between Particle and Nuclear Physics (Lake Louise, Canada, 1986)	86-72018	0-88318-349-8
No. 151	Neural Networks for Computing (Snowbird, UT, 1986)	86-72481	0-88318-351-X
No. 152	Heavy Ion Inertial Fusion (Washington, DC, 1986)	86-73185	0-88318-352-8
No. 153	Physics of Particle Accelerators (SLAC Summer School, 1985) (Fermilab Summer School, 1984)	87-70103	0-88318-353-6
No. 154	Physics and Chemistry of Porous Media—II (Ridge Field, CT, 1986)	83-73640	0-88318-354-4
No. 155	The Galactic Center: Proceedings of the Symposium Honoring C. H. Townes (Berkeley, CA, 1986)	86-73186	0-88318-355-2
No. 156	Advanced Accelerator Concepts (Madison, WI, 1986)	87-70635	0-88318-358-0
No. 157	Stability of Amorphous Silicon Alloy Materials and Devices (Palo Alto, CA, 1987)	87-70990	0-88318-359-9
No. 158	Production and Neutralization of Negative Ions and Beams (Brookhaven, NY, 1986)	87-71695	0-88318-358-7

No. 159	Applications of Radio-Frequency Power to Plasma: Seventh Topical Conference (Kissimmee, FL, 1987)	87-71812	0-88318-359-5
No. 160	Advances in Laser Science–II (Seattle, WA, 1986)	87-71962	0-88318-360-9
No. 161	Electron Scattering in Nuclear and Particle Science: In Commemoration of the 35th Anniversary of the Lyman-Hanson-Scott Experiment (Urbana, IL, 1986)	87-72403	0-88318-361-7
No. 162	Few-Body Systems and Multiparticle Dynamics (Crystal City, VA, 1987)	87-72594	0-88318-362-5
No. 163	Pion–Nucleus Physics: Future Directions and New Facilities at LAMPF (Los Alamos, NM, 1987)	87-72961	0-88318-363-3
No. 164	Nuclei Far from Stability: Fifth International Conference (Rosseau Lake, ON, 1987)	87-73214	0-88318-364-1
No. 165	Thin Film Processing and Characterization of High-Temperature Superconductors	87-73420	0-88318-365-X
No. 166	Photovoltaic Safety (Denver, CO, 1988)	88-42854	0-88318-366-8
No. 167	Deposition and Growth: Limits for Microelectronics (Anaheim, CA, 1987)	88-71432	0-88318-367-6
No. 168	Atomic Processes in Plasmas (Santa Fe, NM, 1987)	88-71273	0-88318-368-4
No. 169	Modern Physics in America: A Michelson-Morley Centennial Symposium (Cleveland, OH, 1987)	88-71348	0-88318-369-2
No. 170	Nuclear Spectroscopy of Astrophysical Sources (Washington, D.C., 1987)	88-71625	0-88318-370-6
No. 171	Vacuum Design of Advanced and Compact Synchrotron Light Sources (Upton, NY, 1988)	88-71824	0-88318-371-4
No. 172	Advances in Laser Science–III: Proceedings of the International Laser Science Conference (Atlantic City, NJ, 1987)	88-71879	0-88318-372-2
No. 173	Cooperative Networks in Physics Education (Oaxtepec, Mexico 1987)	88-72091	0-88318-373-0
No. 174	Radio Wave Scattering in the Interstellar Medium (San Diego, CA 1988)	88-72092	0-88318-374-9
No. 175	Non-neutral Plasma Physics (Washington, DC 1988)	88-72275	0-88318-375-7
No. 176	Intersections Between Particle and Nuclear Physics (Third International Conference) (Rockport, ME 1988)	88-62535	0-88318-376-5